Global Flow of Funds Analysis

Nan Zhang · Yiye Zhang

Global Flow of Funds Analysis

Data, Models, and Applications

Springer

Nan Zhang
Hiroshima Shudo University
Hiroshima, Japan

Yiye Zhang
Cornell University
New York, NY, USA

ISBN 978-981-97-1028-7 ISBN 978-981-97-1029-4 (eBook)
https://doi.org/10.1007/978-981-97-1029-4

This work was supported by the grants-in-aid for scientific research (Japan): [Grant Number Scientific Research (C), 20K01701]

This Springer imprint is published by the registered company Springer Nature Singapore Pte Ltd.
The registered company address is: 152 Beach Road, #21-01/04 Gateway East, Singapore 189721, Singapore

Paper in this product is recyclable.

Preface

The world has undergone significant transformations since 2020, particularly in the realms of politics, economics, healthcare, science and technology, and more. Under the influence of these geopolitical shifts, there has been a noticeable trend in the global economy, transitioning from the process of globalization that began in the 1990s to a phase characterized by deglobalization. The global impact of COVID-19 in 2020 that resulted in a significant loss in the entire generation has further intensified this shift.[1] However, the worldwide pandemic has also accelerated technological and digital transformations. The year 2023 marked a rapid integration of artificial intelligence (AI) with everyday life, making a widespread impact on human cognitive processes worldwide. A common factor underpinning these dramatic changes is the increasing recognition of the importance of information. This heightened focus on analytical comprehension is elevating the significance of data cognition, leading to unprecedented attention and development in the fields of statistics and data science.

This book stands out as the inaugural work dedicated to leveraging data for the observation of global flow of funds (GFF). It has three distinctive features, foremost among them being the integration and advancement of data sources. Grounded in the fundamental principles of the international capital cycle and the dynamics of shifts in domestic and international financial markets, this book brings together the statistical systems of established international institutions. This integration results in the creation of novel data sources that facilitate the harmonization of data essential for international comparisons.

The second is the statistical design of the GFF. Considering the interplay among different financial instruments and the broader financial system, a range of tools have been developed. These include balanced correspondences between countries or regions and the global economy, primary instruments for financial transactions,

[1] World Bank Group, February 01, 2022, We are losing a generation: The devastating impacts of COVID-19, https://blogs.worldbank.org/voices/we-are-losing-generation-devastating-impacts-covid-19.

and the external financial assets and liabilities of a country. Additionally, the 'from-whom-to-whom' statistical matrix model is designed for international observation of transactions between national sectors, as seen in sectoral accounts.

The third feature is the establishment of a theoretical framework for analysis, and the introduction of financial network technology to predict the credit relationship between two or many sides of the transaction, the balance sheets, and financial risk. It links financial flows within the domestic economy to those in the rest of the world, providing a comprehensive analysis of the global financial flow dynamics.

This book makes a significant contribution to financial statistics, international finance, and macro-prudential regulation by introducing a novel analytical perspective and methodology. The proposed statistical framework is built upon the concept of the GFF, leveraging existing metadata to integrate diverse data sources. The resulting GFF statistical matrix is structured as a "who-to-whom" matrix, encompassing both the matrix of external financial assets and liabilities and the matrix of international capital inflow-outflow. These matrices serve to illustrate the relationships between countries in terms of capital operations and balance sheets. The use of matrix data enables the establishment of a financial network, facilitates data visualization, and carries out empirical analysis.

Using GFF as its framework, this book conducts a dynamic analysis of the reciprocal relationship between the current account balance and financial investment in China and the US. The analysis is carried out through the application of a vector error correction model and includes an assessment of the strategic challenges posed by the economic decoupling of the US and China. Furthermore, the book delves into various facets of financial position, credit relationships, interactions, and debt risk within the GFF framework. Notably, it introduces the sectoral from-whom-to-whom financial stock matrix (SFSM) for providing insights into the application of financial networks for evaluating the shock propagation associated with external financial assets and liabilities across different sectors in G-4 economics (China, Japan, the UK, and the US).

This book is divided into five chapters. Chapter 1 introduces the theoretical concept of GFF, statistical framework, data sources, and the method of compiling the GFF statistical matrix, which aims to provide a measurement for the GFF, as discussed in four portions. First, Chap. 1 will define GFF to determine its statistical domains. Second, we set out the ideas and existing data sources published by Bank for International Settlements and International Monetary Funds, etc., and integrate them to measure GFF. Third, the balance sheet approach is used to break down the rest of the world into international investment possion components. An external statistics' matrix (metadata) shows the available external-sector financial data based on the IIP concept. Fourth, we use data science to integrate the data sources, and improve the timeliness of the existing data transmission.

Chapter 2 focuses on the G20 as the subject of observation and endeavors to compile the direct investment matrix, portfolio investment matrix, and cross-border bank credit matrix for the year 2018 using a bilateral approach. Building upon this, a global asset and liability matrix of a composite grouping type is then constructed. Moreover, we employ network theory to discuss an analytical method for the GFF and use countries in the G20 as the research sample to discuss the network centrality, mutual relationships, the financial risk of foreign direct investment, portfolio investment, and cross-border bank credit among the US, Japan, and China. By incorporating network theory into GFF analysis, this book opens a new avenue for measuring and applying GFF. The integration of the financial network into the GFF matrix also facilitates data visualization.

Chap. 3 develops an analytical framework for the external flow of funds to scrutinize the dynamics and challenges associated with the decoupling of China and the US. Initially, the chapter examines the structural relationship between China and the US in terms of savings and investment imbalances from 1980 to 2022. Subsequently, the examination extends to the challenges between China and the US in external financial assets and liabilities, utilizing stock data and concentrating on the external adjustment mode spanning from 2008 to 2022. A vector error correction model is constructed to quantify the relationship between short-term fluctuations and long-term trends of the external flow of funds in China and the US. This analysis assesses the risks associated with China-US economic decoupling and US debt, identifies strategic challenges faced by both parties, and outlines potential countermeasures for the future.

In Chap. 4, which complements the discussions in Chap. 2, we conduct a comparative analysis of the structural changes and debt risks in G20 countries during a unique historical period, focusing on the years 2018–2022. This analysis specifically addresses the China-US economic decoupling and assesses the potential for a debt crisis. We employ stock data to examine the GFF matrix in this extraordinary context. The chapter also delves into the positions of China and the US in the debt securities market, exploring their mutual financing relationships through financial network technology. Furthermore, we provide a statistical estimation of the impact of debt risk transmission. Additionally, Chap. 4 explores the dynamics of China-US external financial assets and liabilities, again utilizing stock data for a detailed examination.

Chapter 5 focuses on improving the sectoral accounts data, establishing the Whom-to-Whom matrix model of the sectors, identifying sectoral interlinkages in G-4 economies, and the statistical estimation of the financial risk shock and spread among G-4 sectors. The SFSM specifically focuses on counterparty national and cross-border exposures of sectors in G-4, designed to create country-specific financial networks, interconnecting each country-level network based on cross-border exposures. Analytical results systematically reveal bilateral exposures among the four countries in the GFF, identifying sectoral interlinkages, characteristics of overseas investment, external shocks, and internal influences. Furthermore, this chapter introduces an eigenvalue and eigenvector decomposition to analyze the effects and provides an analytical description of the shock propagation process.

In 2005, the author (Nan Zhang) published a book titled *Theory and Development of Global Flow of Funds Analysis* by MINERVA. The book centers on the theoretical exploration of GFF and the development of measurement models. However, it did not address the practical challenges associated with compiling GFF statistics, nor did it adopt an "International" perspective. From 2008, the author was employed by the Statistics Department of IMF as Monetary and Financial Statistics Advisor. As a technical assistance expert of the IMF statistical mission, the author was invited to Africa and China many times to participate in financial statistics technical assistance projects and teach advanced courses on Monetary and financial statistics. During this timeframe, the author also presented academic lectures on GFF statistical compilation and analysis methods at events hosted by World Statistics Congress, IARIW General Conference, Annual Conference of the Society for Economic Measurement, and various annual societal meetings in both Japan and China. These engagements significantly contributed to a deeper investigation of financial statistics and GFF statistics.

This book represents an exploration into expanding financial accounts and balance sheets into three-dimensional metadata, coupled with the development of Who-to-Whom analysis. By doing so, this study advances the GFF statistics, allowing for a comprehensive assessment of global financial stability from both national and cross-border sectoral perspectives. The GFF data generated from this study offer valuable insights for analyzing interconnectivity across borders, providing a nuanced understanding of global financial interdependencies.

The GFF analysis requires a multidisciplinary approach, involving expertise in economic statistics, finance, economics, data science, and international relations. The latest data and market dynamics are indispensable for accurate and relevant analysis. Additionally, collaboration with experts in the field and utilizing advanced modeling techniques can enhance the depth and accuracy of the study. This research continues to present numerous avenues for further investigations.

Hiroshima, Japan — Nan Zhang
New York, USA — Yiye Zhang

Acknowledgements

As the authors conclude this manuscript, we extend our heartfelt appreciation to the following professors and colleagues. First and foremost, we express our gratitude to the late Prof. Fengqi Cao of Peking University, who provided long-term support and invaluable guidance. The authors would also like to thank Prof. Tze Leung Lai of Stanford University for his support and encouragement throughout the development of this research.

The authors also like to thank Prof. Tosihisa Toyoda and Prof. Hiroaki Teramoto for their selfless help. In particular, the authors also would like to thank Prof. Itsuo Sakuma of Senshu University, Prof. Kazusuke Tsujimura of Keio University, Prof. Masako Tsujimura of Rissho University, Prof. Satoru Hagino of Reitaku University, and Prof. Kim Jiyoung of Okayama University for their helpful comments in GFF analysis field.

Nan Zhang would like to express my sincere gratitude to Armida San Jose, Mr. Jaroslav Kucera, Ms. Xiuzhe Zhao, and Mr. Artak Harutyunyan of Statistics Department of IMF, who gave him a lot of professional help when he was participating in the IMF statistical missions, so that he has a profound understanding of the practice of statistics. And Mr. Artak Harutyunyan was also the discussant of the paper that Nan Zhang submitted to the 36th IARIW Conference, grateful for Artak's very pertinent and helpful comments.

The authors express their gratitude to Prof. William Barnett, the President of the Society for Economic Measurement, who provided us with three opportunities during the study of GFF statistics to host the Invited Session on GFF Statistics at the 4th to 6th SEM Conferences. These sessions offered a platform to engage in discussions on the establishment of GFF statistics.

The authors extend their gratitude to Mr. Dennis Fixler of the US Bureau of Economic Analysis for providing valuable comments on their paper during the IARIW-OECD special conference in 2015. Additionally, sincere appreciation is expressed to Celestino Giron of the European Central Bank, who served as the discussant for the author's paper at the 35th IARIW Conference in August 2018, for his valuable and enlightening advice.

The initial version of Chap. 2 was presented at the 2021 Spring Annual Meeting of the Japan Society of Monetary Economics, with Prof. Kiyotaka Sato from Yokohama National University serving as the paper's discussant. Professor Takekazu Iwamoto of Kyoto University also acted as the discussant for the paper, specifically the initial version of Chap. 3, during the 81st Annual Meeting of the Japan Association of International Economics in 2022. The authors express their heartfelt thanks to both discussants for their constructive comments.

The authors have previously delivered presentations on GFF statistics at the World Statistics Congress (2013, 2015), IARIW (2012, 2015, 2018, and 2021), the Japanese Joint Statistical Meeting (2017), and the Japan Society of Economic Statistics (2018, 2021–2023). In addition, Nan Zhang served as an invited speaker at the Annual Conference of China Statistics (2017, 2019), the Flow of Funds Statistics Seminar at Keio University (2017, 2018), GFF Analysis Workshop at Fudan University (2017), GFF Statistics Workshop at Tsinghua China Data Center (2019), the Macroeconomic Workshop at the National School of Development, Peking University (2023), and the Monetary Economics Seminar at Kobe University (2023). The authors are grateful for the constructive suggestions and helpful advice extended by the conference hosts: Prof. Dong Qiu (Beijing Normal University), Prof. Qingfu Liu (Fudan University), Prof. Xianchun Xu (Tsinghua University), Prof. Harry Wu (Peking University), and Prof. Yoich Matubayashi (Kobe University). Naturally, any remaining errors are solely the responsibility of the author.

Finally, gratitude to Mr. Yutaka Hirachi from Springer Nature Japan for his invaluable assistance in facilitating the publication of our book.

Hiroshima, Japan
New York, USA
January 2024

Nan Zhang
Yiye Zhang

Contents

About the Authors

Prof. Nan Zhang is a Professor of Statistics at the Faculty of Economic Sciences, Hiroshima Shudo University, Japan. He obtained his Ph.D. in Economics from Ritsumeikan University and has been instructing courses in Statistics, Economic Statistics, and Financial Econometrics at Hiroshima Shudo University since 1995.

He has conducted extensive and in-depth research on Monetary and Financial Statistics, Flow of Fund Analysis, and Global Flow of Funds Statistics and Analysis, resulting in the publication of numerous influential papers and books. Prof. Zhang served as a Special Researcher at the Research Center for Finance and Securities at Peking University from 1997 to 2005, and also held positions as a Visiting Scholar at the East Asian Institute at Columbia University from 2001 to 2002, the Department of Statistics at the UC Berkeley from 2007 to 2008, and the Department of Statistics at Stanford University from 2014 to 2015. Additionally, he acted as an Advisor and Technical Assistance Expert in the Statistics Department of the IMF from 2008 to 2015. In recognition of his contributions, he was honored with the Japan Society of Economic Statistics Award in 2021.

Dr. Yiye Zhang is an Associate Professor at Weill Cornell Medical College of Cornell University and Graduate Faculty in Cornell Systems Engineering. She earned her Ph.D. in Information Systems Management from Carnegie-Mellon University and a Master's in Biostatistics from Columbia University. Her research extensively delves into health information technology, particularly developing methodology and software to analyze large datasets.

Her work predominantly revolves around the effective utilization of electronic health records (EHRs) and clinical decision support systems (CDSS), emphasizing their role in enhancing public health. Her expertise in analyzing vast and complex datasets has been instrumental in advancing data science for generating insights from heterogenous populations. Dr. Zhang's work has been funded by federal agencies in the USA, including the National Institute of Health, Agency for Healthcare Quality and Research, and US Department of Health and Human Services. She was named the Walsh McDermott Scholar in Public Health from 2016 to 2019.

Abbreviations

BIS	Bank for International Settlements
BOP	Balance of Payments
BPM6	BOP and IIP Manual, sixth edition
BSA	Balance sheet approach
CBS	Consolidated banking statistics
CDIS	Coordinated direct investment survey
CNBS	Center for National Balance Sheets of China
CPIS	Coordinated portfolio investment survey
DAL	Domestic assets and liabilities
DI	Direct investment
EAL	External assets and liabilities
EALM	External assets and liabilities matrix
EC	Eigenvector centrality
FA	Financial accounts
FBS	Financial Balance Sheets
FC	Financial corporations
FD	Financial derivatives
FDI	Foreign direct investment
FFA	Flow of funds accounts
FSB	Financial Stability Board
GFC	Great Financial Crisis
GFF	Global flow of funds
GFFM	Global flow of funds matrix
GG	General Government
HH	Household and non-profit institutions serving households
IBS	International Banking Statistics
IFS	International Financial Statistics
IIP	International Investment Position
IMF	International Monetary Fund
LBS	Locational banking statistics
MFS	Monetary and financial statistics

NFC	Non-financial corporations
OE	Other economies
OI	Other investment
PDI	Power of Dispersion Index
PI	Portfolio investment
ROW	Rest of the world
SDI	Sensitivity of Dispersion Index
SFSM	Sectoral from-whom-to-whom financial stock matrix
SNA	System of National Accounts
W-to-W	Who-to-Whom

Chapter 1
Measuring Global Flow of Funds: Statistical Framework, Data Sources, and Approaches

Abstract This chapter aims to provide a measurement for the global flow of funds (GFF), as discussed in four portions. First, the Chapter will define GFF to determine its statistical domains. Second, the document sets out the ideas and existing data sources and integrates them to measure GFF. These concepts and data sources are reflected in the balance of payments, international investment position (IIP), the Coordinated Direct Investment Survey (CDIS), the Coordinated Portfolio Investment Survey, the consolidated banking statistics (CBS), and the rest of the world (ROW) account. Third, the balance sheet approach is used to break down the ROW into IIP components. An external statistics' matrix (metadata) exercise shows the available external-sector financial data based on the IIP concept. As the outcome of the study, this chapter compiled GFF matrix with the pattern of "from-whom-to-whom." Fourth, data science is explored to integrate the data sources, improve the timeliness of the existing data transmission, and illustrate how the GFF matrix operates.

Keywords Global flow of funds · Integrating framework · Data sources · Who-to-whom · Statistics' matrix · Data science

1.1 Introduction

The global flow of funds (GFF) concept is an extension of that for the domestic flow of funds first developed by Copeland (1952). It connects domestic economies with the rest of the world. GFF data can provide valuable information for analyzing interconnectedness across borders and global financial interdependencies. Corresponding to the deregulation of the financial market, researchers began exploring the GFF in the 1990s. Ishida (1993) put forward the idea of GFF analysis, discussed the concept, and then measured the international capital flows among Japan, the United States (U.S.), and Germany. Drawing on this research, Tsujimura and Mizosita (2002a, 2002b and 2003) used the perspective of GFF to analyze European financial Integration. Zhang (2005 and 2008) linked real transactions with financial transactions based on the dynamic flow of funds and established a theoretical framework for GFF analysis

N. Zhang and Y. Zhang, *Global Flow of Funds Analysis*,
https://doi.org/10.1007/978-981-97-1029-4_1

through three factors: domestic savings–investment, foreign trade, and international capital flows. He then also built an econometric model of GFF. Based on the GFF concept, Tsujimura and Tsujimura (2008) conducted pioneering research that used financial matrix methods to test the transmission of financial policy and the effects of the international flow of funds in the Euro area using data from Coordinated Portfolio Investment Survey (CPIS) and Consolidated Banking Statistics (CBS).

Allen et al. (2002) proposed a statistical framework for understanding crises in emerging markets based on the examination of stock variables in the aggregate balance sheet of a country and the balance sheets of its main sectors (assets and liabilities). This framework is consistent with the 2008 System of National Accounts (SNA) and is very instructive for establishing GFF matrix (GFFM) based on "from-whom-to-whom" (W-to-W) format (Cerutti et al., 2017).

In April 2009, the G-20 Finance Ministers and Central Bank Governors Working Group on Reinforcing International Co-operation and Promoting Integrity in Financial Markets called on the International Monetary Fund (IMF) and the Financial Stability Board (FSB) to explore information gaps and provide appropriate proposals for strengthening data collection and reporting back to the Finance Ministers and Central Bank Governors. As a result of the meeting, the IMF and FSB proposed maintenance and expansion of the resultant statistics in October 2009. The principal focus centered on Recommendation 15, as financial and economic crises are characterized by abrupt revaluations or other changes in the capital positions of key sectors of the economy. Recommendation 15 states that, "The Inter-Agency Group on Economic and Financial Statistics (IAG), which includes all agencies represented in the Inter-Secretariat Working Group on National Accounts, to develop a strategy to promote the compilation and dissemination of the balance sheet approach (BSA), Flow of Funds, and sectoral data more generally, starting with the G-20 economies. Data on nonbank financial institutions should be a particular priority."[1] Thus, Recommendation 15 also implies, through its reference to compiling "flow of funds" statistics, a compilation of breakdowns of the financial positions and flows of each economic sector by its counterparty sectors. Datasets providing this kind of information are said to provide "from-whom-to-whom (W-t-W)" financial statistics. In such a situation, we also need to understand and measure the flow of funds between countries, namely the Global Flow of Funds (GFF).

Stone (1966a, 1966b) and Klein (1983) set up the balance sheets of a closed economy in a standard matrix form, distinguishing between financial assets and real assets on the assets side and liabilities side, trying to convert the Use (U) and Make (V) tables of input–output analysis into a Flow of Funds Table by referring to Stone and Klein's method. Their paper considers that Flow of Funds Table can also be a matrix based on the W-to-W format.

There is international awareness of information limitations vis-à-vis the problem that existing data do not describe the risks inherent in a financial system. Previous research has evolved into a discussion of the basic concept of GFF and a proposal

[1] Financial Stability Board and International Monetary Fund (2009). The Financial Crisis and Information aps Gaps- Report to the G-20 Finance Ministers and Central Bank Governors, p. 10.

to establish a statistical framework for GFF. Therefore, the IMF's Statistics Department has organized seven economies with systemically important financial centers to construct a geographically disaggregated GFF mapping of domestic and external capital stocks (Errico et al., 2013, 2014). Those authors delineated key concepts and existing data sources, used the Balance Sheet Approach (BSA) to break down the rest of the world by IIP components. An external statistics' matrix (metadata) shows that external-sector financial data are available by using the IIP concept. The main outcome is a prototype template of stock and flow data, geographically disaggregated by national/regional economies.

Over the past few years, Zhang (IARIW-OECD conference, 2015), Zhang and Zhao (2019), and Zhang (2020, 2021) have focused on three main problems of GFF—its definition, integrating its statistics with a system of national accounts (SNA), and data sources and approaches—in conducting research and pilot compilations of GFF statistics. Using international statistical standards, data on cross-border financial exposures (CPIS, CDIS, IIP, and BIS) can be linked to domestic sectoral account data to build a comprehensive picture of domestic and international financial interconnections. A new challenge for us is to develop a GFF matrix (GFFM) that not only considers risk exposures between countries but also describes debt relationships between counterparty sectors. The GFF project primarily aims to construct a matrix that identifies interlinkages among domestic sectors and with counterparty countries (and possibly counterparty country sectors) to build bilateral financial exposures and support analysis of potential sources of contagion (Zhang, 2022).

Shortly after the 2008 SNA was introduced, the sixth edition of the Balance of Payments and International Investment Position Manual (BPM6) was published by the IMF. Finally, a revised Government Finance Statistics Manual 2014 (GFSM 2014) was made available to the public in 2014. In view of these major developments, it was important to revise the 2000 MFSM and the 2008 MFS Guide to align the methodology used to compute monetary and financial statistics with the new framework. Based on the 2000 and 2008 versions, the IMF published new versions in 2016 in the *Monetary and Financial Statistics Manual and Compilation Guide* (MFSMCG, 2016). From the perspective of economic statistics, since the MFSMCG provides basic data for the national account, flow of funds account (FFA), and balance of payments statistics, the revised international standards on monetary and financial statistics are bound to substantially impact the capital account, balance sheet, FFA, and balance of payments.

The MFSMCG proposes the following approach, shown in Table 1.1, as the basic model for the FFA. In Table 1.1, each sector has two columns, one for net acquisitions of financial assets and the other for net liabilities. To emphasize that transactions in financial instruments are included, net financial investment (NFI) is used (instead of net lending/borrowing) and calculated as the net acquisition of financial assets less the net incurrence of liabilities. For example, the financial assets of the non-financial corporation's sector are 20.4 and its financial liabilities are 26.1, so the NFI of this sector is −5.7. Table 1.1 presents an example of two-dimensional financial statistics for transactions for a single period. In the case of currency and deposits, the total change in financial assets (total domestic minus rest of the world (79.7 − 14.7 =

65.0)) is equal to total changes in liabilities across sectors (29.2 + 35.8 = 65.0). And with the inclusion of the rest of the world (ROW) sector in Table 1.1 the system is closed in the sense that each transaction must have both a creditor and a debtor sector and each financial asset must have a counterparty liability item in another sector's balance sheet.

However, the sectoral accounts dataset presents two main drawbacks which render a comprehensive analysis of counterparty risk exposures difficult. First, the vast majority of countries do not provide detailed information on the counterparty sector of a financial instrument issued by a given sector ("from-whom-to-whom" data). Second, the recent crisis showed that many risks to the global financial system arise from cross-border exposures and in the sectoral accounts data cross-border exposures fall all under the ROW sector without specifying the counterparty country and counterparty sector, it cannot tell us any information with overseas counterparties (Robert Heath & Evrim Bese Goksu, 2017).

The 2008 SNA recommends the development of a detailed FFA. Table 27.4 of 2008 SNA offers a more detailed analysis by showing transactions in assets classified by type of asset and debtor sector in the first part, and by type of liability and creditor sector in the second part.[2] MFSMCG presents the concept of three-dimensional financial statistics, which is shown in Fig. 1.1. It provides information about W-to-W, wherein the breakdowns of financial stocks and flows of each sector include counterpart sectors, addressing the deficiency of two-dimensional financial statistics. By adding the counterpart sector, this three-dimensional approach provides a W-to-W analysis.

Figure 1.1 presents the concept of three-dimensional financial statistics, which provides a framework to present flows and stocks for all categories of instruments and all sectors and subsectors by counterpart sectors. This framework can track who finances whom, the kind of financial instruments used, and the amount of funding involved. Three-dimensional displays show both sides of the transaction and the financial instruments used, which are sometimes referred to as flow of funds statistics. A similar three-dimensional display shows creditors and debtors in each category of financial instruments, sometimes referred to as BSA.

Based on the three-dimensional concept, MFSMCG also presents an idea for establishing a global flow of funds statistics.[3] In analyzing bilateral cross-border flows and stock, the three-dimensional expression is likely to be further expanded by classifying the ROW by country, or even by sector (in other words, who is the basis for domestic and cross-border information by country and sector). Ultimately, this expansion across countries will help establish bilateral financial statistics at the global level. Such "global flow of funds" data have high analytical value, including the ability to measure global liquidity flows and analyze global financial networks. Individual economies can also benefit by identifying possible channels for external shocks to flow into the domestic economy and its sectors.

[2] 2008 SNA, 505.

[3] IMF (2016a, 2016b, 2016c), MFSMCG, 288.

Table 1.1 Basic model for the flow of funds account

Transactions	Nonfinancial corporations		Financial corporations		General government		Households and NPISHs		Total domestic		Rest of the world	
During a period	Changes in financial assets	Changes in liabilities	Changes in financial assets	Changes in liabilities	Changes in financial assets	Changes in liabilities	Changes in financial assets	Changes in liabilities	Changes in financial assets	Changes in liabilities	Changes in financial assets	Changes in liabilities
A. Monetary gold and SDRs			0.4	0.0					0.4	0.0	0	0.4
B. Currency and deposits	16.9	0.1	41.1	29.1	−4.1		25.8		79.7	29.2	−14.7	35.8
C. Debt securities	27.9	−6.7	12.7	−0.8	1.6	15.9	−1.4		40.8	8.4	−35.7	−3.3
D. Loans	1.0	−1.4	−19.8	4.6	−1.7	−7.7	0.0	−19.7	−20.5	−24.2	−5.1	−1.4
E. Equity and investment fund shares	14.4	28.6	9.7	52.4	1.0		−2.2		22.9	81.0	37.6	−20.6
F. Insurance, pension, and standardized guarantee schemes	0.7		−1.6	15.3	0.0		12.8		11.9	15.3	−0.3	−3.7

(continued)

Table 1.1 (continued)

Transactions	Nonfinancial corporations		Financial corporations		General government		Households and NPISHs		Total domestic		Rest of the world	
During a period	Changes in financial assets	Changes in liabilities	Changes in financial assets	Changes in liabilities	Changes in financial assets	Changes in liabilities	Changes in financial assets	Changes in liabilities	Changes in financial assets	Changes in liabilities	Changes in financial assets	Changes in liabilities
G. Financial derivatives and employee stock options	0.0	−2.0	2.4	4.6			1.5	0.9	3.9	3.5	−0.3	0.0
H. Other accounts receivable/payable	−40.5	7.5	12.6	−40.8	5.0	12.0	4.0	3.0	−18.9	−18.3	−0.2	−0.8
Subtotal	20.4	26.1	57.5	64.6	1.8	20.2	40.5	−15.8	120.2	95.1	−18.7	6.4
Net financial investment (net acquisition of financial assets less net incurrence of liabilities)		−5.7		−7.1		−18.4		56.3		25.1		−25.1

Source IMF (2016a, 2016b, 2016c), MFSMCG, Table 8.1, 282
Note NPISHs = non-profit institutions serving households; SDRs = Special Drawing Rights

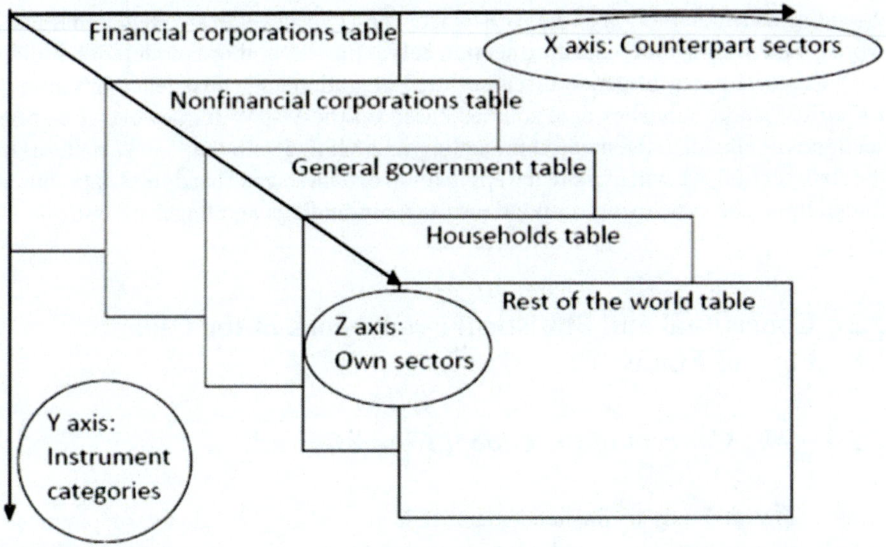

Fig. 1.1 Concept of three-dimensional financial statistics. *Data source* IMF (2016a, 2016b, 2016c) MFSMCG, Fig. 8.2, 288

On the other hand, there is international awareness of the issue that the existing statistical data does not describe the risks inherent in a financial system. Previous research has evolved into a discussion about the basic concept of GFF and a proposal to make a statistical framework for GFF. The recent global crisis showed how easily shocks in one country are transmitted and amplified, and rapid illiquidity in financial markets spreads quickly across national borders. Therefore, IMF's Statistics Department has already organized seven economies with systemically important financial centers to construct a GFF mapping domestic and external capital stocks, geographically broken down, etc.[4] This means that the observation of GFF has not just remained in theoretical research, but has also entered the stage of experiment and statistical application. GFF is the extension of domestic flow of funds. It connects domestic economies with the ROW.[5] GFF data would provide valuable information for analyzing interconnectedness across borders, global liquidity flows, and global financial interdependencies.

However, for GFF statistics creation, integration of data sources and timely collection of data are very important issues. This Chapter referenced "the report of the Financial Crisis and Information Gaps" that was prepared by the FSB and the IMF (2009). The main purpose of the chapter is to measure GFF and apply the result to regular monitoring of the GFF. The composition of this chapter is as follows.

This chapter is structured as follows: The conceptual understanding of economic statistics sets the boundaries for its statistical scope. Consequently, we begin by

[4] Errico et al. (2013).

[5] Zhang (2005).

providing a clear definition of the GFF concept and then develop an integrated framework for measuring GFF. The third section delves into designing a model for building the GFFM. The fourth section discusses the integration and consistency of datasets for GFFM and establishes data sources based on the W-t-W framework. The fifth section will elucidate the method for setting up a GFFM with the W-t-W pattern. In the sixth section, we will explore the application of data science to address data source integration. The conclusion summarizes the main findings and lingering issues.

1.2 Conceptual and Statistical Framework of the Global Flow of Funds

1.2.1 The Concept of the Global Flow of Funds

As the flow of funds in financial markets is related to the balance of payments (BOP), the overseas sector will have an excess fund outflow (net capital outflows) if the current account is in surplus. Conversely, the domestic sector will have net negative inflows. Therefore, when the real economic side of the domestic and overseas economy is analyzed in an open economic system, the balance of savings and investment in the domestic economy will correspond to the current account balance. The transmission mechanism of GFF is shown in Fig. 1.2.

Figure 1.2 illustrates the GFF mechanism among three countries (A, B, and C), an international financial market, and an international organization. The economies of the three countries consist of savings and investment balances, which reflect real

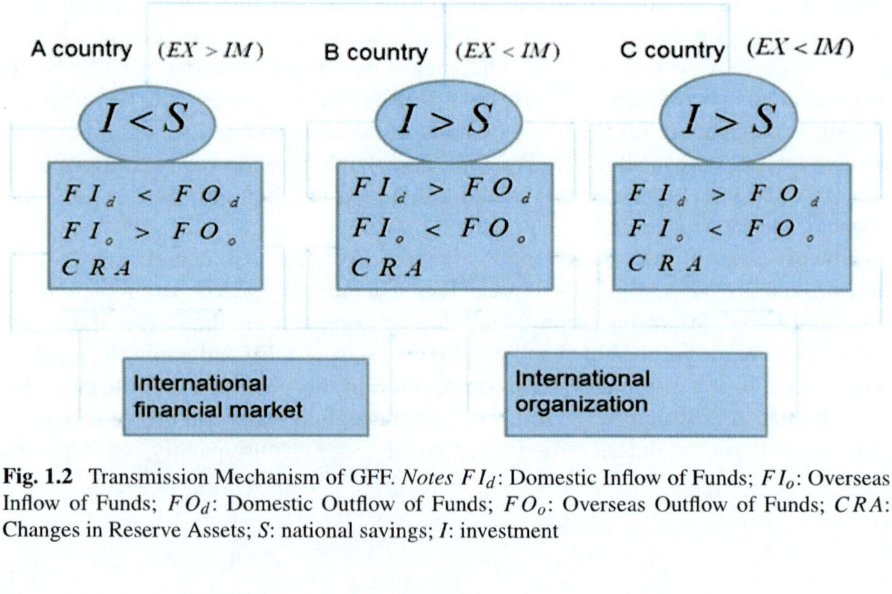

Fig. 1.2 Transmission Mechanism of GFF. *Notes* FI_d: Domestic Inflow of Funds; FI_o: Overseas Inflow of Funds; FO_d: Domestic Outflow of Funds; FO_o: Overseas Outflow of Funds; CRA: Changes in Reserve Assets; S: national savings; I: investment

economic activity, and financial markets, which reflect the financial circulation of funds. Because the spread between each country's domestic and overseas balances (the savings–investments balance) connects the current balance, external fund flows in the financial market are linked to the capital balance. Each country's current and capital transactions are mutually connected, and part of the capital transactions of each country is linked to the international financial market, the IMF, the World Bank, etc., which are parts of GFF.

Figure 1.2 represents the three forms of GFF, which are a capital-exporting country, capital importing country, and key currency country. In capital-exporting countries, such as Country A (e.g., Japan), because savings are greater than investments, the result is a current BOP surplus, which is presented in financial terms as the net increase in financial assets. The broad financial market accepts capital inflows from both home and abroad and simultaneously supplies funds at home and abroad (Zhang & Zhu, 2021).

In the case of a capital importer, such as Country B or C, the current balance deficit is linked to the domestic excess of investment (savings deficit) and the net increase in the financial liability of the financial sector. In the financial market, excess domestic investment leads to an excess of credit, and the current account deficit is financed by the net inflow of funds (capital balance surplus) from overseas. Therefore, regarding the funds account balance of the domestic and overseas sectors in the financial markets of countries B and C, a net outflow of funds occurs in the domestic sector, and a net inflow of funds occurs in the overseas sector. The net inflow of funds from the overseas sector becomes a source of funds for the domestic sector, which attempts to maintain a balance of credit. Moreover, the net outflow of funds from the domestic sector in the financial market causes over-borrowing in the domestic sector, also known as a net increase in financial liability.

In this way, an international capital movement from a country with a surplus current balance to a deficit country arises. The flow of capital moves directly between two nations, from a surplus country to a deficit country or may also arise indirectly in countries through the international financial market, the IMF, the World Bank, etc. These international funds are managed by an agency of a public intergovernmental organization or the World Bank, although most of the funds arise through factors such as the pursuit of interest differential or capital gain and risk aversion through a market mechanism. In any case, from the perspective of the BOP of each country, international capital movement is financed with the balance on the current account, and from a global perspective, it serves as international financial intermediation between a country with excess savings and a country with deficit savings (excess-investment). Moreover, when a capital supplier country is a key currency country, such as the US, the country functions as a supplier of international liquidity. By thoroughly observing the flow of funds, funds mobility (international liquidity and the domestic money supply) can be seen as an integrated system in GFF that connects major power economies because the flow of funds between countries is related to the domestic flow of funds in each of the relevant countries.

Based on the dynamic process of the GFF and the definitional equation of a system of national account, the accounting identity is as follows:

$$B = [EX - IM] + [FI - FO], \quad (1.1)$$

where *B* represents the BOP; *EX* denotes exports; *IM* denotes imports; *FI* denotes fund inflow, and *FO* denotes fund outflow. From the national accounting equation, we derive the following:

$$S - I = EX - IM,$$

where *S* denotes national savings, and *I* denotes investment, which implies that

$$B = [S - I] + [FI - FO], \quad (1.2)$$

where *S* – *I* represents domestic financial markets, and *FI* – *FO* represents international (the ROW[6]) financial markets. However, when we examine the financial relationship between domestic and overseas funds, we find that the domestic net outflow of funds corresponds to the capital account balance. The relationships among the domestic savings–investment balance, financial surplus or deficit, current account, and overseas net funds outflow are expressed in the following structural equations. Using Eq. (1.2), we consider a situation involving two countries, A and B,

$$B_A = [S_A - I_A] + [FI_A - FO_A]$$
$$B_B = [S_B - I_B] + [FI_B - FO_B]$$

Assuming $B_A = B_B = 0$, if for A, $S_A < I_A$, then $FI_A > FO_A$, whereas for B, if $S_B > I_B$, then $FI_B < FO_B$. If $B \neq 0$, then the deficits and surpluses lead to changes in reserve assets, such as currencies, gold, and special drawing rights (SDRs). This is what is illustrated in Fig. 1.1.

If $B \neq 0$, then the deficits and surpluses bring about changes in reserve assets: currencies, gold, and SDRs as shown in Formula (1.3).

$$X - M = (FO - FI) + CRA \quad (1.3)$$

Regard $r_{t-1}FI_{t-1}$ as the interest payments on external debt, and define FRA as the foreign reserve assets. By introducing variable B, we can transform Formula (1.3) into (1.4).

$$(X_t - M_t) - (FI_t - FO_t - r_{t-1}FI_{t-1}) - (FRA_t - FRA_{t-1}) = 0 \quad (1.4)$$

The essence is illustrated as a balance in GFF, that is, the extension of domestic flow of funds, connects the ROW. From the statistical definition of Formulas (1.1)–(1.4), a domestic capital surplus and deficiency in the flow of funds account (FFA) coincide with the current account of the BOP, whereas the overseas flow of funds in the FFA corresponds to the capital account in the BOP. Thus, it is possible to observe

[6] Rest of the world (ROW), which is a sector in Flow of Funds Account.

the systematic process of GFF using FFA and BOP statistics. However, the data about FFA and BOP only provide two-dimensional information, that is, who trades what, but not information about the counterparty, that is, who trades with whom. The 2008 global financial crisis in the US revealed the limitation of this data gap. Therefore, international organizations, such as IMF, proposed the idea of establishing GFF statistics that can provide data about W-t-W).

1.2.2 Statistical Framework

In order to measure financial stress and observe the spread effect of systematic financial crisis through GFF, that needs a new statistical framework which corresponds to the operational structure of GFF. Especially, an integrated framework should be used as the foundation of a statistical monitoring system. When the flow of funds in financial markets is tied up with the balance of payments, the ROW sector will have fund outflow excess (net capital outflows) if the current account is in surplus. Conversely, the domestic sector will have fund inflow excess. Therefore, when the real economic side of the domestic and overseas economy is analyzed under an open economic system, the balance of savings-investment of the domestic economy corresponds to the current account balance. However, domestic net funds outflow corresponds with the capital account balance when we examine the financial relationship between domestic flow of funds and external flow of funds. For this reason, relationships among the domestic savings-investment balance, the financial surplus or deficit, the current account, and the external flow of funds should be expressed in an integrated framework to provide joint routine monitoring of GFF.

A new statistical framework that corresponds to the operational structure of GFF is required to measure financial stress and observe the spread effect of systematic financial crisis through GFF. A statistical monitoring system should be founded based on an integrated framework (PGI 2015). When the flow of funds in financial markets is tied up with the balance of payments, the ROW sector will have excess fund outflows (net capital outflows) if the current account is in surplus. Conversely, the domestic sector will have excess fund inflow. Therefore, when real domestic and overseas economies are analyzed under an open economic system, the saving-investment balance of the domestic economy corresponds to the current account balance. However, an examination of the financial relationship between domestic and external flow of funds shows that domestic net fund outflow corresponds with the capital account balance (Zhang, 2012). Therefore, the relationships between the domestic saving-investment balance, financial surplus or deficit, current account, and external flow of funds should be expressed in an integrated framework to provide joint routine monitoring of GFF (Zhang, 2015).

Table 1.2 shows an integrated framework for GFF based on its definition that should be established to integrate real and financial accounts for measuring GFF. The integrated framework is based on the BSA and uses data stock data (Shrestha et al., 2012). First, for integrating Real and Financial Accounts, we put items in the

Table 1.2 A framework for measuring global flow of funds

Stocks	Flows		Stocks
Opening balance sheet	Transactions	Other changes	Closing balance sheet
Nonfinancial assets	Savings-investment balance		Nonfinancial assets
	Current account balance		
Assets (by financual category)			Assets (by financual category)
Direct investment			Direct investment
Portfolio investment			Portfolio investment
Other investment			Other investment
Reserves assets			Reserves assets
Total assets	Change in financial assets		Total assets
Liabilities (by financial category)			Liabilities (by financial category)
Direct investment			Direct investment
Portfolio investment			Portfolio investment
Other investment			Other investment
Total liabilities	Change in financial liabilities		Total liabilities
Net position	Change in net worth		Net position

non-financial assets, savings-investment balance and current balance categories in a position in the flow diagram to show the structural relationship of real economies and financial economies in GFF. Table 1.2 shows that the financial category includes financial assets, financial liabilities, and net position. External financial positions and flows indicate four aspects to monitor: (1) influence of economic structural changes on current accounts causing saving–investment imbalances, (2) international capital flow risks caused by domestic funds surplus or deficit, (3) international capital flow shocks caused by an imbalance in current accounts and international large-scale capital inflows or outflows and (4) causes of foreign exchange reserve changes and their resulting financial instability pressure. The integrated framework used to construct GFF provides valuable information for analyzing cross-border interconnectedness, global liquidity flows, and financial interdependencies. The framework could also be extended to flow data. For this step, we would break down the data sources by sector and counterpart country (Zhang, 2020).

1.2.3 External Assets and Liabilities Matrix

According to the definition of GFF, and in order to allow for the integration of Real and Financial Accounts for measuring GFF, we must set up an integrated framework for GFF statistic, that is, External Assets and Liabilities Matrix, Financial Instrument Matrix, and Global Flow of Funds Matrix.

In order to measure financial stress and observe the spillover effects of systematic financial crises through GFF, a new statistical framework is needed that corresponds to the operational structure of GFF. It is important that an integrated framework is used as the foundation of a statistical monitoring system. When the flow of funds in financial markets is tied up with the BOP, the ROW has an excess of outflowing funds (net capital outflows) if the current account is in surplus. Conversely, the domestic sector will have an excess of inflowing funds. Therefore, when the real economic side of the domestic and overseas economy is analyzed under an open economic system, the balance of savings–investment corresponds to the current account balance. However, the outflow of domestic net funds corresponds to the capital account balance when we examine the financial relationship between the domestic and external flows of funds. For this reason, relationships among the domestic savings-investment balance, financial surplus or deficit, current account, and external flow of funds should be expressed in an integrated framework to enable comprehensive and regular monitoring of GFF.

The integrated framework is based on the BSA, using stock data. The financial data category includes financial assets, liabilities, and net position, it can be monitored in two aspects of external financial positions and flows. Using the integrated framework to construct GFF statistics would provide valuable information for the analysis of interconnectedness across borders, global liquidity flows, and global financial interdependencies. Furthermore, the framework could also be extended to flow data. For this next step, we then disaggregate the data sources by sector and counterpart country.

As a transitional preparation for producing the GFFM, we need to use External Assets and Liabilities (EAL) matrix. Through Table 1.3, we can connect the relevant information between the ROW sector in flow of funds account with other countries to construct the GFFM. The EAL matrix is also based on the BSA. It depicts for the rest of world sector, the main countries for observation, and all other economies, with each financial instrument/stock of the issuer of liability (the debtor) on the horizontal axis and stocks of the holder of liability (the creditor) on the vertical axis. This table depicts the external flow of funds matrix for the observed countries or regions, where the EAL has been disaggregated into the counterpart country, by the instrument.

The EAL matrix identifies particular sectors, which, like countries, show data for the ROW and how this relates to other economies or regions. Each column corresponds to the balance sheet of the sector in question, with assets and liabilities listed per row by instrument, with counterparty sectors identified for each cell.

Table 1.3 provides a statistical framework for presenting cross-border stocks by counterpart country and sector and instrument. It shows available external-sector financial assets and liabilities' stock data broken down by country. Data in Columns 2–4 of the EAL matrix shows the assets, liabilities, and net assets of county A's external financial, as well as the major financial instruments used by Country A. This is a statistical table of a two-dimensional structure, that is, we can know who did what. The matrix presents external financial asset and liability positions, showing available data by international investment position (IIP) category and instrument:

Table 1.3 External assets and liabilities matrix by balance sheet approach

Country	Country A			Country B			Country …			All other economies			Total of the world		
Financial instrument	A	L	NP	A	L	NP	A	L	NP	A	L	NP	A	L	NP
Direct investment															
Portfolio investment															
Equity securities															
Debt securities															
Long-time debt securities															
Short-time debt securities															
Financial derivatives															
Other investment															
Other equity															
Debt instruments															
Reserve assets															
Total of the world															

Notes A: assets; L: liabilities; NP: Net position
All other economies = Total sum of the World—Total sum of the observed countries

direct investment, portfolio investment equity, and debt securities (the latter displayed separately for long- and short-term debt), other investment (separately for banks and others, using the BIS IBS), and reserve assets. Table 1.1 shows what may be possible in a GFF framework for a country that permits the monitoring of both regional or national and cross-border (by country and sector) financial positions. However, we haven't been known the W-t-W by what instruments, which is as a statistical matrix of the three-dimensional structure.

Table 1.4 Financial Instrument Matrix on a W-to-W Basis

Counterpart countries (investment from)	Counterpart countries (investment in)					
	Country A	Country B	…	All other economies	Total of the world	
Country A						
Country B						
…						
All other economies						
Total of the world						

Note Table 1.4 was made with reference to Table 1.2 of Shresthas' paper[7]

1.2.4 Financial Instrument Matrix

Although Table 1.3 is modeled after a traditional account format, it cannot show the inter-sectoral W-to-W relationships needed to measure financial positions and flows. Therefore, in order to know "who is financing whom, in what amount, and with which type of financial instrument," we constructed the GFFM on a W-to-W basis. Table 1.4 reflects this approach and shows the financial instrument categories.

Table 1.4 is based on a specific analysis, namely the matrix of a financial instrument designed in accordance with the W-to-W form. According to the specific analytical purpose, the statistical scope can cover only certain relevant countries or regions as the observation object. The columns show a country's funds used by other countries (assets), and the rows show if a country should raise funds from other countries (liabilities). Table 1.4 accurately reflects the relationship between empirical data and the underlying structure. By setting up a sector as the other economies, the relationship of a financial instrument and the GFF is as follows: other economies = the total for all countries in the world—the total for all countries being analyzed. We can use Table 1.4 to speculate the corresponding input coefficient, observe the impact of changes in the financial instruments on the financial markets, and determine the extent of the impact on other related countries.

According to analytical need, a GFFM resulting from the from-whom-to-whom table can be created to illustrate the country vis-à-vis the country through each financial instrument. These instruments show the connections between financial positions, such as direct investment and portfolio investment. Likewise, every financial instrument can be disaggregated within the matrix on a from-whom-to-whom basis. Instruments located in the rows of the table describe a country relative to the counterpart country's assets, while instruments located in the columns describe a country relative to the counterpart country's liabilities. If all the financial instruments are totaled, that amount will equal the sum total of external financial assets and liabilities in the given country. In this way, EAL will have been disaggregated into the counterpart country, as well as by main instruments, based on the IIP.

7 Shrestha et al. (2012).

1.3 A Model for Building Global Assets and Liabilities Matrix

Table 1.5 is a version update for the GFF theoretical framework (Zhang & Zhao, 2019) consistent with IIP statistical standards and based on a structure in which W-t-W data are used to establish the GFF statistical framework according to the double-entry principle. According to the Balance of Payments and International Investment Position Manual, Sixth Edition (BPM6), IIP can be set as foreign financial assets and external debt. Each "column" represents the assets of the respective country, while each "row" corresponds to the liabilities associated with individual financial instruments, with the counterparty country for each cell being identified.[8]

Table 1.5 provides a statistical framework for deriving the GFFM. Assets are subdivided into five parts: direct investment (DI), portfolio investment (PI), financial derivatives, other investments (OI), and reserve assets. Liabilities are divided into four parts: DI, PI, financial derivatives, and OI. The net financial position is external financial assets plus reserve assets minus liabilities; this measure is consistent with the statistical framework of IIP. Using this statistical framework, the GFF statistics can reflect stock information of financial assets and liabilities of the transactions between a region and other countries at a particular time. Importantly, the GFF statistics are consistent with IIP statistics standards and exhibit unique methodological characteristics that can be summarized as follows.

(1) In order to reflect the relationship between W-to-W, GFF statistics use the parallel processing method wherein transactions and countries (sectors) are rows, namely, by putting the transaction items that direct investments, securities investments, financial derivatives, and other investments to countries (sectors) in the rows, whereas each country (sector) is in the columns. Accordingly, we can determine the dual relationship of a transaction item in countries (sectors), which can show the scale of the position item and reflect from-whom-to-whom-by-what relationships in a two-way format. For example, a5–a8 (see column a and row 5–8, direct investment can be represented as a5, portfolio investment as a6, financial derivatives as a7 and other investment as a8) in the table shows Country A transactions in the columns by showing which financial instruments are used for transactions bringing how much funds to country B. As this can provide two-way information about the financing structure of Country A with Country B, we also can identify and understand the financing scale and corresponding information on counterparties. At the same time, we can also capture information about where country A is located in the row vectors from other countries to raise funds. We can also acquire relevant information on country B in the row vectors on its fund-raising from Country A, Country C, etc.
(2) To reflect the actual situation of international capital in a country or a region, and in order to establish the GFFM table for the application analysis, we set

[8] Depending on the purpose of the analysis, we can also set the column as a liability and the row as an asset. See Chap. 3 for a detailed explanation.

Table 1.5 A statistical template for global assets and liabilities matrix

		a	b	c	d	e	f	g	h	i	
Creditor by country		Country A	Country B	Country C	…	All other economies	Total liabilities of financial instruments	Total liability	Difference (A > L)	Total	
Debtor by country and financial instrument											
Country A	Direct investment										1
	Portfolio investment										2
	Financial derivatives										3
	Other investment										4
Country B	Direct investment										5
	Portfolio investment										6
	Financial derivatives										7
	Other investment										8
Country C	Direct investment										9
	Portfolio investment										10

(continued)

Table 1.5 (continued)

	Financial derivatives										11
	Other investment										12
……	……										13
All other economies	Direct investment										14
	Portfolio investment										15
	Financial derivatives										16
	Other investment										17
Total asset of financial instruments	Direct investment										18
	Portfolio investment										19
	Financial derivatives										20
	Other investment										21
Total asset											22
Difference (L > A)											23
Total											24
Financial net worth											25

(continued)

Table 1.5 (continued)

Reserve assets										26
Monetary gold										27
Special drawing rights										28
Reserve position in the fund										29
Other reserve assets										30
Adjustment item										31
Net financial position										32

Notes (i) Net worth is the difference between assets and liabilities (2008SNA, P29)

(ii) Adjustment item is an item for balancing the net worth, reserve assets, and net financial position in Global Flow of Funds Matrix (GFFM), and put it in row 27. It is derived from the net worth of each county by:

a. Adjustment item = Net Financial Position—Financial Net Worth—Reserve assets, and
b. Net Financial Position = Financial Net Worth + Reserve assets + Adjustment item

countries (sectors) in rows and columns by the principle of W-to-W tabulating. We also designed an "all other economies" sector (see column e and rows 9–12 that can be represented as e9, e10, e11, e12). The relationship of these "all other economies" and the world total can be expressed as follows: "liabilities of all other economies" = total liabilities—liabilities of the total for specific countries. That is, e9 = f9 − (a9 + b9 + c9 + d9), ..., e12 = f12 − (a12 + b12 + c12 + d12).

(3) Each "column" shows a country how to use funds by transaction item, namely, who outputs how much funds by what item; each "row" represents how a country raises funds through four financial instruments, namely, who inputs how much funds by what item. The difference between the total of the row and column in row 23, shows the balance between the use of external funds financing for a certain country at a particular point in time, that is, the net output of funds. For instance, Country A's net worth equals country A's total assets minus its total liabilities, that is, a23 = a22 − (g1 + g2 + g3 + g4). Were, the sum of all countries' assets equals the sum of all countries' liabilities.

(4) To maintain symmetry in the W-t-W matrix, the difference term is reset—"difference (L > A)" is set in row 23, and "difference (A > L)" is set in column h. If L > A, put the number of the net liability in row 23 of the country; otherwise, use 0; and if A > L, put the number of the net asset in column h of the country; otherwise, use 0. This way, the total liabilities of a country in the row plus the difference is equal to the country's total assets in the column plus the difference. Thus, "Total" set in a column and "Total" put in a row will balance.[9]

(5) Corresponding to the various transaction instruments of various countries rows 26–30 show part of the reserve assets, specifically monetary gold, special drawing rights, reserve positions in the fund, and other reserve assets. Denoting reserve assets as an instrument in Table 1.5 shows a balanced relationship between net worth and net financial position and the components thereof. For example, country A's component of reserve assets can be shown as a26 = a27 + a28 + a29 + a30. In order to correspond to the reserve assets of each country, the financial net worth item is set in row 25, with net assets represented as positive and net liabilities as negative.

(6) The bottom row in Table 1.5, namely row 32, reflects net IIP, corresponding to Table 1.5's Net Financial Position obtained for each country. These data are taken from IIP and reflect the overall equilibrium conditions of national external financial positions. Theoretically, adding Reserve assets to the Financial Net Worth of a country should reveal the external net financial position of

[9] This is a theoretical setting of the statistical framework, but there are biases in practice. Because total global assets will not equal to total global liabilities even if we had perfect data sources, with the difference generated by the fact that monetary gold does not have counterpart liability. Another source of inconsistency are the countries' assets and liabilities vis-à-vis international organizations, as these are not residents of any country.

the country.[10] For example, a32 = a25 + a26, and b32 = b25 + b26…, etc. However, since there are factors, like the non-compatibility of IIP data and other datasets and the difficulty in selecting the financial investment item, the actual external net financial investment figures are inconsistent with the above theoretical relationship. Therefore, in order to attain balance when adding the net worth in row 25 to the reserve assets in row 26 so they are equal to the financial position in row 32 of Table 1.5, we need to set up an adjustment item for balancing the net worth, the reserve assets and net financial position in GFFM, and put is in row 31. The net financial position of each country is calculated using net worth, i.e., financial net worth plus reserve assets and adjustment item is equal to net financial position, such as a32 = a25 + a26 + a31, b32 = b25 + b26 + b31, …, e32 = e25 + e26 + e31.

(7) Because the main purpose of compiling the GFFM table is to observe cross-border capital positions, the diagonal line elements in the matrix are zero. Each position is the result of financial investment between the domestic and foreign countries and does not include a country's internal financial investments.

(8) In the bold blue line box at the top half of Table 1.5, if the financial instruments of each country in the rows are merged, we derive a square matrix with the same number of rows as columns. Therefore, we can use this square matrix to make statistical inferences regarding real-life cases.

(9) In addition, deriving the GFFM based on the W-t-W can improve the quality and consistency of data, thereby providing more opportunities for cross-checking and balancing information. The GFFM, built using stock data, can be extended to flow data to quantify the bilateral flow of funds. Using Table 1.5, we find that the statistical information can answer the following synthesis problems: "What is the main trading partner on bilateral financing?" "What financial instruments are used?", and "What is the structure and scale of bilateral financing?" We can get the three-dimensional financial statistics, that is, global who-to whom matrices with cross-border interlinks.

1.4 Integration and Consistency of Datasets

The GFF data should be based on existing statistical data and thus share many similarities in approach with the existing statistical data. GFF data sources include not only ROW sectoral data, from national accounts, but also monetary and financial statistics (IMF, 2016a, 2016b, 2016c), IIP statistics, BIS locational banking statistics, and OECD financial accounts (FA). The prototype template for the main data is shown in Fig. 1.3, with two data sources for measuring GFF: (1) data sources for establishing the External assets and liabilities matrix (EALM); (2) data sources for operationalizing the Sectoral from-whom-to-whom stock matrix (SFSM). The two

[10] When discussing reserve assets, it should be clarified that these are included also on the liabilities side in the IIP data within the relevant functional categories of the relevant countries (except monetary gold).

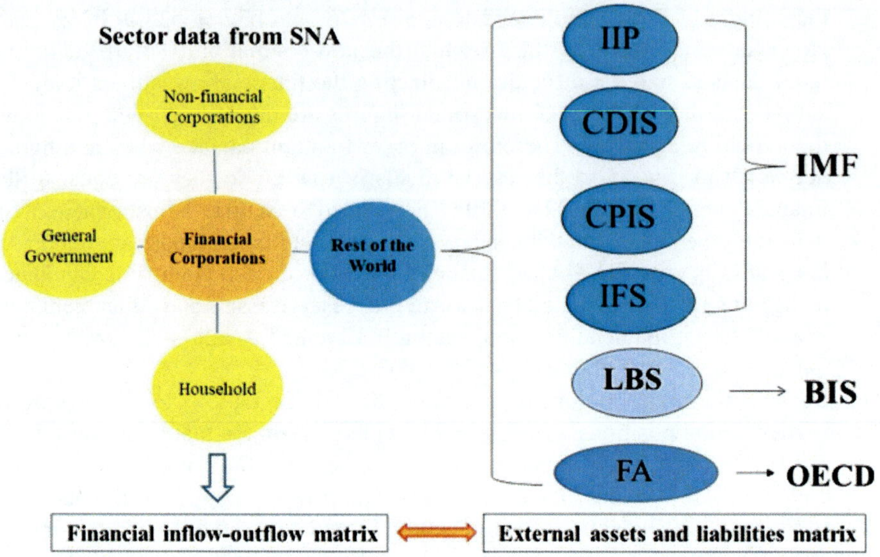

Fig. 1.3 Prototype template for measuring GFF

matrices can be linked to reflect counterparty country sectoral debt relationships between countries and cross-border sectors and extended to flow data.

1.4.1 Datasets for Measuring GFF

The EALM is based on the BSA, with Rest of world (ROW) data drawn from national accounts and IIP. The EALM presents data on whatever external-sector financial stock data are available by IIP category, drawing on IMF and BIS data sources. The IIP is the link between domestic and external matrices. For a detailed description of the EALM and SFSM data sources, see our previous paper (Zhang & Zhao, 2019, and Zhang, 2022), here we focus on EALM data sources and integrate them with the sectoral data to establish the SFSM. And we will discuss the data sources for measuring SFSM in Chap. 5.

1.4.2 Data Sources for Measuring GFF

Data from IMF's Monetary and Financial Statistics, IIP, and National Accounts are used to derive the BSA matrix. The BSA matrix can provide information about a country's or region's financial corporations' stock positions for residents and nonresidents. In the EALM, the datasets with bilateral counterpart country details are collected by the IMF and BIS as follows:

(1) Coordinated Direct Investment Survey (CDIS): The CDIS (IMF, 2015) provides bilateral counterpart country details on inward direct investment positions (i.e., direct investment into the reporting economy) cross-classified by the economy of immediate investors. It also provides data on outward direct investment positions (i.e., direct investment abroad by the reporting economy), cross-classified by the economy of immediate investment, as well as mirror data[11] for all economies (see Errico et al., 2013). In the CDIS data, outward investments are netted out in the sense that there exist negative values, which render the calculation of country shares in foreign investments impossible, and the CDIS is only available after 2009.
(2) Coordinated Portfolio Investment Survey (CPIS): The CPIS data show countries' cross-border portfolio investments broken down by counterparty country and instrument type (debt securities and equities). CPIS provides bilateral counterpart country details covering holdings of asset stock positions by reporting economies and derived (mirror) liabilities information for all economies. The CPIS's purpose is to improve statistics on holdings of portfolio investment assets in the form of equity, long-term debt, and short-term debt. It is also used to collect comprehensive information, including geographical details on the issuer's country of residence, the stock of cross-border equities, long-term bonds and notes, and short-term debt instruments, for use in the compilation or improvement of IIP statistics on portfolio investment capital.
(3) Other investment: Another investment is a residual category that includes positions and transactions other than those included in direct investment, portfolio investment, financial derivatives, employee stock options, and reserve assets.[12] Other investments include (a) other equity; (b) currency and deposits; (c) loans (including use of IMF credit and IMF loans); (d) nonlife insurance technical reserves, life insurance and annuity entitlements, pension entitlements, and provisions for calls under standardized guarantees; (e) trade credit and advances; (f) other accounts receivable/payable; and (g) Special Drawing Rights (SDR) allocations (SDR holdings are included in reserve assets). In order to reflect the bilateral counterpart country for loans, deposits, and other assets and liabilities, this paper uses the related dataset with BIS International Banking Statistics (IBS) instead of IIP statistics.
(4) The BIS statistics provide information on banks' total foreign claims not broken down by instruments but broken down by counterparty country and recently also by counterparty sector (banks, private nonbank, public). The BIS compiles and publishes two sets of statistics on international banking activity, namely the Locational Banking Statistics (LBS) and Consolidated Banking Statistics

[11] The term "mirror data" refers to the same data seen from different perspectives. For instance, banks' loans to households could be called mirror data of household debt to banks.

[12] IMF, *Balance of Payments Manual,* 6th edition (BPM6), 111.

(CBS). This paper uses data on cross-border claims and liabilities from LBS[13] as our main source, because these statistics provide information about the currency composition of banks' balance sheets and the geographical breakdown of their counterparties. The LBS data capture outstanding claims and liabilities of internationally active banks located in reporting countries against counterparties residing in more than 250 countries and regions.[14] Banks record their positions on an unconsolidated basis, including intragroup positions between offices of the same banking group. The data is compiled following the residency principle that is consistent with the BOP statistics, and compatible with IIP, CDIS, and CPIS. In this regard, the major advantage of the BIS' LBS data, compared to the banking flows collected from the BOP statistics, is the detailed breakdown of the reported series by counterparty countries. However, for the geographical breakdown of loan exposures the BIS banking statistics are used as a proxy and it is assumed that the whole economy loan foreign claims follow the same geographical breakdown as banks' total foreign claims. This feature enables us to identify changes in the supply factors of banking flows from changes in demand for bank credit in counterparty countries.

(5) For data on reserve assets, we use the IIP as the basic data source and can reference the Currency Composition of Official Foreign Exchange Reserves (COFER). The IIP dataset complements the CPIS and BIS datasets by providing sectoral information on who is holding foreign assets and who is issuing liabilities held by nonresidents. To supplement data on reserve assets, International Financial Statistics (IFS), which includes World Total Reserves, World Gold, World Reserve Position in the Fund, World SDR Holdings, and World Foreign Exchange, can also be used. IIP data have been used to supplement the data for constructing the EAL matrix. The IIP is a subset of the national balance sheet, the net IIP plus the value of nonfinancial assets equaling the net worth of the economy, which is the balancing item of the national balance sheet. The IIP relates to a point in time, usually at the beginning of the period (opening value) or the end of the period (closing value).

(6) The data sources of EALM are statistical tables that reflect external assets and liabilities between countries, but it can be extended to sectoral tables between sector by sector between countries. That is, make EALM link a sectoral table reflecting financial flows in the domestic sector for getting more detailed information on W-to-W basis.

GFF Statistics can construct a statistical framework if concepts, definitions, and classifications underlying these statistics are standardized across economies. Fortunately, we can get these standards from *2008SNA*, the IMF (2016a, 2016b, 2016c) *Monetary and Financial Statistics Manual and Compilation Guide*, *Balance*

---

[13] The BIS locational banking statistics (LBS) are reported by banking offices located in selected countries, including many offshore financial centers, and exclude the assets and liabilities of banking offices outside of these countries. The number of LBS-reporting countries increased from 14 in 1977 to 47 in 2017.

[14] BIS, https://stats.bis.org/statx/srs/table/a6.2 on 11/4/2023 11:04: AM.

of Payments Manual (BPM6), and the BIS's *Guidelines for Reporting the BIS International Banking Statistics.*

1.5 Creating the GFFM

1.5.1 A Matrix Model for Measuring a Financial Instrument

According to the framework shown in Table 1.4, in order to meet the special tracking analysis of a financial investment, first, we created a matrix for measuring a financial instrument, namely the matrix of foreign direct investment, portfolio investment, and international banking credit, we set up Tables 1.6, 1.7, and 1.8 on a W-t-W basis to describe the three matrixes. This kind of matrix can not only show bilateral stocks by financial instrument, but also illustrate the exact situation of each instrument according to its use and source among main countries.

Table 1.6 uses the data of geographic breakdown of Outward Direct Investment Positions published with the IMF (CDIS, Table 1.6), which includes G20 countries and "Other Economies". It is a matrix based on a W-to-W benchmark: the columns show assets, and the rows represent liabilities. The matrix is a square matrix, with the same number of rows as columns, which is an orthogonal matrix. We can use the matrix to make various statistical estimates for meeting the needs.

Table 1.6 has the following four characteristics. First, by using the form W-to-W, we can observe and analyze the bilateral relations of relevant countries in direct investments; the elements on the diagonal are zero, which means that the matrix does not include domestic financial investment. Second, we can understand the structure of the global direct investments market, and the proportion and influence of relevant countries in the direct investments market. Third, using the direct investments assets located in a column and subtracting the liabilities in each row, we can see the net assets and the relevant information of the counterparty. Fourth, Table 1.6 shows the balance position on assets and liabilities for each country and the global market in direct investments.

In the bottom half of Table 1.6, a differential item is introduced. When the differential item represents a situation where liabilities are less than assets (L < A), the net assets figure is positive, and a zero is displayed in the corresponding row, indicating the net assets of the respective country. Conversely, when the differential item represents a situation where liabilities are greater than assets (L > A), the net assets figure is negative, and a zero is displayed in the corresponding column, indicating the net liabilities of the respective country. Following this procedure, we can observe the balance. In general, the total of each row does not match the total of each column. However, by incorporating the difference item, the sum of the rows in the matrix equals the sum of the columns.

Using the same methodology, we also employed CPIS data and LBS data to create the International Portfolio Investment Matrix and the Cross-border Bank Credit

Table 1.6 International direct investment matrix for G20 (as of end-2020, US dollars, millions)

	AR	AU	BR	CA	CN	FR	DE	IN	ID	IT	JP	KR	LU	MX
AR		515	4614	2679	1111	2206	3412	26	4	900	813	380	1742	2123
AU			873	35554	34158	9307	16736	301	27	696	101508	6172	6973	
BR	−43	1748		12993	1634	31125	10622	110	234	8179	18179	5374	55083	11299
CA	−147	13759	11879		18892	13748	14175	−65	77	2628	25527	2392	44684	1780
CN	10	10165	354	12122		28147	86684	639	715	9191	193338	82693	7594	118
FR	2	1059	189	8737	4004		105610	301	19432	30127	13339	1569	176728	
DE		3820	−378	7613	5727	61132		293	13	34954	33134	7948	220284	221
IN	0	994	20	2737	1407	15609	16643		46	2796	45890	12063	3183	189
ID	0	3811	6	4230	8004	1665	2134	154		103	25594	6627	1035	0
IT	812	726	289	1011	1254	87515	43309	45	3		3931	1213	83031	−23
JP		2742	25	1463	2515	30774	1683	303	154	1211		8963	7164	3
KR	5	1410	−18	4218	8609	7377	10575	479	35	866	57792		1684	108
LU	−164	5378	4686	135711	11393	17014	149583	−895	−138	25852	8868	949		−5302
MX	454	616	2763	35664	873	6128	19662	495	1	2361	10531	3827	11029	
NL	69	4134	−44793	76549	20462	192100	332293	6117	1363	21820	116616	8479	435166	42427
RU	3	55	0	105	2188	19462	18115	611	0	4811	2403	3362	4823	10
SA				33	2931	3594	1376	126		5466	4523	−123	295	
SG	16	16126	26	45653	55064	15119	14652	25285	23768	1677	104863	20226	82055	59
ZA	24	2659	36	804	6732	1664	7128	226	0	631	3893	210	713	87
ES	831	20	5186	3167	3492	85982	91250	195	−10	45546	8365	1363	123887	20993
CH	33	418	−10346	−124	6760	54570	15914	3552	35	3185	35141	644	321451	394
TR		35	25	115	1069	3586	9360	58	0	2981	1267	321	5291	

(continued)

Table 1.6 (continued)

	AR	AU	BR	CA	CN	FR	DE	IN	ID	IT	JP	KR	LU	MX
GB	5	15323	984	35152	4207	108297	65052	12134	16	9695	99481	2004	151078	17163
US	627	97963	6857	490769	37995	285078	411308	1874	461	31631	647718	63668	316778	20850
Others	4460	70560	21956	168533	606674	347277	373478	32322	−18598	103681	212531	42263	732142	−6206
Total assets	6995	254037	5235	1085488	847154	1428474	1820753	84687	27639	350987	1775246	282588	2793892	106292
Net liabilities	77324	536618	558659	0	2366961	0	0	488172	212869	118289	0	0	849766	439319
Total	84319	790655	563894	1085488	3214115	1428474	1820753	572859	240507	469277	1775246	282588	3643658	545612

	NL	RU	SA	SG	ZA	ES	CH	TR	GB	US	Others	Total liabilities	Net assets	Total
AR	9997	16		370	34	11453	4114		2134	20015	15661	84319	0	84319
AU	40660	50		30745	2339	1375	8948	−17	95093	151171	247988	790655	0	790655
BR	112472	3	13	11603	188	52753	11698	183	24065	105314	89065	563894	0	563894
CA	115046	1382		1035	896	10246	39298	2	54153	355306	86498	813192	272297	1085488
CN	40920	266	2090	148721	515	4754	18579	275	25480	86907	2453839	3214115	0	3214115
FR	115715	818	532	1960	751	31838	127866	231	123648	72184	127161	963801	464673	1428474
DE	206592	4430	1978	3227	3346	20245	91434	1946	78964	115320	227657	1129900	690852	1820753
IN	53512	261	166	77535	107	2047	32291	172	85407	99752	120032	572859	0	572859
ID	31554	0	52	57994	1	241	2217	187	8146	31859	54892	240507	0	240507
IT	97810	677	87	−116	−50	14595	28612	465	44684	13774	45622	469277	0	469277
JP	20886	65	28	35629	1	712	−22	4	14466	62748	40798	232313	1542932	1775246
KR	17753	31	3084	18652	−9	507	5159	18	17300	35053	42686	233376	49211	282588

(continued)

Table 1.6 (continued)

	NL	RU	SA	SG	ZA	ES	CH	TR	GB	US	Others	Total liabilities	Net assets	Total
LU	316884	−13313	6676	48002	−551	1589	66951	−1493	521976	992638	1351363	3643658	0	3643658
MX	112657	15	4	922	5	88430	15025	23	25423	184911	23793	545612	0	545612
NL		33526	36747	108525	127775	8202	209153	15745	433383	1190180	1136407	4512447	0	4512447
RU	38484		108	3655	27	563	13029	2001	32275	3896	299060	449047	0	449047
SA	10553	0			0	1209	1317	31	6574	11386	4590	53881	19012	72892
SG	44685	11203	1808		8	617	54758	−110	73647	367455	666784	1625443	0	1625443
ZA	25759	4	202	183		825	2184	0	41329	8846	28987	133127	49566	182692
ES	127584	4126	−87	3860	12		36429	28	112270	38765	152460	865714	0	865714
CH	377514	12033			5873	6985		253	70413	174798	345369	1424863	109215	1534078
TR	23590	3685	649	863	8	4950	3122		13083	2416	48449	124923	0	124923
GB	240175	15460	458	10596	5961	57391	101921	−9		642767	624599	2219910	695396	2915305
US	483991	4326	6262	27257	3474	88423	300276	2578	486879		809409	4626452	897879	5524331
Others	877686	160494	12036	222987	31981	122382	359719	20007	524513	756872		9176491	3263420	12439911
Total assets	3542479	239558	72892	814207	182692	532333	1534078	42522	2915305	5524331	12439911	38709775		
Net liabilities	969968	209490	0	811236	0	333381	0	82401	0	0	0			
Total	4512447	449047	72892	1625443	182692	865714	1534078	124923	2915305	5524331	12439911			

Source CDIS and Data extracted from IMF Data Warehouse on 11/3/2023 14:24:15 PM

Table 1.7 International portfolio investment matrix for G20 (as of end-2020, US dollars, millions)

	AR	AU	BR	CA	CN	FR	DE	IN	ID	IT	JP	KR	LU	MX
AR		428	88	1379	4	403	653	0	1	1977	474	92	6319	18
AU	0		29	30180	9939	36480	55829	4	131	6828	189649	17464	51284	1
BR	411	4131		14165	1543	7004	5673	1	0	1333	9971	9140	37645	2426
CA	72	36633	110		5536	30582	80153	53	99	5038	91850	12708	86535	81
CN	30	20336	55	44105		14575	11759	618	306	2578	34749	23940	122714	144
FR	1	28428	273	49206	9341		539709	158	2	199100	300213	28568	550413	109
DE	92	32891	166	39424	14398	254294		757	35	92262	136322	11107	422864	130
IN	0	8390	2	20765	1369	14018	6518		457	370	20752	4118	60971	18
ID	0	2103	0	6027	1072	3175	9894	92		2154	11966	1835	29582	26
IT	0	5513	80	10194	2332	241352	148437	0	15		87588	3147	200032	4
JP	1	55929	17	86329	16510	111966	50758	50	107	18363		29081	176858	50
KR	0	17098	682	27495	5472	14956	15149	6	59	1965	26327		60258	20
LU	96	11294	4088	13099	13450	515248	781542	1217		841195	134690	38825		880
MX	10	3224	2024	10064	809	10040	19495	0	0	7096	20502	1765	44356	
NL	4	26955	238	38700	3859	316444	340963	97	2	87762	123986	10682	246782	128
RU	10	1595	0	3856	583	1401	7485	0	0	1238	3669	1048	23071	11
SA	0	561		289	227	1050	4177	4	5	1358	3069	2105	6085	1
SG	0	11192	20	7391	6547	2843	8857	131	16597	1005	22900	6068	22963	7
ZA	2	2757	2	6605	517	2318	5052	0		1360	5377	1088	22979	
ES	11	6192	575	10075	1277	213587	181807	1	5	135501	73482	4547	124417	346
CH	3		1491	41161	4348	39169	79721	5	0	11939	37090	8349	126471	72
TR	0	1398	5	2289	264	1817	4364	0	0	1908	4424	317	13717	7

(continued)

Table 1.7 (continued)

	AR	AU	BR	CA	CN	FR	DE	IN	ID	IT	JP	KR	LU	MX
GB	6	91811	768	96035	21535	256348	212449	1035	177	54027	185243	38654	436775	128
US	33089	436311	21685	1473898	178439	427335	584011	5448	949	164475	2072497	345031	1618024	20202
Others	43	220798	10368	220903	600483	793872	1210569	1498	4000	390750	1476896	105955	1378062	52939
Total	33882	1025968	42767	2253633	899852	3310276	4365025	11174	22946	2031581	5073686	705631	5869178	77749
Net Liabilities	8115	272735	366665	0	997554	960483	0	668499	222957	0	0	147519	0	316719
Total Assets	41997	1298703	409432	2253633	1897405	4270758	4365025	679672	245902	2031581	5073686	853150	5869178	394468

	NL	RU	SA	SG	ZA	ES	CH	TR	GB	US	Other	Total liabilities	Net assets	Total
AR	1128	68	8	534	30	299	342	0	1438	18285	8029	41997	0	41997
AU	34494	126	1011	35725	621	2889	26441	3	64552	381048	353973	1298703	0	1298703
BR	13439	130	519		715	3000	4868	36	39516	170654	83113	409432	0	409432
CA	34393	297	282	20466	356	3711	44427	14	63399	1195724	462621	2175140	78493	2253633
CN	22399	60	14652	198743	1863	287	12404	25	89566	290122	991377	1897405	0	1897405
FR	240118	998	15554		1463	83486	89459	24	191788	701012	1241336	4270758	0	4270758
DE	265049	2707	9467	37773	843	38571	97499	64	169429	526281	1314347	3466770	898255	4365025
IN	13809	185	539	67574	613	221	5166	10	32643	237809	183356	679672	0	679672
ID	11248	237	1786	25933	57	345	2917	0	12226	68655	54573	245902	0	245902
IT	23587	61	1418		1615	157623	13189	6	53428	147658	451708	1548988	482593	2031581
JP	65737	103	17334		890	11308	51026	0	167009	1281840	1111255	3252519	1821167	5073686
KR	21302	64	6827	79769	197	973	17646	2	43436	301505	211941	853150	0	853150

(continued)

Table 1.7 (continued)

	NL	RU	SA	SG	ZA	ES	CH	TR	GB	US	Other	Total liabilities	Net assets	Total
LU	141126	16537	8064	30489	14735	246623	294113	64	208635	191693	1231799	4739501	1129677	5869178
MX	13033	1448	0		77	4697	6319	2	19853	159283	70369	394468	0	394468
NL		7624	0	15083	7904	54190	69069	22	107866	625067	625646	2709073	0	2709073
RU	6371		478		106	141	3548	27	14635	72448	57865	199586	0	199586
SA	2025	74			20	120	1867	2	4541		28576	56156	273749	329904
SG	9503	602	1294		538	282	5307	0	19952	100079	131255	375333	1205632	1580965
ZA	11529	99	380			106	2564	2	16266	76685	44801	200490	0	200490
ES	54034	164	4531		101		15378	5	42326	163581	358102	1390044	0	1390044
CH	34365	706	6521	17451	10153	8873		6	70969	632833	237817	1369514	317282	1686796
TR	5108	939	7648		41	249	1735		7162	23864	20550	97805	0	97805
GB	128979	15399	23323	48709	60129	36197	96681	125		1422401	1747549	4974481	0	4974481
US	692130	14040	114212	425437	27296	78687	407153	1039	1244770		9070690	19456847	0	19456847
Others	625773	41743	94058	577280	53141	258698	417679	423	1240781	5569209		20419672	4746728	25166400
Total	2470679	104414	329904	1580965	183503	991577	1686796	1901	3926186	14357736	25166400	76523406		
Net Liabilities	238394	95171	0	0	16987	398467	0	95905	1048296	5099111	0			
Total Assets	2709073	199586	329904	1580965	200490	1390044	1686796	97805	4974481	19456847	25166400			

Data Source CPIS and Data extracted from IMF Data Warehouse on 11/2/2023 5:42:16 PM
Notes "0" Indicates a value less than US$ 500,000, "c" Indicates data are confidential, (-) Indicates that a figure is zero

Table 1.8 Cross-border bank credit matrix for G20 (as of end-2020, US dollars, millions)

	AR	AU	BR	CA	CN	FR	DE	IN	ID	IT	JP	KR	LU	MX
AR		2				667				32		26	60	152
AU	33		93	2614	21690	14950	9333	1135	1567	311	11344	866	1992	34
BR	4	11		704	851	8770	1245	11		213	1296	456	3865	4
CA	41	6511	495		17218	9451	2206	557	98	185	5843	280	11441	11955
CN		27167	5	12858		19658				1523		30804	15565	1
FR	161	22414	1117	3942	32555		228119	1922	500	100140	38613	479	239706	4351
DE		4083	632	1430		120231				39275		4888	92683	213
IN		2479				3130				243		4133	896	0
ID		715				3015				78		5934	60	
IT	158	866	185	157	3069	103473	52006	32	41		2550	14	32110	129
JP		25704	244			265979				454		6290	3319	42
KR	14	199	24	559	20651	5208	3716	9	516	27	7594		236	16
LU	47	227	617	963	13760	91653	79230	42	46	22035	2970	501		270
MX	0	2	9	275	132	174	1977			4	493	179	13	
NL	93	2152	309	4284		66517	99535	31		9695	5383	732	15769	45
RU		5		4		6088				5325		739	4676	9
SA		64		37		6287				172		3482	1322	
SG		32772	1	6671		26108				310		4882	1039	56
ZA	1	59	6	77	2096	725	566	116	1	26	217	10	112	1
ES	1022	408	1174	332	1950	70824	45842	26	8	10881	316	55	9803	2621
CH	3689	1095	1329	3290	3393	69418	52638	559	271	21945	2489	76	39315	5992
TR		6				8898				4625		2244	643	

(continued)

Table 1.8 (continued)

	AR	AU	BR	CA	CN	FR	DE	IN	ID	IT	JP	KR	LU	MX
GB	174	99666	10752	154904	94904	334310	378878	20194	1721	47242	117594	10119	117214	8153
US	13487	38744	21948	212389	141598	153984	118710	24299	7539	20243	242919	39957	95332	73829
Others	2064	25651	−5756	403839	649475	778245	533341	12981	−486	22445	370411	64830	−151319	−90010
Total assets	20987	290999	33184	809327	1003341	2167764	1607343	61914	11823	307430	810032	181977	535854	17864
Net liabilities	0	0	89601	15036	0	335812	0	78830	15853	64239	667027	0	0	0
Total	20987	290999	122785	824363	1003341	2503576	1607343	140744	27675	371670	1477059	181977	535854	17864

	NL	RU	SA	SG	ZA	ES	CH	TR	GB	US	Others	Total liabilities	Net assets	Total
AR	1842					914	486		462	5395	5532	15570	5417	20987
AU	9880	81	186	26192	578	308	3092	108	43757	49246	68067	267459	23540	290999
BR	8253		1	33	35	4091	94	1	5159	43668	44020	122785	0	122785
CA	5944	57		2598	72	412	4565	405	142475	536413	65140	824363	0	824363
CN	1285				636	1083	2424		45464	30331	620702	809504	193837	1003341
FR	110742	22698	9982	20688	3164	55704	73251	443	738089	395073	399723	2503576	0	2503576
DE	58199				2722	29806	28884		395959	62136	408648	1249789	357554	1607343
IN	801				94	689	1113		19307	13686	94173	140744	0	140744
ID	204					62	431		2325	2824	12028	27675	0	27675
IT	7527	959	3348	250	144	23978	2832	249	39212	4809	93572	371670	0	371670
JP	3210				29	356	7215		327354	521516	315347	1477059	0	1477059
KR	781	471	215	26231	2	524	546	98	5553	18348	61204	152743	29234	181977

(continued)

Table 1.8 (continued)

	NL	RU	SA	SG	ZA	ES	CH	TR	GB	US	Others	Total liabilities	Net assets	Total
LU	22294	2318	304	5120	113	15287	45861	418	58288	10251	136499	509114	26740	535854
MX	12	0		4	0	1142	21		115	3211	922	8685	9179	17864
NL				18903	313	26112	34236	2131	94331	73622	185834	640028	147663	787691
RU	855				5	944	2184		10300	112	33873	65119	68633	133752
SA	329				4	638	3605		32722	4685	38773	92120	86656	178776
SG	44550				310	3524	12385		94461	70540	449780	747388	0	747388
ZA	150	4	5	109		14	177	7	6625	1550	9739	22393	17852	40245
ES	14235	2482	7429	612	68		7579	180	29728	12441	67169	287183	110416	397599
CH	24699	15723	6182	15850	1384	13605		6405	208075	22137	300166	819724	0	819724
TR	5705				6	1309	4646		22069	1063	41965	93178	0	93178
GB	175045	14758	79360	82758	20381	44672	178405	11932		1341773	1341094	4686003	0	4686003
US	59732	15100	13203	74508	6336	23992	47788	4899	842667		1451818	3745021	0	3745021
Others	231419	59100	58560	429953	3849	148433	5598	11542	1436709	107343		3731223	1137569	4868792
Total assets	787691	133752	178776	703809	40245	397599	467416	38820	4601205	3332173	4868792	23410117		
Net liabilities	0	0	0	43579	0	0	352308	54358	84798	412848	0			
Total	787691	133752	178776	747388	40245	397599	819724	93178	4686003	3745021	4868792			

Source BIS, LBS, A6.2 By country (residence) of counterparty and location of reporting bank, outstanding at end-December 2021, in millions of US dollars

Matrix. These two tables are presented as Tables 1.7 and 1.8, respectively. As the creation of Table 1.8 involves transforming data from an account form to a matrix, we will provide a comprehensive explanation of the procedure and methodology for constructing the LBS matrix in Chap. 2.

1.5.2 A Matrix of Multiple Financial Instruments

According to the structure presented in Table 1.5, when we merge the data from Table 1.6 with that in Tables 1.7 and 1.8, we can create a comprehensive map of bilateral relationships between national and regional economies using CDIS, CPIS, BIS, and IIP data, that is, Table 1.9. These matrices have the potential to be expanded to include flow data, allowing us to quantify gross bilateral flows in terms of: (i) transactions; (ii) alterations in the value of financial assets/liabilities; and (iii) other changes in asset/liability volumes.

Table 1.9 shows what may be possible in a GFF framework for a country to enable monitoring of financial positions at both region/nation and cross-border levels through financial instruments. Table 8.8 is also based on W-to-W benchmark, the "column" as Assets, and "row" represents liabilities. The matrix here has the same number of rows as columns too, which is a square matrix.

Table 1.9 is an illustration of the GFFM as of the end of December 2021. Each row of the matrix has two statistical groupings, including countries and three financial instruments for showing the source of funds, that is, direct investment (DI), portfolio investment (PI), and other investment (OI), covering the main structural elements of external financial liabilities. Financial assets are listed by country in the columns to show fund uses, with the counterparty sectors identified for each cell. The columns of the matrix delineate 25 sectors: 24 for countries, All other economies, Total liability of Financial Instruments, Total liabilities, Difference (A > L), and Total. The total of all sector's assets is equal to the total of all sector's liabilities. The columns of the matrix are configured to understand the external assets of many countries, thereby displaying both national and regional perspectives. Each column corresponds to the balance sheet of the sector in question; which countries or regions should appear in the matrix depends on the specific purpose of the analysis. The data in Table 1.9 are derived from IMF Data Warehouse and BIS' IBS. But Financial Derivatives (FD) data are not used in Table 1.9 because many countries lack such data.

We used data from CDIS, CPIS, and LBS instead of OIs to compile the GFF matrices for each country. Table 1.9 shows cross-border liabilities of debtors (rows) and cross-border claims of asset holders (columns). The GFFM reveals structural equilibrium relationships as follows. First, we can determine both the distribution and scale of EAL for a country and show the basic structure of its external investment position. By analyzing the rows of the matrix, we can determine the sources of inward financial investment to a country (debtor), and thorough analysis of the columns of the matrix, we can also identify the destinations of outward financial investments from a country (creditor). At the same time, we also know that the rows in the matrix

Table 1.9 External Asset and Liability Matrix for G20 (as of end-2020, USD millions)

Holder of claim (creditor) / Issuer of liability (debtor)	Financial Instruments	AR	AU	BR	CA	CN	FR	DE	IN	ID	IT	JP	KR	LU
AR	Direct investment		515	4614	2679	1111	2206	3412	26	4	900	813	380	1742
	Portfolio investment		428	88	1379	4	403	653	0	1	1977	474	92	6319
	Other investment		2	0	0	0	667	0	0	0	32	0	26	60
	Direct investment	0		873	35554	34158	9307	16736	301	27	696	101508	6172	6973
AU	Portfolio investment	0		29	30180	9939	36480	55829	4	131	6828	189649	17464	51284
	Other investment	33		93	2614	21690	14950	9333	1135	1567	311	11344	866	1992
	Direct investment	−43	1748		12993	1634	31125	10622	110	234	8179	18179	5374	55083
RB	Portfolio investment	411	4131		14165	1543	7004	5673	1	0	1333	9971	9140	37645
	Other investment	4	11		704	851	8770	1245	11	0	213	1296	456	3865
CA	Direct investment	−147	13759	11879		18892	13748	14175	−65	77	2628	25527	2392	44684

(continued)

Table 1.9 (continued)

Holder of claim (creditor) / Issuer of liability (debtor)	Financial Instruments	AR	AU	BR	CA	CN	FR	DE	IN	ID	IT	JP	KR	LU
	Portfolio investment	72	36633	110		5536	30582	80153	53	99	5038	91850	12708	86535
	Other investment	41	6511	495		17218	9451	2206	557	98	185	5843	280	11441
CN	Direct investment	10	10165	354	12122		28147	86684	639	715	9191	193338	82693	7594
	Portfolio investment	30	20336	55	44105		14575	11759	618	306	2578	34749	23940	122714
	Other investment	0	27167	5	12858		19658	0	0	0	1523	0	30804	15565
FR	Direct investment	2	1059	189	8737	4004		105610	301	19432	30127	13339	1569	176728
	Portfolio investment	1	28428	273	49206	9341		539709	158	2	199100	300213	28568	550413
	Other investment	161	22414	1117	3942	32555		228119	1922	500	100140	38613	479	239706
DE	Direct investment	0	3820	−378	7613	5727	61132		293	13	34954	33134	7948	220284

(continued)

Table 1.9 (continued)

Holder of claim (creditor) / Issuer of liability (debtor)	Financial Instruments	AR	AU	BR	CA	CN	FR	DE	IN	ID	IT	JP	KR	LU
	Portfolio investment	92	32891	166	39424	14398	254294		757	35	92262	136322	11107	422864
	Other investment	0	4083	632	1430	0	120231		0	0	39275	0	4888	92683
IN	Direct investment	0	994	20	2737	1407	15609	16643		46	2796	45890	12063	3183
	Portfolio investment	0	8390	2	20765	1369	14018	6518		457	370	20752	4118	60971
	Other investment	0	2479	0	0	0	3130	0		0	243	0	4133	896
ID	Direct investment	0	3811	6	4230	8004	1665	2134	154		103	25594	6627	1035
	Portfolio investment	0	2103	0	6027	1072	3175	9894	92		2154	11966	1835	29582
	Other investment	0	715	0	0	0	3015	0	0		78	0	5934	60
IT	Direct investment	812	726	289	1011	1254	87515	43309	45	3		3931	1213	83031

(continued)

Table 1.9 (continued)

Holder of claim (creditor) / Issuer of liability (debtor)	Financial Instruments	AR	AU	BR	CA	CN	FR	DE	IN	ID	IT	JP	KR	LU
	Portfolio investment	0	5513	80	10194	2332	241352	148437	0	15		87588	3147	200032
	Other investment	158	866	185	157	3069	103473	52006	32	41		2550	14	32110
JP	Direct investment	0	2742	25	1463	2515	30774	1683	303	154	1211		8963	7164
	Portfolio investment	1	55929	17	86329	16510	111966	50758	50	107	18363		29081	176858
	Other investment	0	25704	244	0	0	265979	0	0	0	454		6290	3319
KR	Direct investment	5	1410	−18	4218	8609	7377	10575	479	35	866	57792		1684
	Portfolio investment	0	17098	682	27495	5472	14956	15149	6	59	1965	26327		60258
	Other investment	14	199	24	559	20651	5208	3716	9	516	27	7594		236
LU	Direct investment	(164)	5378	4686	135711	11393	17014	149583	−895	−138	25852	8868	949	

(continued)

Table 1.9 (continued)

Holder of claim (creditor) / Issuer of liability (debtor)	Financial Instruments	AR	AU	BR	CA	CN	FR	DE	IN	ID	IT	JP	KR	LU
	Portfolio investment	96	11294	4088	13099	13450	515248	781542	1217	0	841195	134690	38825	
	Other investment	47	227	617	963	13760	91653	79230	42	46	22035	2970	501	
MX	Direct investment	454	616	2763	35664	873	6128	19662	495	1	2361	10531	3827	11029
	Portfolio investment	10	3224	2024	10064	809	10040	19495	0	0	7096	20502	1765	44356
	Other investment	0	2	9	275	132	174	1977	0	0	4	493	179	13
NL	Direct investment	69	4134	−44793	76549	20462	192100	332293	6117	1363	21820	116616	8479	435166
	Portfolio investment	4	26955	238	38700	3859	316444	340963	97	2	87762	123986	10682	246782
	Other investment	93	2152	309	4284	0	66517	99535	31	0	9695	5383	732	15769
RU	Direct investment	3	55	0	105	2188	19462	18115	611	0	4811	2403	3362	4823

(continued)

Table 1.9 (continued)

Holder of claim (creditor) / Issuer of liability (debtor)	Financial Instruments	AR	AU	BR	CA	CN	FR	DE	IN	ID	IT	JP	KR	LU
	Portfolio investment	10	1595	0	3856	583	1401	7485	0	0	1238	3669	1048	23071
	Other investment	0	5	0	4	0	6088	0	0	0	5325	0	739	4676
SA	Direct investment	0	0	0	33	2931	3594	1376	126	0	5466	4523	−123	295
	Portfolio investment	0	561	0	289	227	1050	4177	4	5	1358	3069	2105	6085
	Other investment	0	64	0	37	0	6287	0	0	0	172	0	3482	1322
SG	Direct investment	16	16126	26	45653	55064	15119	14652	25285	23768	1677	104863	20226	82055
	Portfolio investment	0	11192	20	7391	6547	2843	8857	131	16597	1005	22900	6068	22963
	Other investment	0	32772	1	6671	0	26108	0	0	0	310	0	4882	1039
ZA	Direct investment	24	2659	36	804	6732	1664	7128	226	0	631	3893	210	713

(continued)

Table 1.9 (continued)

Holder of claim (creditor) / Issuer of liability (debtor)	Financial Instruments	AR	AU	BR	CA	CN	FR	DE	IN	ID	IT	JP	KR	LU
	Portfolio investment	2	2757	2	6605	517	2318	5052	0	0	1360	5377	1088	22979
	Other investment	1	59	6	77	2096	725	566	116	1	26	217	10	112
ES	Direct investment	831	20	5186	3167	3492	85982	91250	195	(10)	45546	8365	1363	123887
	Portfolio investment	11	6192	575	10075	1277	213587	181807	1	5	135501	73482	4547	124417
	Other investment	1022	408	1174	332	1950	70824	45842	26	8	10881	316	55	9803
CH	Direct investment	33	418	−10346	−124	6760	54570	15914	3552	35	3185	35141	644	321451
	Portfolio investment	3	0	1491	41161	4348	39169	79721	5	0	11939	37090	8349	126471
	Other investment	3689	1095	1329	3290	3393	69418	52638	559	271	21945	2489	76	39315
TR	Direct investment	0	35	25	115	1069	3586	9360	58	0	2981	1267	321	5291

(continued)

Table 1.9 (continued)

Holder of claim (creditor) / Issuer of liability (debtor)	Financial Instruments	AR	AU	BR	CA	CN	FR	DE	IN	ID	IT	JP	KR	LU
	Portfolio investment	0	1398	5	2289	264	1817	4364	0	0	1908	4424	317	13717
	Other investment	0	6	0	0	0	8898	0	0	0	4625	0	2244	643
GB	Direct investment	5	15323	984	35152	4207	108297	65052	12134	16	9695	99481	2004	151078
	Portfolio investment	6	91811	768	96035	21535	256348	212449	1035	177	54027	185243	38654	436775
	Other investment	174	99666	10752	154904	94904	334310	378878	20194	1721	47242	117594	10119	117214
US	Direct investment	627	97963	6857	490769	37995	285078	411308	1874	461	31631	647718	63668	316778
	Portfolio investment	33089	436311	21685	1473898	178439	427335	584011	5448	949	164475	2072497	345031	1618024
	Other investment	13487	38744	21948	212389	141598	153984	118710	24299	7539	20243	242919	39957	95332
Others	Direct investment	4460	70560	21956	168533	606674	347277	373478	32322	(18598)	103681	212531	42263	732142

(continued)

Table 1.9 (continued)

Holder of claim (creditor) / Issuer of liability (debtor)	Financial Instruments	AR	AU	BR	CA	CN	FR	DE	IN	ID	IT	JP	KR	LU
	Portfolio investment	43	220798	10368	220903	600483	793872	1210569	1498	4000	390750	1476896	105955	1378062
	Other investment	2064	25651	(5756)	403839	649475	778245	533341	12981	(486)	22445	370411	64830	(151319)
Total asset of Financial Instruments	Direct investment	6995	254037	5235	1085488	847154	1428474	1820753	84687	27639	350987	1775246	282588	2793892
	Portfolio investment	33882	1025968	42767	2253633	899852	3310276	4365025	11174	22946	2031581	5073686	705631	5869178
	Other investment	20987	290999	33184	809327	1003341	2167764	1607343	61914	11823	307430	810032	181977	535854
Total assets		61864	1571004	81186	4148449	2750347	6906514	7793121	157775	62407	2689998	7658963	1170196	9198924
Difference (L > A)		85438	809353	1014925	15036	3364514	1296295	0	1235501	451678	182529	667027	147519	849766
Total		147302	2380357	1096111	4163485	6114861	8202808	7793121	1393276	514085	2872527	8325991	1317715	10048690
Financial net worth		−85438	−809353	−1014925	−15036	−3364514	−1296295	1946661	−881088	−343687	122680	5227465	116374	2672533
Reserve assets		39387	46445	355620	89687	3356529	224503	268777	586178	135897	210956	1387860	441907	1188
Monetary gold		3758	4000	4101	0	118246	148389	204808	36966	4758	149342	46390	4795	137
Special drawing rights		1354	4492	4234	8886	11495	11554	17125	1510	1605	8381	20214	3371	361

(continued)

Table 1.9 (continued)

Holder of claim (creditor) / Issuer of liability (debtor)	Financial Instruments	AR	AU	BR	CA	CN	FR	DE	IN	ID	IT	JP	KR	LU
Reserve position in the fund		386	2532	4578	4722	10765	7483	9951	5420	1135	5715	11277	3625	424
Other reserve assets		33889	35421	342707	76079	3216023	57077	36894	542282	128398	47519	1309979	430117	265
Adjustment item		167498	−1840	105065	840752	2294783	201319	404733	−57177	−72184	−292519	−3193190	−71086	−2631637
Net Financial Position		121447	−764748	−554240	915404	2286798	−870472	2620172	−352087	−279975	41116	3422134	487195	42084

MX	NL	RU	SA	SG	ZA	ES	CH	TR	GB	US	Others	Total liability of Financial Instruments	Total liabilities	Difference (A > L)	Total
2123	9997	16	0	370	34	11453	4114	0	2134	20015	15661	84319	141885	5417.015	147302
18	1128	68	8	534	30	299	342	0	1438	18285	8029	41997			
152	1842	0	0	0	0	914	486	0	462	5395	5532	15570			
0	40660	50	0	30745	2339	1375	8948	−17	95093	151171	247988	790655	2356817	23540.311	2380357
1	34494	126	1011	35725	621	2889	26441	3	64552	381048	353973	1298703			
34	9880	81	186	26192	578	308	3092	108	43757	49246	68067	267459			
11299	112472	3	13	11603	188	52753	11698	183	24065	105314	89065	563894	1096111	0	1096111
2426	13439	130	519	0	715	3000	4868	36	39516	170654	83113	409432			
4	8253	0	1	33	35	4091	94	1	5159	43668	44020	122785			

(continued)

Table 1.9 (continued)

MX	NL	RU	SA	SG	ZA	ES	CH	TR	GB	US	Others	Total liability of Financial Instruments	Total liabilities	Difference (A > L)	Total
1780	115046	1382	0	1035	896	10246	39298	2	54153	355306	86498	813192	3812695	350790	4163485
81	34393	297	282	20466	356	3711	44427	14	63399	1195724	462621	2175140			
11955	5944	57	0	2598	72	412	4565	405	142475	536413	65140	824363			
118	40920	266	2090	148721	515	4754	18579	275	25480	86907	2453839	3214115	5921024	193837	6114861
144	22399	60	14652	198743	1863	287	12404	25	89566	290122	991377	1897405			
1	1285	0	0	0	636	1083	2424	0	45464	30331	620702	809504			
0	115715	818	532	1960	751	31838	127866	231	123648	72184	127161	963801	7738136	464673	8202808
109	240118	998	15554	0	1463	83486	89459	24	191788	701012	1241336	4270758			
4351	110742	22698	9982	20688	3164	55704	73251	443	738089	395073	399723	2503576			
221	206592	4430	1978	3227	3346	20245	91434	1946	78964	115320	227657	1129900	5846460	1946661	7793121
130	265049	2707	9467	37773	843	38571	97499	64	169429	526281	1314347	3466770			
213	58199	0	0	0	2722	29806	28884	0	395959	62136	408648	1249789			
189	53512	261	166	77535	107	2047	32291	172	85407	99752	120032	572859	1393276	0	1393276
18	13809	185	539	67574	613	221	5166	10	32643	237809	183356	679672			
0	801	0	0	0	94	689	1113	0	19307	13686	94173	140744			
−0	31554	0	52	57994	1	241	2217	187	8146	31859	54892	240507	514085	0	514085

(continued)

Table 1.9 (continued)

MX	NL	RU	SA	SG	ZA	ES	CH	TR	GB	US	Others	Total liability of Financial Instruments	Total liabilities	Difference (A > L)	Total
26	11248	237	1786	25933	57	345	2917	0	12226	68655	54573	245902			
0	204	0	0	0	0	62	431	0	2325	2824	12028	27675			
−23	97810	677	87	−116	−50	14595	28612	465	44684	13774	45622	469277	2389934	482593	2872527
4	23587	61	1418	0	1615	157623	13189	6	53428	147658	451708	1548988			
129	7527	959	3348	250	144	23978	2832	249	39212	4809	93572	371670			
3	20886	65	28	35629	1	712	−22	4	14466	62748	40798	232313	4961892	3364099	8325991
50	65737	103	17334	0	890	11308	51026	0	167009	1281840	1111255	3252519			
42	3210	0	0	0	29	356	7215	0	327354	521516	315347	1477059			
108	17753	31	3084	18652	−9	507	5159	18	17300	35053	42686	233376	1239270	78445	1317715
20	21302	64	6827	79769	197	973	17646	2	43436	301505	211941	853150			
16	781	471	215	26231	2	524	546	98	5553	18348	61204	152743			
−5302	316884	−13313	6676	48002	−551	1589	66951	−1493	521976	992638	1351363	3643658	8892273	1156417	10048690
880	141126	16537	8064	30489	14735	246623	294113	64	208635	191693	1231799	4739501			
270	22294	2318	304	5120	113	15287	45861	418	58288	10251	136499	509114			
	112657	15	4	922	5	88430	15025	23	25423	184911	23793	545612	948765	9179	957943
	13033	1448	0	0	77	4697	6319	2	19853	159283	70369	394468			

(continued)

Table 1.9 (continued)

MX	NL	RU	SA	SG	ZA	ES	CH	TR	GB	US	Others	Total liability of Financial Instruments	Total liabilities	Difference (A > L)	Total
	12	0	0	4	0	1142	21	0	115	3211	922	8685			
42427		33526	36747	108525	127775	8202	209153	15745	433383	1190180	1136407	4512447	7861548	147663	8009211
128		7624	0	15083	7904	54190	69069	22	107866	625067	625646	2709073			
45		0	0	18903	313	26112	34236	2131	94331	73622	185834	640028			
10	38484		108	3655	27	563	13029	2001	32275	3896	299060	449047	713752	68633	782385
11	6371		478	0	106	141	3548	27	14635	72448	57865	199586			
9	855		0	0	5	944	2184	0	10300	112	33873	65119			
0	10553	0		0	−0	1209	1317	31	6574	11386	4590	53881	202157	379416	581573
1	2025	74		0	20	120	1867	2	4541	0	28576	56156			
0	329	0		0	4	638	3605	0	32722	4685	38773	92120			
59	44685	11203	1808		8	617	54758	−110	73647	367455	666784	1625443	2748164	1205632	3953796
7	9503	602	1294		538	282	5307	0	19952	100079	131255	375333			
56	44550	0	0		310	3524	12385	0	94461	70540	449780	747388			
87	25759	4	202	183		825	2184	0	41329	8846	28987	133127	356009	67418	423427
0	11529	99	380	0		106	2564	2	16266	76685	44801	200490			
1	150	4	5	109		14	177	7	6625	1550	9739	22393			

(continued)

Table 1.9 (continued)

MX	NL	RU	SA	SG	ZA	ES	CH	TR	GB	US	Others	Total liability of Financial Instruments	Total liabilities	Difference (A > L)	Total
20993	127584	4126	(87)	3860	12		36429	28	112270	38765	152460	865714	2542940	110416	2653357
346	54034	164	4531	0	101		15378	5	42326	163581	358102	1390044			
2621	14235	2482	7429	612	68		7579	180	29728	12441	67169	287183			
394	377514	12033	0	0	5873	6985		253	70413	174798	345369	1424863	3614100	426498	4040598
72	34365	706	6521	17451	10153	8873		6	70969	632833	237817	1369514			
5992	24699	15723	6182	15850	1384	13605		6405	208075	22137	300166	819724			
0	23590	3685	649	863	8	4950	3122		13083	2416	48449	124923	315907	0	315907
7	5108	939	7648	0	41	249	1735		7162	23864	20550	97805			
0	5705	0	0	0	6	1309	4646		22069	1063	41965	93178			
17163	240175	15460	458	10596	5961	57391	101921	(9)		642767	624599	2219910	11880394	695396	12575790
128	128979	15399	23323	48709	60129	36197	96681	125		1422401	1747549	4974481			
8153	175045	14758	79360	82758	20381	44672	178405	11932		1341773	1341094	4686003			
20850	483991	4326	6262	27257	3474	88423	300276	2578	486879		809409	4626452	27828320	897879	28726199
20202	692130	14040	114212	425437	27296	78687	407153	1039	1244770		9070690	19456847			
73829	59732	15100	13203	74508	6336	23992	47788	4899	842667		1451818	3745021			
(6206)	877686	160494	12036	222987	31981	122382	359719	20007	524513	756872		9176491	33327386	9147717	42475103

(continued)

Table 1.9 (continued)

MX	NL	RU	SA	SG	ZA	ES	CH	TR	GB	US	Others	Total liability of Financial Instruments	Total liabilities	Difference (A > L)	Total
52939	625773	41743	94058	577280	53141	258698	417679	423	1240781	5569209		20419672			
(90010)	231419	59100	58560	429953	3849	148433	5598	11542	1436709	107343		3731223			
106292	3542479	239558	72892	814207	182692	532333	1534078	42522	2915305	5524331	12439911				
77749	2470679	104414	329904	1580965	183503	991577	1686796	1901	3926186	14357736	25166400				
17864	787691	133752	178776	703809	40245	397599	467416	38820	4601205	3332173	4868792				
201905	6800849	477724	581573	3098981	406440	1921508	3688290	83243	11442696	23214240	42475103				
756038	1208362	304661	0	854815	16987	731849	352308	232664	1133094	5511959	0				
957943	8009211	782385	581573	3953796	423427	2653357	4040598	315907	12575790	28726199	42475103				
−512084	2276109	−128411	260808	167744	73421	−287521	402903	−277165	−2413682	−3236100	−4390453				
199055	53723	595772	453641	362181	54,992	81319	1083436	93272	177218	627306					
7286	36934	138754	433	0	7,623	17152	63306	43241	18593	493605					
4054	7068	6995	8392	1106	2,157	3642	4904	1407	14320	52942					
3541	2960	5527	3637	1639	961	3217	2078	162	7263	36370					
184174	6761	444495	441178	359432	44,252	57307	1013148	48461	137041	44389					
−220986	−1225355	49370	−115233	439494	−16174	−962350	−618893	−201609	1586166	−12098630					
−534015	1104477	516731	599217	969419	112239	−1168552	867446	−385502	−650297	−14707424					

Sources IMF's CPIS (2016b), CDIS (2016a), IIP; BIS's LBS (2016c)

Notes (1) There is a clear criterion to distinguish direct and PIs (i.e., investment of 10% or more of the voting power in DI enterprises). The IMF's CPIS and CDIS strictly follow this criterion. Therefore, there is no overlapping between these two datasets. Moreover, the data on "Other Investment" in Table 1.9 are from LBS. Because the data of LBS are consistent in concept and scope with those of IIP, CDIS, and CPIS, LBS should be selected instead of CBL. The data on BIS's LBS overlaps with the CPIS data, so to prevent double counting, we selected data from LBS, which covers all instruments, out of which loans and deposits are used to compile Table 1.9

(2) The data of financial derivatives are not included in Table 2 because of the absence of statistics on financial derivatives in many G20 countries.

will always sum to the columns; that is, total global assets = total global liabilities. Second, put a difference item after the item of Total assets or Total liabilities, which shows the point on a row "a country held the total assets of financial instruments ≠ total liabilities of the country"; and from the point on column "a country held the total liabilities of financial instruments ≠ total assets of the country." Therefore, we can observe the structure of EAL for a country. Third, from the balance of external financial assets and liabilities, we can get the balanced relationship between "total liabilities of a country − total assets of a country = the country's net financial assets," which can reveal the balance between domestic and foreign financial assets and liabilities.

We can discern the relationship between items such as Financial Net Worth, Reserve Assets, Adjustment Items, and Financial Position in Table 1.9 by referring to the descriptions of these items in Table 1.5.

Table 1.9 can further indicate the scope of external financing conditions, such as (1) the proportion of and relationship with the international financial market; (2) the risk of imbalance in external financial assets and liabilities; an (3) transmission route of impacts from the outbreak of a financial crisis in a country or region as well as a country to enable implementation of an effective financial policy in terms of the impacts arising from other countries. For brevity, we focus on G-20 to trace the effects of external financing such as DI, PIs, and bank credit funds.

1.6 Data Science for Measuring GFF

Big data refers to the conventional software tools in a certain period to crawl the content management and the data set. Big Data Technology (BDT) refers to the various types of data, and the ability to quickly obtain valuable information. It is suitable for big data technology, including database, data mining, distributed file system, distributed database, cloud computing platform, the Internet, and scalable storage system. BDT can also be used to compile monetary and financial statistics, especially to measure the Global Flow of Funds (GFF). This section explores how to use BDT to integrate the data sources, and improve the timeliness of the existing data transmission. Applying BDT to measure GFF can provide an important basis for policy authorities to guard against financial risks.

When we read Table 1.9, we can know that it is not easy to compile this table. It needs to use many different data sources with different statistical criteria, and some of which had a long lag time. Through the above research of constructing statistical framework and arranging data sources, we can conclude that the key problem for establishing GFF statistics is the benchmark of data sources and timeliness of data reporting. Some data are compiled by IMF and BIS which are both based on the BPM6, but some parts of the data are overlapping. For example, CPIS is compiled by IMF, which mainly consists of securities statistics, while banking statistics are made by BIS, but the banking credit business also includes some securities trading. That is, data collected from different sources have some overlapping and omitted. If we can make the same benchmark for data sources, it will facilitate in data collection and improve the quality of data. If we improve the timeliness of reporting data, it will be easy to data compilation, thereby ensuring the timeliness of data publication.

1.6.1 BDT for GFF Measurement

We need to solve two issues for establishing GFF: to clear benchmarks on data sources and to use BDT to solve the standardization of data transmission, reduce statistical errors, and improve data publication timeliness to reflect financial risk changes (Hilbert 2016). Table 1.9 shows that the lag of the data published by CDIS is more than one year, and the lag period of CPIS publication is six months, which cannot meet the needs of financial regulation. In the following sections, we will focus on applying BDT to solve data transmission standardization and improve data timeliness.

Big data is a term for data sets that are so large or complex that traditional data processing application software is inadequate to deal with them. Challenges include capture, storage, analysis, data curation, search, sharing, transfer, visualization, querying, updating, and information privacy. Big data can be described by the following five basic characteristics (Global Pulse, 2012).

Big data is a term for datasets that are large and complex that traditional data processing application software is inadequate to deal with them. Challenges include capture, storage, analysis, data curation, search, sharing, transfer, visualization, querying, updating, and information privacy. Big data can be described by the following five basic characteristics:

(1) **Volume** The quantity of generated and stored data is the size of the data that determines the value and potential insight and whether the data can be considered big.
(2) **Variety** The type and nature of data aid in data analysis to effectively use the resulting insight.
(3) **Velocity** In this context, the speed at which data are generated and processed to meet the demands and challenges depends on the path of growth and development.
(4) **Variability** Dataset inconsistency can hamper data management.
(5) **Veracity** The quality of captured data can vary greatly, affecting the accuracy of the analysis.

Based on the above general interpretation of big data, the preparation of GFF data and monetary and financial statistics also characterize big data; BDT can be used to handle GFF data and application analysis. International institutions, such as the IMF, can propose to member countries to establish a network of data transfer agreements concerning the submission of direct investment, securities investment, financial derivative products, and international banking. Hence, certain international institutions can popularize data transmission standardization and improve data transmission timeliness. Timely GFF monitoring helps track massive behavioral data on international capital flows through the Internet and mining analysis, reveals the regularity of GFF, and establishes research conclusions and countermeasures.

1.6.2 Data Sources Integrate of CDIS, CPIS, and IIP

In the financial big data era, numerous financial products and trading activities are accessed through the network, including fixed networks and mobile networks. Among them, mobile networks will gradually become a significant channel of big data financial transactions. With the improvement in law and regulatory policies and continuous development of BDT, rich financial products and transactions will continue to rise, and financial information gathering through the network is also increasingly becoming convenient.

BDT application in the financial field, including GFF, has three levels: integrating Internet data sources, generalizing statistical standards of different data sources, and establishing the subject classification and standard coding system. As noted above, the data sources of GFF are the IMF and BIS. IMF data source can be divided into CDIS, CPIS, and IIP, but the statistical methods of CDIS and CPIS differ from those of IIP. CDIS and CPIS have similar stock data, including a cross-border matrix, which reflects the situation of counterparts. IIP also provides stock data on direct investment, securities investment, financial derivatives, other investments, and reserve assets. However, IIP data only reflect each country's respective external financial positions and do not include the information of counterparts. Therefore, the data source of CDIS, CPIS, and IIP must be integrated to make an instrument of a country in IIP consistent with CDIS and CPIS. For instance, assume that the total assets of country A's direct investment in a table is equal to the assets of direct investment of the same country in IIP, then the total liabilities of country A's direct investment in a table equals the liabilities of direct investment of the same country in IIP. This condition can ensure that IIP has the same statistical range as CDIS and CPIS and can avoid double calculations and omissions.

1.6.3 Statistical Standards Consistency: Treatment of Other Investments

The fourth section of this Chapter is a conceptual introduction that explains the concept of other investment instruments and its data sources. The additional investment covers other equity; currency and deposits; loans; insurance, pension, and standardized guarantee schemes; trade credit and advances; other accounts receivable and payable-other; and SDRs.[15] However, other investments have not been compiled in a matrix form, such as CDIS or CPIS. Therefore, as an alternative method and data collection, we adopt the location of banks' offices (LBS) data, which belongs to the BIS statistics. Thus, we need to solve the statistical standard consistency issue of different data sources. The BIS publishes two sets of statistics on the activity of internationally active banks. First, locational statistics detail the currency and geographical composition of banks' balance sheets according to the LBS. Second, consolidated statistics describe banks' country risk exposures according to the nationality of banking groups (CBS). The BIS also publishes three sets of statistics on money issuance and bond markets: international debt securities (IDS), domestic debt securities (DDS), and total debt securities (TDS). These statistics are harmonized in the Handbook on Securities Statistics,[16] an internationally agreed framework for classifying debt security issues. In other words, the data that we need to use comes from different data sources and publishing agencies. Therefore, coordinating with international agencies is necessary to develop standard benchmarks and statistical ranges and thus avoid double calculation and data omission.

Through an international statistical benchmark, we can use BDT to measure GFF using the Internet of things and statistical techniques to collect data, compile GFF statistics, and release relevant information. To achieve this goal, we need to improve the environment of data transmission and prepare for three aspects. First, the creation of a new international agreement on data transmission is required in conjunction with relevant countries and international organizations that formulate relevant international agreements on data transmission. Second, data transmission improvement based on a statistical benchmark is required. According to relevant international agreements and uniform statistical benchmarks, participating countries timely report relevant data to the IMF and other international organizations. Third, the IMF's coordination and leadership in data transmission management must be strengthened. International organizations collate data through established procedures and publish all data that meet the online statistical benchmark within the time limit.

1.6.4 Impacts of BDT Application

Compared with traditional statistics, innovation through BDT is based on intelligent sensor. Information acquisition technology, such as software devices, builds a system of information standardization in accordance with the requirements for extensive systems interconnection. This standardization includes data transmission, statistics processing, and an official information announcement, such as statistical information standardization. This standardization will have the following effect on establishing GFF statistics.

---

[15] IMF, Balance of payments and international investment position compilation guide, 2017.

[16] IMF, BIS, and ECB (2015), Handbook on Securities Statistics.

(1) **Enhanced Data Quality and Financial Supervision** The rapid development of information technology has not only made it easier for countries to report their data to international organizations, such as the IMF, but also greatly expanded the amount of data held by international organizations. Also, data from the IMF and other international organizations can be collated, and timely feedback is provided to the world to meet statistical information demand at all levels. BDT can reduce the statistical error of data transmission and processing and improve data quality. The application of high-quality GFF statistics can increase the controllability of financial risks and timely discover and solve possible financial risks. Thus, determining the regularity of financial risk can be more accurate, and the financial supervision level of policymakers improved.

(2) **Reduced Financial Information Asymmetry** The transmission of financial data in the international financial market was slow due to time, place, physical border, and institutional constraints; financial information demand is far greater than its supply. However, in the financial network era, financial data transmission will improve, and data processing will be faster. Thus, financial information demand and supply are balanced.

(3) **High efficiency for Measuring GFF** BDT is efficient in financial data compilation; many online processes and actions have been completed, and some actions are automatic. At the right time and place, necessary financial information is provided to users appropriately. At present, the lagged period of some financial information disclosure is long. For example, the CDIS data released a lag period of approximately one year; the CPIS and LBS data released a lag period of about six months. If we use BDT in financial statistic compilation, the efficiency of data transmission and consolidation will improve. Furthermore, strong data analysis ability enables high-efficiency analysis of financial transactions and the market.

1.6.5 Data Science Applications for GFF

Managing the flow of funds effectively requires robust database management systems. These systems utilize various models and techniques to ensure the accurate, efficient, and secure processing of financial data. Data science applications for the global flow of funds are a multifaceted discipline, ranging from database design, statistical analysis, data visualization, and machine learning. Below we briefly introduce several concepts.

(1) **Data Models and Financial Transactions**

Essential for mapping the flow of funds, database concepts such as relational database management systems (RDBMS) and entity-relationship (ER) diagrams provide a structured approach to manage financial data (Harrington, 2016). A relational database structures data into tables where each table represents a different type of entity. It has been widely used for storing and managing data from across the sectors such as banking, retail, healthcare, and more. The key feature of a relational database is its ability to establish relationships among tables through common fields or keys. Using ER diagrams, these models facilitate the representation of complex relationships among different types of financial data, aiding in a comprehensive understanding of fund flows and dependencies. This structure not only facilitates efficient data retrieval through sophisticated querying languages like Structured Query Language (SQL) but also ensures data integrity and consistency. Relational databases, using languages like SQL, follow the ACID (Atomicity, Consistency, Isolation, Durability) framework. This framework is critical in maintaining the integrity and security of financial transactions. Atomicity ensures transactions are entirely processed or not at all; consistency maintains transaction rules; isolation ensures transactions are processed independently; and durability guarantees transactions are completed even in case of system failures.

Contrary to relational databases, NoSQL, standing for "Not Only SQL," refers to a category of database management systems that differ from traditional RDBMS in some significant ways. NoSQL databases are particularly known for their ability to handle large volumes of structured, semi-structured, and unstructured data. NoSQL is often used in big data and real-time web applications for their speed, scalability, and ease of integration with distributed computing platforms. While we managed GFF data in this book in the relational format, graph databases can be used to handle data whose relationships are well represented as a graph and are as important as the data itself.

(2) **Data Science and Analytics**

Databases can be separated into operational database to store transactional level data, and data warehouses to store aggregate and predictive data. Availability of rich, time-series data allows for various analytics. For example, statistical models such as linear regression, time-series analysis, and logistic regression can be used for analyzing historical data and identifying trends. We may perform Eigenvalue and Eigenvector computation to understand system stability, stress points, and shock propagation in the GFF. Principal Component Analysis (PCA) can be used to find the principal components that capture the most variance in the data. Machine learning models including unsupervised algorithms such as clustering, and supervised algorithms such as deep neural networks can be used for predictive analytics and pattern recognition. For example, clustering techniques can be used to identify underlying clusters in the data that may exhibit strong associations.

There are also graphical models, which can be considered both statistical and machine learning for its ability to infer distributions and predict future events, that are suitable in the analysis of GFF data. We can format the GFF matrix as a Bayesian network, a probabilistic graphical model, that allows us to model complex systems of conditional dependencies to understand and predict the behavior of financial markets. In addition, network analysis provides the ability to visualize the complex flow of funds across sectors, allowing viewers to digest and interpret the data more easily. We can view the global flow of funds as a network of nodes and edges, and use network metrics such as degree distribution, shortest paths, and community detection to analyze the strength and natural clusters in the global sectors.

These aforementioned analyses can be commonly performed using Open-source software in Python and R. In the rest of the book, we use Gephi, an open-source graph visualization software to visualize the GFF. Thus, it is possible to create analytical platforms to manage, monitor, and analyze GFF. By leveraging various data models, understanding transaction lifecycles, utilizing both basic and full database models, and adhering to stringent security protocols and frameworks like ACID in relational databases, financial institutions can effectively manage and track the global flow of funds.

1.7 Concluding Remarks

This chapter reviewed the definition of GFF, clarified the integrated framework for measuring GFF, and compiled GFF for external financial positions and flows on a from-whom-to-whom basis. Also, the chapter addressed essential data gaps in currently available macroeconomic statistics. The paper elaborates on the main attributes of the integrated macroeconomic accounts and the GFF matrix, which allow it to serve as the framework for compiling sector accounts, including financial positions and flows on a from-whom-to-whom basis. Notably, the GFF integrated framework ensures three consistency rules as follows:

The core statistical structure of the GFFS for external financial positions and flows focuses on showing not only who does what but also who does what with whom. This Chapter recommends that the GFF should become a part of the SNA in the future to incorporate the from-whom-to-whom relationship as the central underlying principle for compiling and disseminating external financial positions and flows.

The advantage of using IMF and BIS data to compile a GFF matrix within the integrated SNA framework (as opposed to using fragmented data from different sources) is that such a framework ensures data consistency for all entities and economic flows and positions. Thus, a systematic understanding is developed on the relationships between economic flows in the real and financial spheres; financial interconnectedness; and linkages between the domestic economic and external economic matrices (e.g., between saving–investment, financial surplus or deficit, the balance of payment, and international capital flows).

This chapter discussed the establishment of a GFF statistical framework and data collection sources, compared and integrated different data sources, and analyzed the possibility of using BDT to compile GFF statistics. The paper also discussed how to use BDT in financial statistics, including GFF statistics, and its effects.

The chapter has explored the issues of statistical agencies, such as IMF and BIS, and measurement of CDIS, CPIS, FD, CBS, and reserve assets that correspond to national accounts, the balance of payments statistics, international investment position, and financial statistics. Thus,

BDT is used to compile GFF statistics. Innovations in data collection, compilation, or reporting also need to be addressed as well as how BDT is used to formulate a new statistical benchmark for GFF measurement in macroeconomic and financial statistics. The policy recommendations of this paper are as follows:

(1) With the development of the Internet, international financial transactions will be rapid, and transaction volume will increase. Establishing a GFF statistical system is necessary to prevent financial risks.
(2) Establishing data transmission standards is required to use BDT between international organizations, such as the IMF and BIS, and participating countries.
(3) Based on the standard statistical benchmarks, a network data transmission agreement for measuring GFF must be established between the member states and international organizations, such as the IMF. This agreement includes direct investment, portfolio investment, financial derivatives, other investments, and foreign currency reserve assets. The deadline for national data reporting and the timing of IMF data release must also be determined in the form of W-to-W, and a system of information sharing must be implemented.

Finally, according to the statistical framework, this paper provided an example, clarified the GFF matrix methodologies, and outlined specific data sources. Countries are likely to face difficulties in compiling GFF accounts. Thus, this paper suggests that steps may be implemented depending on current statistical development status, resource requirements, and analytical and policy needs. As GFF statistics are established and perfected in the future, the following steps should also be taken:

(1) Data source integration of CDIS, CPIS, IIP, IFS, and BIS statistics is required to establish GFF statistics following the SNA creation standard. There is also a need to set up the GFF account to connect with the Flow of Funds account in the SNA. However, this requires additional external financial positions in new data collection systems, as described above for GFFS databases.
(2) For the rest of the world sector, further details for the main observed countries by subsectors and other economic flows may also be considered. From-whom-to-whom external financial positions flow for subsectors of the main observation countries, and possibly other economic flows should be considered.
(3) Sectors (subsectors) and specific instruments (loans, deposits, direct investment, portfolio investment, other investment banks, reserve position in the Fund, and foreign exchange) of financial positions and flows on a from-whom-to-whom basis should move from an aggregated subsector and instrument details to a disaggregated subsector and instrument details.
(4) Lastly, the BSA and external matrices could potentially be extended to flow data to identify transactions, revaluation changes, and other asset and liability volume changes. This may be a challenging task because flow data needs to be broken down by counterpart country.
(5) Of course, we also should note that in compiling the financial inflow-outflow tables, very strong assumptions are used. For example, it is assumed that the

financing structure for different sectors "can be considered to be relatively stable." This is a string assumption to work with, particularly in the context of cross-border flows, which can be very volatile even in very short term. The same applies when extending the calculations to compile country and sector W-to-W matrices.

References

Allen, M., Rosenberg, C., Keller, C., Setser, B., & Roubini, N. (2002). A balance sheet approach to financial crisis. IMF working paper WP/02/210 (pp. 44–47). Washington, DC. https://www.imf.org/external/pubs/ft/wp/2002/wp02210.pdf

Bank for International Settlements (2013) *Guidelines for reporting the BIS international banking statistics.*

Cerutti, E., Claessens, S., & Rose, A. K. (2017). How important is the global financial cycle? Evidence from capital flows, IMF Working Paper WP/17/193, Washington, D.C.

Copeland, M. A. (1952). A study of money flows in the united states. In *National Bureau of Economic Research* (pp. 103–285).

Errico, L., Walton, R., Hierro, A., AbuShanab, H., & Amidžic, G. (2013). Global flow of funds: Mapping bilateral geographic flows. In *Proceedings 59th ISI World Statistics Congress*, Hong Kong (pp. 2825–2830).

Errico, L., Harutyunyan, A., Loukoianova, E., Walton, R., Korniyenko, Y., Amidžic, G., AbuShanab, H., & Shin, H. S. (2014). Mapping the shadow banking system through a global flow of funds analysis. IMF Working Paper WP/14/10. Washington, DC. https://www.imf.org/en/Publications/WP/Issues/2016/12/31/Mapping-the-Shadow-Established Principal Global Indicators (PGI) Website (2015). http://www.principalglobalindicators.org/default.aspx

Financial Stability Board and International Monetary Fund. (2009). The financial crisis and information gaps. Report to the G-20 finance ministers and central bank governors. http://www.imf.org/external/np/g20/pdf/102909.pdf

Global Pulse. (2012). Big data for development: Opportunities and challenges (White p. by E. Letouzé). United Nations, Retrieved 4 July, 2017, from http://www.unglobalpulse.org/projects/BigDataforDevelopment

Harrington, J. L. (2016). *Relational database design and implementation.* Morgan Kaufmann.

Hilbert, M. (2016). Big data for development: A review of promises and challenges. *Development Policy Review*, *34*(1), 135–174.https://doi.org/10.1111/dpr.12142. June 20, 2017

IMF. (2016). *Update of the monetary and financial statistics manual (MFSM) and the monetary and financial statistics compilation guide (MFSCG).*

IMF. (2016a). Coordinated direct investment survey (CDIS) data. http://data.imf.org/?sk=40313609-F037-48C1-84B1-1F1CE54D6D5&ss=1393552803658

IMF. (2016b). *Coordinated portfolio investment survey (CPIS) data.* http://data.imf.org/?sk=B981B4E3-4E58-467E-9B90-9DE0C3367363

IMF, BIS and ECB. (2015). *Handbook on securities statistics.* Retrieved 10 March, 2017 from http://www.imf.org/external/np/sta/wgsd/pdf/hss.pdf

Ishida, S. (1993). *Flow of funds in Japanese economy.* Toyo Keizai Shinpo-Sha (pp. 170–205).

Klein, L. R. (1983). *Lectures in econometrics* (pp. 1–46). North-Holland.

Robert Heath, & Evrim Bese Goksu (2017) Financial stability analysis: what are the data needs? IMF Working Paper, WP/17/153.

Shrestha, M., Mink, R., & Fassler, S. (2012). An integrated framework for financial positions and flows on a from-whom-to-whom basis: Concepts, status, and prospects. IMF Working Paper WP/12/57.

Stone, R. (1966a). The social accounts from a consumer's point of view. *Review of Income and Wealth, 12*, 1–33.

Stone, R. (1966b). *Input-output and demographic accounting: A tool for educational planning.*

Tsujimura, K., & Mizosita, M. (2002a). *European financial integration in the perspective of global flow of funds.* Keio Economic Observatory Discussion Paper No.72, Tokyo.

Tsujimura, K., & Mizosita, M. (2002b). *Flow-of-funds analysis: fundamental technique and policy evaluation.* Keio University Press.

Tsujimura, K., & Mizosita, M. (2003). *How to become a big player in the global capital market: A flow of funds approach.* Keio Economic Observatory Discussion Paper No. 84, Tokyo.

Tsujimura, K. and M. Tsujimura (2008) International Flow-of-Funds Analysis: Techniques and Applications, Keio University Press, 3–59.

Zhang, N. (2005) Global Flow of Funds Analysis in Theory and Development, Kyoto, Japan, Minerva, 75–100.

Zhang, N. (2008). *Global-flow-of-funds analysis in a theoretical model, quantitative economic analysis* (pp. 103–119). International Trade and Finance, Hakata, Kyushu University Press.

Zhang, N. (2012). New frameworks for measuring global-flow-of-funds: Financial stability in China. In *The 32nd general conference of the international association for research in income and wealth (IARIW).*

Zhang, N. (2015). *Measuring global flow of funds and integrating real and financial accounts.* Working paper, 2015 IARIW-OECD conference: W(h)ither the SNA? April 16–17, 2015. http://www.iariw.org/c2015oecd.php

Zhang, N., & Zhao, X. Z. (2019). *Measuring global flow of funds: A case study on China* (Vol. 31, No.1). Japan and the United States, Economic Systems Research.

Zhang, N. (2020). *Flow of funds analysis: Innovation and development* (pp. 257–281). Springer.

Zhang, N. (2021). Measuring global flow of funds: Who-to-whom matrix and financial network. In *36th annual virtual general conference.* https://iariw.org/wp-content/uploads/2021/07/Zhang_Paper.pdf

Zhang, N., & Zhu, L. (2021). Global flow of funds as a network: The case study of the G20. *Japanese Journal of Monetary and Financial Economics, 9*, 21–56.

Zhang, N. (2022). Measuring global flow of funds: Who-to-whom matrix and financial network. *Japanese Journal of Statistics and Data Science, 5*, 899–942.

Chapter 2
Global Flow of Funds as a Network: Cross-Border Investment in G20

Abstract This study measures global financial stability by constructing a global flow of funds (GFF) matrix model based on its inherent market mechanisms. We discuss the basic concept of GFF, integrate the data sources, establish a GFF statistical matrix, which can be used to evaluate the financial risks and influences among its members, and estimate bilateral exposures between countries for three different financial instruments within and across the G20 economies. Then, we use financial network analysis to construct the financial relationships between countries. Moreover, we employ the network theory to discuss an analytical method for the GFF and use countries in the G20 as the research sample to discuss the network centrality, mutual relationships, and the financial risk of foreign direct investment, portfolio investment, and cross-border bank credit among the United States, Japan, and China. This study establishes a GFF statistical matrix and introduces the network theory into the GFF analysis, which opens a new field for measuring and applying GFF.

Keywords Global flow of funds · Data sources · Who-to-whom matrices · Financial networks

2.1 Introduction

The global flow of funds (GFF) concept is an extension of the concept of the domestic flow of funds developed by Copeland (1952). It connects domestic economies with the rest of the world. Due to the deregulation of the financial market, researchers began exploring the GFF in the 1990s. Ishida (1993) proposed the idea of GFF analysis, discussed the concept, and then measured international capital flows among Japan (JP), the United States (US), and Germany (DE). The GFF is a domestic and international capital flow of funds. In particular, the GFF refers to the flow of international capital due to financing and current account imbalances caused by the savings–investment gap. Therefore, the GFF includes the flow of all domestic funds due to investment and savings, links to current balances, and connects international capital flows.

N. Zhang and Y. Zhang, *Global Flow of Funds Analysis*,
https://doi.org/10.1007/978-981-97-1029-4_2

A GFF analysis demonstrates the characteristics and structure of the flow of funds; it involves two aspects, which are (1) the relationship between the real and financial economy and (2) the relationship between domestic funds and international capital flows. The key to solving these two relationships is to get the balance of savings–investment flows, trade flows, and foreign capital flows. A GFF analysis is related to the domestic savings–investment gap and external financial surplus or deficit and considers international capital flows caused by changes in the current account. A GFF analysis is a broader extension of the flow of funds analysis and an expansion of the analysis of the domestic flow of funds and international capital flows.

As the flow of funds in financial markets is related to the balance of payments (BOP), the overseas sector will have an excess fund outflow (net capital outflows) if the current account is in surplus. Conversely, the domestic sector will have net negative inflows. Therefore, when the real economic side of the domestic and overseas economy is analyzed in an open economic system, the balance of savings and investment in the domestic economy will correspond to the current account balance.

In this way, an international capital movement from a country with a surplus current balance to a deficit country arises. The flow of capital moves directly between two nations, from a surplus country to a deficit country or may also arise indirectly in countries through the international financial market, the IMF, the World Bank, etc. These international funds are managed by an agency of a public intergovernmental organization or the World Bank, although most of the funds arise through factors such as the pursuit of interest differential or capital gain and risk aversion through a market mechanism. In any case, from the perspective of the BOP of each country, international capital movement is financed with the balance on the current account, and from a global perspective, it serves as international financial intermediation between a country with excess savings and a country with deficit savings (excess-investment). Moreover, when a capital supplier country is a key currency country, such as the US, the country functions as a supplier of international liquidity. By thoroughly observing the flow of funds, funds mobility (international liquidity and the domestic money supply) can be seen as an integrated system in GFF that connects major power economies because the flow of funds between countries is related to the domestic flow of funds in each of the relevant countries.

From the statistical definition of Eqs. (1.1–1.4) in Chap. 1, a domestic capital surplus and deficiency in the flow of funds account (FFA) coincide with the current account of the BOP, whereas the overseas flow of funds in the FFA corresponds to the financial account in the BOP. Thus, it is possible to observe the systematic process of GFF using FFA and BOP statistics. However, the data about FFA and BOP only provide two-dimensional information, that is, who trades what, but not information about the counterparty, that is, who trades with whom. The 2008 global financial crisis in the US revealed the limitation of this data gap. Therefore, international organizations, such as IMF, Bank for International Settlements (BIS), and Financial Stability Board (FSB), proposed the idea of establishing GFF statistics that can provide data about from-whom-to-whom (W-t-W).

There is international awareness of data limitations as the existing data do not describe the risks inherent in a financial system (Robert, 2013). Previous research has focused on the basic concept of GFF and the proposal to establish a statistical framework for GFF. Therefore, the IMF's Statistics Department has organized seven economies with systemically important financial centers to construct a geographically disaggregated GFF mapping of domestic and external capital stocks and presented an approach to understand the US shadow banking system using a new GFF conceptual framework (Errico et al., 2013, 2014). The authors delineated the key concepts and existing data sources and used the balance sheet approach (BSA) to categorize the rest of the world according to the International Investment Position (IIP) components. The main outcome of this study is a prototype template of stock and flow data, geographically disaggregated by national/regional economies. The GFF data can provide valuable information for analyzing interconnectedness across borders and global financial interdependencies.

Castrén and Rancan (2014) developed a financial network, the "macro-network," that depicts the connections between the main financial and non-financial sectors in the various financial instruments of the euro area. Antoun de Almeida (2015) used sectoral accounts data and data from the Coordinated Portfolio Investment Survey (CPIS), IIP, and BIS to estimate bilateral exposures between financial and non-financial sectors in three different financial instruments within and across G-4 economies. European Central Bank (ECB) applied analytical theories and methods of financial networks to GFF and observed the financial risks.[1] Moreover, Girón et al. (2018) studied the Propagation of Quantity Shocks in "who-to-whom networks." The study focused on banking, and most of the nodes in the network represented government or banks and other institutions. This is considerably different from our network architecture, which uses the country as the node.

Based on previous studies, we present a new statistical approach to measuring GFF and provide an empirical example. Zhang (2016) discussed related problems, such as GFF's data sources, its statistical framework, and the analysis method. He constructed a statistical model of 11 countries, with China (CN), JP, and the US as the main object of observation for measuring GFF, to observe the capital credit relationship among these countries as well as the influence and sensitivity of each country in the GFF (Zhang & Zhao, 2019). Moreover, Zhang (2020) has published a collection of scholarly works on the flow of funds analysis. The book also includes GFF statistics and analysis of theoretical methods and applications of the academic monograph. Zhang and Zhu (2021) introduce the financial network method into GFF analysis and make a meaningful attempt.

It is necessary to strengthen research on the GFF analysis method, use GFF to measure financial risks, observe the spillover effect of systematic financial crises, and observe the situation that triggers an international financial crisis. Using GFF statistics, we can observe interlinkages of counterparties and transmission channels of cross-border capital flows and use them to analyze financial vulnerabilities, risk accumulation, and the causes and effects of imbalances.

In this study, we apply the GFF statistical matrix using G20 as the subject of observation to further explore and develop the GFF analysis method. Based on the

---

1 . ECB website for journalists: www.euro-area-statistics.org.

theoretical framework of the analysis, to reveal the relationship between countries' foreign financial investment and their financial stability, we must first discuss the basis of statistical framework and data sources for measuring GFF. In view of the existing works that have been conducted in this domain and the gaps therein, we aim to present a new statistical approach to measure GFF and include an empirical example to illustrate its operational potential. Research that measures financial risks and observes triggers and spillovers of systematic financial crises using a GFF analysis is required to strengthen the literature on GFF statistical methods.

The novel contributions of this study are as follows. It integrates relevant international financial statistics, creates a GFF matrix model table using W-t-W analysis, and compiles a GFF statistical matrix table that includes G20 countries. Moreover, a network theory is introduced to conduct a comparative international cross-border capital analysis with CN, JP, and the US as the main observation objects. It opens a new field of analysis based on the benchmark of W-t-W, and objectively reveals the status of G20 countries and counterparties in the GFF and the main existing problems.

The rest of this paper is as follows. Section 2.2 improves the GFF statistical framework and reduces statistical discrepancies, discusses the integration and consistency of data sources, such as enhancing the consistency between the IIP, CPIS, the Coordinated Direct Investment Survey (CDIS), BIS statistics, financial account (FA) of OECD Stat, and financial instruments BOP/ROW consistency, and discusses the methodology for preparing counterpart country tables. In Sect. 2.3, we establish the statistical matrix of GFF and clarify the structure and equilibrium relationship of the statistical matrix. Then, the statistical matrix of major financial instruments, such as direct investment (DI), portfolio investment (PI), and international bank credit, is derived. Section 2.4 conducts a financial network analysis of GFF and uses the power of dispersion index (PDI) and the sensitivity of dispersion index (SDI)[2] to show the financial position and degree centrality among the G20. Moreover, we conduct an empirical analysis on CN, JP, and the US.

2.2 Data Sources

In this study, we aim to map external capital stocks to show the characteristics and structure of the external flow of funds, including interlinked international capital stocks and flows. Using GFF statistics, we observe interlinkages between counterparties and transmission channels of cross-border capital flows and use them to analyze the vulnerabilities of financial positions, risk build-up, and causes and effects of imbalances. This can provide basic information for decision making by financial policy authorities. According to Zhang and Zhao (2019), a GFF matrix derived from

[2] It was Rasmussen (1956) who invented the dispersion indices for the input–output analysis. While the PDI is the mean-normalized column sum, the SDI is the mean-normalized row sum of the Leontief inverse.

the W-t-W table can be created to illustrate investment relationships between countries through each type of financial instrument. These instruments show the connections between financial positions, including direct and PIs. Likewise, each financial instrument can be disaggregated within the matrix on a W-t-W basis. Instruments in the rows of the table describe a country's investment (lending) relative to the counterpart country's assets, whereas instruments located in the columns describe a country's assets relative to the counterpart country's liabilities. If all the financial instruments are totaled, that amount will equal the sum of external financial assets and liabilities in the country.

In this analysis, the W-t-W table such as Table 1.5 in Chap. 1 can be used to create a GFF matrix to illustrate how the financial instruments of a country relate to those of another country. These instruments show the connections between financial positions, such as DI and PIs. Moreover, every financial instrument can be disaggregated within the matrix on a W-t-W basis. Instruments in the rows of the table describe a country relative to the counterpart country's assets, whereas instruments in the columns describe a country relative to the counterpart country's liabilities. If all the financial instruments are totaled, the amount will be equal to the sum of external financial assets and liabilities in the country. Thus, based on the IIP, the external assets and liabilities have been disaggregated into the counterpart countries and by the main instruments.

The statistical framework delineated in Table 1.5 in Chap. 1, and the corresponding data sources can provide information about fundraising. It can indicate financial stability, comparability across GFF within a country and across countries, and the spillover effect for taking corresponding financial policies in domestic and global financial markets. Using W-t-W, we can reveal some special needs of financial supervision, which can be used to compile a separate matrix for measuring each financial instrument.

2.2.1 Data Sources from IMF

The GFF data are based on existing statistical data and thus both share many similarities (IMF, 2006). The GFF data sources include not only the rest of the world account of national accounts but also monetary and financial statistics (IMF, 2016), the IIP statistics, BIS locational banking statistics, and OECD's financial accounts. In this chapter, we focus on the database from IMF and BIS. OECD's financial accounts are mainly used to compile sectoral accounts of GFF, so we will discuss it in Chap. 5.

Data from the IMF's Monetary and Financial Statistics, the IIP, and National Accounts are used to derive the BSA matrix. The BSA matrix can provide information about the stock positions of a country or region's financial corporations of both residents and nonresidents. For the external assets and liabilities matrix, datasets with bilateral counterpart country details are collected by the IMF and BIS as follows:

(i) Foreign direct investment (FDI): The CDIS provides bilateral counterparty details on inward DI positions (i.e., DI into the reporting economy), cross-classified by the economy of the immediate investors. It also provides data on outward DI positions (i.e., the reporting economy's DI abroad), cross-classified by the economy of the immediate investment, and mirrors data for all economies.

(ii) PI: The CPIS provides bilateral counterparty details on the stock holding positions of the reporting economies and the derived (mirror[3]) liabilities of all economies. The CPIS's purpose is to improve statistics on the holdings of PIs in the form of equity and long- and short-term debts. It also collects comprehensive information, including the issuer's country of residence, stock of cross-border equities, long-term bonds and notes, and short-term debt instruments, to compile or improve the IIP statistics on cross-border PIs.

(iii) Other investment: Other investment is a residual category that includes positions and transactions other than those included in DI, PI, financial derivatives, employee stock options, and reserve assets.[4] Other investment includes (a) other equity; (b) currency and deposits; (c) loans (including IMF credit and loans); (d) non-life insurance technical reserves, life insurance and annuity entitlements, pension entitlements, and provisions for calls under standardized guarantees; (e) trade credit and advances; (f) other accounts receivable/payable; and (g) SDR allocations (SDR holdings are included in reserve assets). To reflect the loans, deposits, and other assets and liabilities of bilateral counterparties, we use a related dataset of BIS International Banking Statistics instead of the IIP statistics.

2.2.2 Data Sources from BIS

The BIS compiles and publishes two sets of statistics—Locational Banking Statistics (LBS) and Consolidated Banking Statistics (CBS)—on international banking activity. These statistics cover the balance sheets of internationally active banks. The LBS provides information about the geographical and currency composition of banks' assets and liabilities, including intragroup business. The CBS measures banks' country risk exposures on a worldwide consolidated basis. Both data sets are collected under the auspices of the Committee on the Global Financial System and reported to the BIS at a country, rather than by individual banks.

This study uses LBS data on cross-border claims and liabilities as the main source of data because these statistics provide information about the currency composition of banks' balance sheets and the geographic breakdown of their counterparties. At end-2023, The LBS data captures outstanding claims and liabilities of internationally active banks located in the reporting countries against counterparties in more than

[3] The term "mirror" refers to the same data seen from different perspectives. For instance, banks' loans to households could be called mirror data of household debt to banks.

[4] IMF (2013), *Balance of Payments Manual*, 6th edition (BPM6), p. 111.

229 countries and regions.[5] Banks record their positions on an unconsolidated basis, including intragroup positions between offices of the same banking group. However, we know that CDIS[6] data and CPIS[7] data are represented in matrix form, while LBS data is published in account form. Thus, in order to integrate these data into a statistical framework of GFF in the form of W-t-W, it is necessary to transform LBS data[8] into matrix form. The steps and methods for converting LBS data from account form to matrix form are described below.

(1) LBS includes Global tables and the Country tables. We first select A6.2 in the Country table, that is, A6.2 By country (residence) of counterparty and location of reporting bank.
(2) Select A6.2 Location of reporting bank from the BLS. Select some countries of interest as survey target countries and regions such as G20 countries.
(3) As an illustration, we use Canadian data from A6.2 by Table 2.1.
(4) The Table 2.1 structure, selected from the Country tables (A6.2) of BIS International LBS, is divided from left to right into three parts: Cross-border positions by the location of the banking office, Claims, and Liabilities. Consequently, for countries whose data is not included in the table, it is necessary to convert the data into matrix form using mirror data instead. There are three key points.

a. The relationship between claims and liabilities. Claims refer to Canadian assets in these reporting countries, the countries or territories listed on the left side of Table 2.1, while Liabilities denote the liabilities of Canada to the reporting countries in Table 2.1.
b. Make up for missing data. The country tables of banks' cross-border positions encompass 30 countries and regions in Table 2.1, but they do not include Argentina, China, India, Indonesia, Russia, Saudi Arabia, Singapore, and Turkey in Table 2.1. Consequently, to construct the matrix table for G20, it is essential to incorporate the data of the aforementioned countries into the list of countries on the left side of Table 2.1, necessitating the utilization of mirror data.
c. Avoid double counting. Since banks in numerous countries are involved in securities activities, the measurement of claims or liabilities is restricted to specific categories, such as "Of which: loans and deposits," encompassing all instruments. This approach mitigates the risk of double counting with international securities statistics (CPIS), ensuring a more precise measurement of global financial flows. However, if the research focus is solely on observing the capital assets and liabilities of international banks, the research scope should encompass all instruments.

---

[5] BIS, https://stats.bis.org/statx/srs/table/a6.2 on 12/31/2023 11:07: AM.

[6] Such as Table 6-o: Outward Direct Investment Positions by All Reporting Economies Cross-classified by Counterpart Economies, which from CDIS, IMF.

[7] Such as Table 11: Geographic Breakdown of Total Portfolio Investment Assets: Total Portfolio Investment, which from CPIS, IMF.

[8] BIS, https://stats.bis.org/statx/toc/LBS.html on 11/3/2023 16:47: AM.

Table 2.1 Banks' cross-border positions on residents of Canada

Banks' cross-border positions on residents of Canada								
Outstanding at end-December 2022, in millions of US dollars								
Canada	Claims				Liabilities			
Q4 2022	All sectors		Of which: non-banks		All sectors		Of which: non-banks	
	All instruments	Of which: loans and deposits	All instruments	Of which: loans and deposits	All instruments	Of which: loans and deposits	All instruments	Of which: loans and deposits
Cross-border positions	819,839	573,624	282,629	180,146	506,859	466,868	133,912	127,886
By location of banking office								
Australia	13,980	6925	2345	1719	5268	1956	797	570
Austria	2143	780	1317	590	223	128	108	108
Belgium	8850	1964	2633	143	451	341	196	195
Brazil	40	22	0	0	1584	1571	0	0
Chile	2341	24	…	…	5675	3877	26	26
Chinese Taipei	5326	1166	947	204	1064	840	395	371
Denmark	396	146	91	17	414	207	56	55
Finland	2740	14	582	10	351	20	20	20

(continued)

Table 2.1 (continued)

Banks' cross-border positions on residents of Canada								
Outstanding at end-December 2022, in millions of US dollars								
Canada	Claims				Liabilities			
Q4 2022	All sectors		Of which: non-banks		All sectors		Of which: non-banks	
	All instruments	Of which: loans and deposits	All instruments	Of which: loans and deposits	All instruments	Of which: loans and deposits	All instruments	Of which: loans and deposits
France	47,391	14,852	16,415	12,595	21,527	5398	2774	2720
Germany	43,183	\	\	\	5528	\	\	\
Greece	7	…	7	…	291	291	291	291
Guernsey	3821	1445	1215	35	304	124	56	56
Hong Kong SAR	34,258	15,504	11,280	1374	3160	2656	1729	1601
Ireland	2685	1426	862	1301	9463	8299	264	250
Isle of Man	42	11	11	11	194	193	193	193
Italy	812	117	292	61	195	179	160	160
Japan	60,774	\	43,357	\	22,889	\	17,546	\

(continued)

Table 2.1 (continued)

Banks' cross-border positions on residents of Canada								
Outstanding at end-December 2022, in millions of US dollars								
Canada	Claims				Liabilities			
Q4 2022	All sectors		Of which: non-banks		All sectors		Of which: non-banks	
	All instruments	Of which: loans and deposits	All instruments	Of which: loans and deposits	All instruments	Of which: loans and deposits	All instruments	Of which: loans and deposits
Jersey	428	256	125	14	218	218	218	218
Korea	4151	2611	2823	2371	980	577	212	209
Luxembourg	7677	1489	3581	653	1290	1186	211	211
Macao SAR	1595	1254	73	35	202	16	16	16
Mexico	1292	1292	1	1	958	958	17	17
Netherlands	11,596	7449	4893	\	\	\	\	\
Philippines	381	325	2	2	109	105	107	103
South Africa	318	316	82	82	76	71	35	32
Spain	3908	2894	2392	2212	1034	363	228	228
Sweden	1400	1054	1010	901	167	133	38	37

(continued)

Table 2.1 (continued)

Banks' cross-border positions on residents of Canada								
Outstanding at end-December 2022, in millions of US dollars								
Canada	Claims				Liabilities			
Q4 2022	All sectors		Of which: non-banks		All sectors		Of which: non-banks	
	All instruments	Of which: loans and deposits	All instruments	Of which: loans and deposits	All instruments	Of which: loans and deposits	All instruments	Of which: loans and deposits
Switzerland	11,241	1424	5137	931	3365	2354	1600	1043
United Kingdom	210,000	173,787	49,859	31,835	131,872	131,872	28,538	28,538
United States	307,463	304,775	111,596	110,325	250,266	248,339	61,071	59,250

Sources BIS, https://stats.bis.org/statx/srs/table/A6.2?c=CA&p=20224, on 23/10/2023

For detailed instructions on converting LBS's account data into matrix form, please consult the supplementary annex located at the end of this chapter and Table A.5 Cross-border banking credit matrix.

2.2.3 *Data Preparation for China*

Since the flow data sometimes have negative numbers, we adjusted the data in order to obtain the ratio coefficients of all positive numbers and then compile the financial matrix table. That is, the negative data of the liabilities in the flow of funds table is moved to the corresponding asset side, which makes it positive data, and the negative data of the asset side is moved to the corresponding liability side, making it become positive data. This adjustment method is both in accordance with the principle of the double-entry accounting, and also the meaning of the actual economy, which is the data of the amount of capital flow that is positive.

As for the stock data, the People's Bank of China published the 2017–2020 stock data of financial asset-liability for the first time in 2022, and Li and Zhang (2020) compiled the stock data of asset-liability for 2000–2019. This data is based on the basic concepts of 2008SNA, the framework of transaction items and sector classification, and is consistent with the accounting principles of 2008SNA. We use the above stock data to estimate the PDI and the SDI, and the results show that the two coefficients are distributed in the same quadrant, so it can be considered that the integration and accuracy of these data are basically the same. Next, based on the analytical framework of GFF, use the above cross-border asset-liability to prepare the asset-liability matrix.

2.3 Develop a Cross-Border Asset-Liability Matrix

When transforming financial account-type data into matrix-format, it is essential to consult Stone and Klein's prior research literature. Before constructing the GFF matrix, it is important to first summarize the theoretical contributions of Stone and Klein related to the transformation of account form and matrix data within the System of National Accounts. Their theory proves valuable in the development of both asset-approach and liability-approach matrices.

2.3.1 *Stone Formula and Klein Formula*

In the international standard flow of funds statement, the rows represent each item (i.e., m financial instruments) and the columns represent each institutional sector (i.e., n institutional sectors), within which assets and liabilities are listed separately according to double-entry bookkeeping. This Chapter focuses on the method of establishing the Y matrix (sector × sector).

Two methods for converting T-type accounts into Y matrices were proposed by Stone (1966) and Klein (1983) respectively. When Stone (1966) presided over the revision of the 1968SNA, he prepared the financial matrix table reflecting the asset-liability relationship of various sectors by referring to Table U and Table V (see Table 2.2 for details). According to the rows and columns, the matrix is divided into four parts: institutional sector, financial transaction items, physical assets, and total.

The n rows and n columns of the original matrix Y on the left correspond to each institutional sector, each row represents the assets of each sector, and each column represents the liabilities of each sector, reflecting the relationship from who-to-whom. M rows and m columns on the second block show the various financial instruments, which constitute two matrices with n institutional sectors, namely, "sector × instrument" matrix (A_{jk}) and "instrument × sector" matrix (L'_{jk}). Among them, the A_{jk} shows the j sector held the k class financial instrument assets, and L'_{jk} shows the j sector held the k class financial instrument liability. Next, according to the row and column respectively embedded in real assets and accumulate savings, respectively for e_j and z'_j. The last row and column describe the total of rows w_j, l_k, ξ and the total of columns x'_j, a'_k, ε. The financial matrix table reflects the basic structure of input–output table of U-V type. According to this framework, the flow of funds account of T-shaped type can be converted into Y matrix (sectors × instruments) and X matrix (instruments × instruments).

Table 2.2 A matrix of sectoral assets and liabilities

	n sectors	m financial claims	Real asset/accumulated saving	Row totals
n sectors	**Y**	A_{jk}	e_j	w_j
m financial claims	L'_{jk}	**X**		l_k
Real asset/accumulated saving	z'_j			ξ
Column totals	x'_j	a'_k	ε	

Source Stone (1966) 19–24

Notes (a) A capital letter denotes a matrix

(b) Small Roman letters denote vectors. These are written as column vectors: a row vector is written with a prime superscript, as is the transposition of a matrix. Diagonal matrices are denoted by a symbol for a vector surmounted by a circumflex accent. The letter i is used to denote the unit vector.

(c) Small Greek letters denote scalars.

(d) The subscript of a letter with j denotes the sector and k denotes the claim.

(e)A_{jk}, I_{jj}, and i'_k denote, respectively, a matrix whose rows relate to sectors and whose columns relate to claims; the unit matrix of order equal to the number of sectors; and the unit row vector with elements equal in number to the number of claims.

(f) Y represents the matrix of sectors by sectors (n × n) and X represents the matrix of financial instruments by instruments (m × m).

According to the definition of Stone (1966), the asset input coefficient a'_{jk} is calculated by the vector and row sum of the asset transpose matrix, while the liability input coefficient l_{jk} is calculated by the liability matrix and column sum. If known the w and k in Table 2.1, the two input coefficients are as follows:

$$a'_{jk} = A'_{jk}\widehat{w}_j^{-1} \tag{2.1}$$

$$l_{jk} = L_{jk}\hat{l}_k^{-1} \tag{2.2}$$

Among them, the $\widehat{w}_j$ for vector w_j diagonal matrix, $\hat{l}_k$ for vector l_k diagonal matrix, the top right corner-1 and apostrophes are the symbols for inverse matrix and transposed, respectively.

Klein (1983) also put forward the research idea of linking the fund flow statement with the national income account and the input–output table in a matrix representation. His analytical approach is similar to Stone's (1966), but with a different definition of the input coefficient. Klein calculated each proportion by dividing each item of assets and liabilities with the total of the other party, focusing on the fact that the use of funds of each subject depends on the funds raised, and then monitoring the relationship between the fundraising and use of each department.

Continuing to take Table 2.2 as an example, Klein (1983) defined the input coefficient of the flow of funds matrix as follows.

$$d'_{jk} = L'_{jk}\widehat{w}_j^{-1} \tag{2.3}$$

$$c_{jk} = A_{jk}\hat{l}_k^{-1} \tag{2.4}$$

where, d'_{jk} represents the liability ratio of various financial instruments in the total assets held by each sector, and c_{jk} represents the asset ratio which each sector in the total liabilities of each financial instrument.

Therefore, Stone formula using the right side of the T-shaped account (debt) as its basis into Y matrix, expressed as Y^S, and Klein formula using T-shaped account on the left (asset) for calculating standard into Y matrix, expressed as Y^K. Before calculating the input coefficient matrix, the flow of funds statement of T-shaped account is first divided into two matrix tables by the using and raising of funds. Table E reflects the use of funds and table R reflects the raising of funds. For details, refer to Tsujimura and Tsujimura (2018) and Zhang (2020). E table and R table for m × n matrix, the matrix elements e_{kj} and r_{kj} respectively.

Based on Stone's formula and Klein's formula, with flow (or stock) data which get from R table and E table, can find out the corresponding input coefficient matrix C^S (Stone method) and C^K (Klein method), and then establish Y^S and Y^K matrix with W-to-W form. On this basis, we can investigate the capital ripple effect from the asset side and the liability side, calculate the total amount of economic transactions generated by the use or raising of funds in a certain sector, and deduce the limit effect of net induced economic transactions by applying the principle of Leontief inverse matrix.

Through E table and R table, further define the diagonal matrix $\widehat{T}, \widehat{T}^{E}$ and $\widehat{T}^{R}.\widehat{T}$ as a n × n matrix, the diagonal elements is t_j, everything else is 0. Similarly, $\widehat{T}^{E}$ and $\widehat{T}^{R}$ are m × m diagonal matrix, respectively by t_k^E and t_k^R as elements. In addition, the elements of vectors ε and ρ are ε_j and ρ_j, respectively. The details are as follows[9]:

$$t_k^E = \sum_j e_{kj}; \quad t_k^R = \sum_j r_{kj} \tag{2.5}$$

$$t_j = \max\left(\sum_k e_{kj}, \sum_k r_{kj}\right) \tag{2.6}$$

$$\varepsilon_j = t_j - \sum_k e_{kj} \geq 0; \quad \rho_j = t_j - \sum_k r_{kj} \geq 0 \tag{2.7}$$

As you can see, formulae t_k^E and t_k^R respectively the kth term's total financial assets and total financial liabilities, theory all the sector (including foreign) into account, can get $t_k^E = t_k^R$; Formula (2.6) the t_j shows the larger value between total assets and total liabilities in j sector; Formula (2.7) ε_j and ρ_j respectively fund surplus and deficiency in j sector.

We use the method of table U and table V in the input–output model to deal with table E and table R, and the superscript S and K represent Stone's formula and Klein's formula respectively, as follows:

$$U^S \equiv \text{R}; \quad V^S \equiv E' \tag{2.8}$$

$$U^K \equiv \text{E}; \quad V^K \equiv R' \tag{2.9}$$

where the apostrophe denotes transpose. Further, the use of matrix U^S, V^S, U^K, V^K each element of the divided by column sum (row sum), to define the coefficient matrix B^S, D^S, B^K, D^K. In Stone formula, D^S for E (assets) transposed matrix input coefficient, B^S for R (debt) matrix input coefficient. In Klein formula, D^K for R (debt) transposed matrix input coefficient, B^K for E (assets) input coefficient matrix. Then, we can get the matrices with W-to-W form, namely Y^K, and the corresponding coefficient matrix C^S and C^K in the following manner:

$$C^S = D^S B^S; \quad Y^S = C^S \hat{T}, \tag{2.10}$$

$$C^K = D^K B^K; \quad Y^K = C^k \hat{T}. \tag{2.11}$$

[9] See Tsujimura and Tsujimura (2018, 161–162).

When the economy is in a state of unbalanced growth, the financial risks in a certain sector have different ripple effects on the use and raising of funds. By observation of Stone formula from debt as a benchmark flow of funds matrix Y^S, column represents a fundraising (debt), row represents fund use (assets). If a column total of Y^S matrix is greater than the corresponding row total, the net debt of this sector increases, thus the possibility of debt default risk can be observed. This situation will affect the external impact of other sectors' assets, and then affect the overall flow of funds system. As the same, the asset side as a benchmark to observe matrix Y^K, the column represents fund use (assets), and row show the raising (liabilities). Such as Y^K a column total is less than the corresponding row total, the net assets reduce the sector, which can be viewed as the department's net worth fell to form a liquidity risk. This change will also affect the overall flow of funds system. Because $t_k^E = t_k^R$, so Y^S and Y^K these two matrices are a diagonal symmetry, namely $Y^K = (Y^S)'$, each sector of the asset-liability changes in flow of funds system is a completely symmetrical transmission mechanism.

2.3.2 Creating the GFF Matrix for G20

As at 2018, the G20 members were Argentina (AR), Australia (AU), Brazil (BR), Canada (CA), China (CN), the European Union (EU), France (FR), Germany (DE), India (IN), Indonesia (ID), Italy (IT), Japan (JP), Mexico (MX), Russia (RU), Saudi Arabia (SA), South Africa (ZA), Korea (KR), Turkey (TR), the United Kingdom (GB), and the United States (US). Singapore (SG) is a permanent guest invitee. Due to G20 restrictions, Switzerland (CH), Spain (ES), Luxembourg (LU), and the Netherlands (NL) were selected to represent the EU (aside from FR, DE, and IT, which are also EU members); therefore, the observations and analysis in this study include 24 countries.

Using the layout of Table 1.5, we established a GFF matrix of G20, which is shown in Table 2.3 that includes 24 countries and other economies. This updated GFF matrix makes it possible for a country to use a GFF framework to monitor financial positions at region/nation and cross-border levels through financial instruments.

Table 2.3 is also based on the W-t-W benchmark, the column represents assets, and the row represents liabilities. The matrix has the same number of rows and columns, i.e., a square matrix. Table 2.3 illustrates the GFF matrix as at the end of December 2018. Each row of the matrix has two statistical groupings, including countries and three financial instruments, to show the source of funds, whether DI, PI, or other investment (OI), of the main structural elements of external financial liabilities. Financial assets are listed by country in the columns to show what the funds are used for, and the counterparty sectors of each cell are identified. The improvements in the updated version of the GFF matrix are as follows. We used data from CDIS, CPIS, IIP, and LBS instead of OIs to compile the GFF matrices for each country.

Table 2.3 GFF matrix for G20 (as of end−2018, in millions of USD)

		Holder of claim (creditor)												
		Financial Instruments	AR	AU	BR	CA	CN	FR	DE	IN	ID	IT	JP	KR	LU
Issuer of liability (debtor)	AR	Direct investment		0	4908	2010	1010	2156	2388	6	0	1594	0	378	4004
		Portfolio investment		252	364	1452	234	753	2625	0	2	2951	1060	178	15,552
		Other investment		1	513	3	0	683	468	0	0	71	579	57	61
	AU	Direct investment	0		176	29,235	24,988	18,453	13,370	276	608	2694	65,321	11,764	18,231
		Portfolio investment	0		11	25,525	9319	34,286	48,098	2	455	6788	146,000	13,018	45,831
		Other investment	23		9	5516	17,500	12,173	20,021	4888	780	1010	91,087	4774	4957
	RB	Direct investment	467	3100		9177	3052	27,089	13,986	260	136	13,250	21,266	6274	58,000
		Portfolio investment	471	3695		15,068	1638	8160	5047	52	0	1290	12,906	10,723	45,461
		Other investment	17	515		0	911	17,943	3329	7	0	502	11,416	2507	10,771
	CA	Direct investment	0	23,772	−2503		11,081	9176	16,733	555	35	3199	17,070	4135	102,405
		Portfolio investment	1	24,546	121		5126	31,076	60,762	1	3	4518	67,610	8304	72,932
		Other investment	52	9216	42		11,285	17,913	31,249	249	131	958	48,331	2128	9122
	CN	Direct investment	0	9812	918	9963		24,017	91,014	831	17,602	11,720	121,165	77,643	11,581
		Portfolio investment	0	13,089	8	24,615		11,170	5829	536	1086	735	22,982	14,822	54,441
		Other investment	0	41,568	34	7017		34,012	22,713	0	0	2309	57,687	46,319	18,753
	FR	Direct investment	1	2228	776	7190	6360		98,798	111	15,841	33,901	15,516	1065	143,383
		Portfolio investment	0	26,766	220	39,837	6023		403,500	22	6	171,721	258,069	23,457	418,172

(continued)

Table 2.3 (continued)

		Holder of claim (creditor)												
		Financial Instruments	AR	AU	BR	CA	CN	FR	DE	IN	ID	IT	JP	KR	LU
		Other investment	188	19,798	5662	5559	8250		215,538	2308	193	84,719	184,912	1846	116,045
	DE	Direct investment	0	0	82	8345	11,988	81,426		395	1	43,767	22,467	4409	148,667
		Portfolio investment	60	30,047	221	31,757	10,583	186,620		1	15	77,313	121,766	7759	351,651
		Other investment	524	9125	473	5589	18,925	102,405		1558	359	73,985	80,881	5061	93,509
	IN	Direct investment	0	1145	−13	1788	2156	6333	26,329		88	6959	24,280	12,443	2756
		Portfolio investment	0	6868	4	17,531	983	7091	4255		1857	314	19,352	3060	53,070
		Other investment	0	3072	0	1419	0	6584	10,263		0	322	22,315	4388	262
	ID	Direct investment	0	1611	2	2317	8835	318	2792	242		855	31,155	7735	981
		Portfolio investment	0	0	0	5677	999	1432	7439	62		1320	10,134	1428	29,282
		Other investment	0	1399	0	0	0	4601	3627	0		240	20,221	6495	259
	IT	Direct investment	1	0	757	305	1857	104,196	38,717	85	0		4141	909	78,163
		Portfolio investment	0	4412	65	8282	1769	227,970	135,159	7	28		54,498	1853	169,275
		Other investment	141	448	139	0	684	330,215	76,909	30	18		30,108	536	36,481
	JP	Direct investment	0	927	221	6423	3160	28,803	10,225	57	**−8**	2135		7089	6634
		Portfolio investment	0	40,476	25	62,038	10,130	148,094	32,835	19	133	10,434		21,858	144,482
		Other investment	21	15,833	528	34,966	23,890	187,625	34,083	2062	1396	3361		7620	6230

(continued)

Table 2.3 (continued)

		Holder of claim (creditor)												
		Financial Instruments	AR	AU	BR	CA	CN	FR	DE	IN	ID	IT	JP	KR	LU
	KR	Direct investment	0	765	0	1812	6128	5619	10,225	463	**−285**	1717	38,372		1608
		Portfolio investment	0	12,229	1288	18,222	5652	10,511	8298	5	120	623	22,025		41,998
		Other investment	13	3517	169	1101	22,764	14,014	7305	904	315	51	29,939		755
	LU	Direct investment	0	0	8289	75,148	14,316	60,091	182,689	99	**−0**	44,692	12,698	6183	
		Portfolio investment	298	11,792	1707	11,676	12,044	463,586	625,880	264	1963	650,383	110,059	24,929	
		Other investment	79	1123	2825	31,524	11,271	191,802	186,655	107	37	30,695	152,636	752	
	MX	Direct investment	1172	601	1270	19,505	744	5588	14,187	108	0	2446	11,914	3047	25,911
		Portfolio investment	11	2776	1148	10,011	872	11,314	17,000	0	6	4303	19,572	1274	40,655
		Other investment	0	185	209	0	100	3738	3115	0	0	492	11,631	3157	237
	NL	Direct investment	1571	5424	**−17,746**	26,762	18,129	177,372	173,772	13,015	**−1033**	64,234	116,814	8399	789,281
		Portfolio investment	0	22,044	377	24,881	3027	283,453	276,843	6	7036	60,339	112,967	8058	202,031
		Other investment	364	6535	1590	4257	0	120,658	176,064	0	0	10,866	61,991	1350	24,188
	RU	Direct investment	0	179	6	748	6578	20,885	20,511	88	0	12,163	1523	3052	12,166
		Portfolio investment	3	1374	1	3082	1570	2547	4217	0	0	1058	3230	881	20,930
		Other investment	0	10	0	4	0	9184	8108	0	0	6590	4185	1494	4316
	SA	Direct investment	0	0	0	1	484	5795	1912	238	0	5300	5203	210	124
		Portfolio investment	0	127	0	250	52	1656	1085	5	67	126	1003	495	1921
		Other investment	0	56	0	176	0	7157	2300	0	0	200	3137	4203	2666
	SG	Direct investment	0	17,852	231	6985	35,970	10,249	15,518	15,280	30,161	940	71,573	16,182	44,443

(continued)

Table 2.3 (continued)

		Holder of claim (creditor)													
		Financial Instruments	AR	AU	BR	CA	CN	FR	DE	IN	ID	IT	JP	KR	LU
		Portfolio investment	0	11,375	22	5688	5986	2202	6715	53	664	686	16,481	5124	20,026
		Other investment	0	32,060	1	6224	0	34,415	31,497	0	0	244	156,478	5757	7194
	ZA	Direct investment	0	759	93	630	6000	2049	6430	389	107	2036	7104	223	8302
		Portfolio investment	0	2532	1	6343	579	1626	4314	2	0	1199	7507	1024	21,555
		Other investment	3	286	2	35	2238	3254	2588	602	1	359	5608	226	997
	ES	Direct investment	342	34	7579	5864	829	57,169	67,328	191	7	41,468	7358	955	81,935
		Portfolio investment	66	4802	1661	9519	1103	177,367	145,493	0	10	102,194	52,062	2814	109,685
		Other investment	959	269	8445	0	4914	118,213	68,704	15	8	44,390	27,178	318	9917
	CH	Direct investment	80	258	**−21,644**	7090	4748	57,366	42,151	2945	28	9661	5909	444	418,379
		Portfolio investment	0	12,404	1461	29,534	3938	28,131	54,502	8	9	9401	31,243	5163	83,951
		Other investment	3328	2220	23	783	1346	68,344	61,495	227	158	5080	26,530	660	42,052
	TR	Direct investment	0	0	167	947	1602	3941	9624	41	3	7706	0	1195	5630
		Portfolio investment	0	497	5	2682	306	2662	4132	0	40	1897	5069	334	17,234
		Other investment	0	13	0	706	0	11,527	22,147	0	0	10,292	4314	1904	1796
	GB	Direct investment	0	83,217	6406	74,194	16,542	148,105	144,983	4721	368	25,292	154,551	10,352	730,851
		Portfolio investment	39	67,516	518	87,458	15,927	239,266	194,598	35	9	60,938	170,246	31,099	362,506

(continued)

Table 2.3 (continued)

		Holder of claim (creditor)													
		Financial Instruments	AR	AU	BR	CA	CN	FR	DE	IN	ID	IT	JP	KR	LU
		Other investment	372	147,320	4502	65,088	70,742	516,543	332,537	22,394	706	78,085	273,357	5290	68,032
	US	Direct investment	4666	91,697	20,371	459,192	67,038	237,198	304,209	11,808	320	41,987	487,940	90,626	523,722
		Portfolio investment	28,484	333,372	16,422	988,562	132,022	298,069	387,734	4475	3669	130,317	1,515,981	207,054	1,097,152
		Other investment	13,970	51,080	8614	399,548	124,076	287,595	254,255	9159	6267	50,335	1,132,671	35,098	40,798
	Others	Direct investment	33,928	253,632	198,084	230,148	1,728,676	414,532	343,595	30,689	8786	175,185	325,424	109,309	1,261,168
		Portfolio investment	214	159,776	15,238	170,081	268,075	544,008	862,239	270	4918	276,874	1,286,953	70,274	1,019,581
		Other investment	1518	102,095	51,653	184,701	431,866	535,303	551,050	28,236	7934	105,225	1,240,012	80,099	459,626
	Total asset of Financial Instruments	Direct investment	42,228	497,011	208,431	985,781	1,982,270	1,507,926	1,651,488	82,893	72,765	554,900	1,568,766	384,021	4,478,326
		Portfolio investment	29,649	792,767	40,889	1,599,772	497,957	2,723,051	3,298,601	5825	22,094	1,577,724	4,068,775	464,979	4,439,373
		Other investment	21,572	447,744	85,433	754,216	750,762	2,635,901	2,126,020	72,746	18,303	510,381	3,677,204	222,039	959,024
	Total assets		93,449	1,737,522	334,753	3,339,769	3,230,989	6,866,878	7,076,109	161,464	113,162	2,643,005	9,314,745	1,071,039	9,876,723
	Difference (L > A)		94,936	363,502	883,179	0	392,317	0	0	881,088	343,687	0	0	0	0
	Total		188,385	2,101,024	1,217,932	3,339,769	3,623,307	6,866,878	7,076,109	1,042,552	456,849	2,643,005	9,314,745	1,071,039	9,876,723
	Financial net worth		**−94,936**	**−363,502**	**−883,179**	139,946	**−392,317**	862,688	1,881,454	**−881,088**	**−343,687**	122,680	5,227,465	116,374	2,672,533
	Reserve assets		65,779	43,257	374,715	83,931	3,167,992	166,628	198,230	396,116	120,654	152,510	1,265,290	403,575	939
	Monetary gold		2256	1316	2781	0	76,331	100,525	139,055	21,690	3230	101,177	31,409	4795	92
	Special drawing rights		4354	193	4046	7941	10,690	11,284	16,453	1463	1553	7712	18,484	3427	344
	Reserve position in the fund		372	776	2344	2706	8479	4692	6315	3164	1096	3451	10,955	2021	322
	Other reserve assets		58,797	40,972	365,544	73,284	3,072,492	50,128	36,406	369,800	114,776	40,170	1,204,442	393,333	181
	Adjustment item		93,909	**−404,693**	**−86,889**	313,099	**−668,172**	**−1,538,457**	355,768	51,231	**−94,018**	**−373,382**	−3,412,374	−83,906	−2,635,521
	Net Financial Position		64,752	**−724,938**	**−595,353**	536,976	2,107,502	**−509,141**	2,435,451	**−433,740**	**−317,051**	**−98,192**	3,080,381	436,043	37,951

(continued)

Table 2.3 (continued)

		Holder of claim (creditor)															
		Financial Instruments	MX	NL	RU	SA	SG	ZA	ES	CH	TR	GB	US	Others	Total liability of Financial Instruments	Total liabilities	Difference (A > L)	Total
Issuer of liability (debtor)	AR	Direct investment	1631	13,804	11	0	0	57	24,869	3067	0	0	9522	12,538	83,953	188,385	0	188,385
		Portfolio investment	14	2481	55	154	0	13	625	734	3	2433	34,430	13,940	80,305			
		Other investment	132	2209	0	0	0	0	1464	739	0	4063	6384	6700	24,127			
	AU	Direct investment	0	80,673	455	0	0	8343	1505	6592	−12	47,867	163,999	45,537	540,075	2,101,024	0	2,101,024
		Portfolio investment	0	28,909	90	1473	27,729	986	2083	24,963	0	64,851	332,553	300,910	1,113,879			
		Other investment	0	0	91	2206	24,559	474	719	4353	25	58,040	46,598	147,267	447,070			
	RB	Direct investment	8158	180,722	0	0	0	544	60,307	10,433	0	13,334	79,032	42,433	551,021	1,217,932	0	1,217,932
		Portfolio investment	3100	12,732	75	3772	0	544	0	4601	16	26,443	168,692	86,709	411,195			
		Other investment	2618	7359	0	0	68	261	11,409	1966	1	16,949	61,973	105,194	255,716			
	CA	Direct investment	1169	257,820	1669	0	0	271	11,690	28,304	16	35,472	368,498	38,096	928,665	3,199,823	139,946	3,339,769
		Portfolio investment	45	27,984	257	287	18,535	1525	2260	35,351	10	55,360	981,173	370,549	1,768,335			
		Other investment	127	8732	74	241	2428	139	1885	7226	94	89,839	207,529	53,833	502,823			

(continued)

Table 2.3 (continued)

	Holder of claim (creditor)															
	Financial Instruments	MX	NL	RU	SA	SG	ZA	ES	CH	TR	GB	US	Others	Total liability of Financial Instruments	Total liabilities	Difference (A > L)	Total
CN	Direct investment	−46	25,615	254	0	0	120,425	3554	22,628	191	16,665	109,332	780,047	1,454,931	3,623,307	0	3,623,307
	Portfolio investment	119	12,516	28	12,357	111,724	1105	187	6860	9	57.967	159,127	666,232	1,177,542			
	Other investment	33	4625	0	0	0	537	5391	3529	0	71.739	34,340	640,228	990,834			
FR	Direct investment	0	185,172	2979	0	0	1056	22,763	63,888	99	98,704	68,035	92,240	860,105	6,004,191	862,688	6,866,878
	Portfolio investment	168	192,320	426	7188	0	1041	69,381	74,126	8	176,736	556,593	986,285	3,412,065			
	Other investment	237	77,774	19,491	14,576	18,251	1206	65,461	58,279	326	526,711	121,665	183,025	1,732,020			
DE	Direct investment	−127	333,825	8125	0	0	3931	24,687	59,647	1727	38,573	137,148	130,349	1,059,432	5,194,655	1,881,454	7,076,109
	Portfolio investment	179	226,997	2490	5664	0	552	28,142	78,708	33	146,754	398,767	1,164,484	2,870,564			
	Other investment	217	86,634	8756	2817	13,293	1322	34,910	44,826	3726	347,916	75,670	252,177	1,264,658			
IN	Direct investment	180	25,379	121	0	0	1211	2921	6743	201	18,457	42,444	132,028	313,948	1,042,552	0	1,042,552

(continued)

Table 2.3 (continued)

	Holder of claim (creditor)																
	Financial Instruments	MX	NL	RU	SA	SG	ZA	ES	CH	TR	GB	US	Others	Total liability of Financial Instruments	Total liabilities	Difference (A > L)	Total
	Portfolio investment	5	9941	32	2198	50,851	631	0	3917	9	30,004	176,016	161,604	549,593			
	Other investment	0	1064	0	0	0	475	426	1545	0	31,937	16,938	78,001	179,011			
ID	Direct investment	0	22,262	5	0	0	50	84	1528	−49	6782	10,240	25,108	122,854	456,849	0	456,849
	Portfolio investment	22	7923	63	2457	23,866	182	141	2415	3	10,667	66,617	44,913	217,044			
	Other investment	0	0	0	0	0	0	121	493	0	4579	2662	72,254	116,951			
IT	Direct investment	0	172,061	2775	0	0	820	12,262	18,221	140	20,228	33,080	34,600	523,317	2,520,326	122,680	2,643,005
	Portfolio investment	19	35,168	63	2361	0	1274	129,386	9427	1	46,271	115,191	330,081	1,272,560			
	Other investment	6	10,395	1850	3397	124	34	72,747	6225	812	76,032	4491	72,627	724,449			
JP	Direct investment	0	37,312	55	0	0	68	471	16,914	3	8934	113,254	26,147	268,824	4,087,280	5,227,465	9,314,745

(continued)

Table 2.3 (continued)

	Holder of claim (creditor)															
	Financial Instruments	MX	NL	RU	SA	SG	ZA	ES	CH	TR	GB	US	Others	Total liability of Financial Instruments	Total liabilities	Difference (A > L)	Total
	Portfolio investment	8	45,039	40	31,930	0	442	4386	32,440	1	145,272	1,007,631	744,695	2,482,409			
	Other investment	137	2958	106	202	72,820	204	503	7623	68	344,607	375,144	214,060	1,336,047			
KR	Direct investment	0	22,176	33	0	0	0	1881	3909	2	7154	39,021	12,145	152,746	954,665	116,374	1,071,039
	Portfolio investment	14	14,622	19	6131	47,323	31	424	12,396	1	33,042	213,376	151,090	599,438			
	Other investment	4	0	103	21	18,557	0	87	2902	2	16,337	17,730	65,891	202,481			
LU	Direct investment	0	319,537	**−12,113**	0	0	2129	9838	193,568	535	149,011	726,121	648,112	2,440,945	7,204,189	2,672,533	9,876,723
	Portfolio investment	1629	108,943	18,582	1074	24,052	14,783	168,673	222,249	45	118,562	138,919	941,392	3,673,484			
	Other investment	7	27,206	1560	315	7823	299	11,248	56,397	95	136,969	62,380	175,956	1,089,761			
MX	Direct investment		114,681	6	0	0	163	42,007	5536	−8	12,908	95,873	8296	365,953	844,319	0	844,319

(continued)

Table 2.3 (continued)

	Holder of claim (creditor)																
	Financial Instruments	MX	NL	RU	SA	SG	ZA	ES	CH	TR	GB	US	Others	Total liability of Financial Instruments	Total liabilities	Difference (A > L)	Total
	Portfolio investment		9091	483	3117	0	0	5853	6825	2	19,661	146,286	55,580	355,839			
	Other investment		1169	0	0	2	2	30,591	3001	0	4537	40,271	20,090	122,527			
NL	Direct investment	26,137		40,415	0	0	3859	**−4744**	154,576	17,573	189,099	810,238	590,920	3,204,069	6,342,857	2,276,109	8,618,966
	Portfolio investment	94		4841	1806	12,532	542	38,320	62,913	12	104,557	452,853	489,618	2,169,151			
	Other investment	52		0	0	14,672	1481	44,398	22,463	3161	251,442	41,244	182,862	969,638			
RU	Direct investment	0	94,345		0	0	1389	1020	23,300	695	13,483	14,071	182,015	408,217	671,882	0	671,882
	Portfolio investment	16	9683		1603	0	106	0	2994	16	10,898	56,599	48,677	169,486			
	Other investment	5	2591		0	0	9	983	2085	0	17,355	168	37,091	94,178			
SA	Direct investment	0	11,751	0		0	**−1**	1089	1135	**−47**	5727	10,884	6328	56,136	181,911	260,808	442,719
	Portfolio investment	5	101	10		0	0	0	268	1	1986	7584	16,218	32,961			

(continued)

Table 2.3 (continued)

	Holder of claim (creditor)																
	Financial Instruments	MX	NL	RU	SA	SG	ZA	ES	CH	TR	GB	US	Others	Total liability of Financial Instruments	Total liabilities	Difference (A > L)	Total
	Other investment	0	0	0		0	5	611	3389	0	30,927	3917	34,071	92,815			
SG	Direct investment	0	57,274	3471	0		358	673	25,363	32	11,713	254,670	127,353	746,290	1,739,019	167,744	1,906,763
	Portfolio investment	0	9411	389	666		535	128	4337	0	18,860	88,738	119,080	317,169			
	Other investment	100	46,285	0	0		260	1349	18,754	0	88,901	47,528	198,513	675,560			
ZA	Direct investment	0	24,181	35	0	0		1199	2506	8	13,250	7313	11,976	94,591	355,686	73,421	429,107
	Portfolio investment	0	9420	56	2384	0		0	2700	0	17,438	91,279	32,065	202,024			
	Other investment	16	656	4	5	51		235	554	55	21,601	1780	17,915	59,071			
ES	Direct investment	11,395	168,339	6441	0	0	323		10,958	207	92,871	35,425	63,668	660,688	2,214,219	0	2,214,219
	Portfolio investment	464	47,179	15	5766	0	91		12,084	15	39,075	138,743	280,519	1,130,725			
	Other investment	728	15,522	2363	11,809	676	3		4719	282	53,192	6915	43,267	422,806			
CH	Direct investment	465	483,950	17,760	0	0	4250	12,414		205	54,608	253,253	123,604	1,477,925	3,062,169	402,903	3,465,072

(continued)

Table 2.3 (continued)

	Holder of claim (creditor)																
	Financial Instruments	MX	NL	RU	SA	SG	ZA	ES	CH	TR	GB	US	Others	Total liability of Financial Instruments	Total liabilities	Difference (A > L)	Total
	Portfolio investment	76	24,589	109	6154	0	5663	6105		7	58,207	458,052	172,386	991,095			
	Other investment	30	26,819	15,583	10,696	13,936	436	7232		7191	212,861	16,315	69,804	593,149			
TR	Direct investment	0	22,773	8229	0	0	2	5760	2309		8189	3903	27,447	109,467	373,175	0	373,175
	Portfolio investment	9	4660	714	1112	0	58	333	1658		7463	27,911	20,585	99,362			
	Other investment	0	10,937	0	0	0	55	7653	4853		29,943	1931	56,275	164,346			
GB	Direct investment	13,372	669.120	6378	0	0	23,556	118,636	81.713	4114		796,564	365,805	3,478,840	11,427,365	0	11,427,365
	Portfolio investment	239	118,662	4235	13,549	37,267	55,246	32,660	77,034	32		1,359,579	1,440,965	4,369,623			
	Other investment	397	279.121	17,991	87,360	73,132	14,174	123,509	146,812	10,647		689,270	551,521	3,578,902			
US	Direct investment	65,555	875.817	7332	0	0	15,903	99,592	296,364	1815	400,970		346,053	4,450,174	23,773,814	0	23,773,814

(continued)

Table 2.3 (continued)

		Holder of claim (creditor)																
		Financial Instruments	MX	NL	RU	SA	SG	ZA	ES	CH	TR	GB	US	Others	Total liability of Financial Instruments	Total liabilities	Difference (A > L)	Total
		Portfolio investment	21,461	493,217	3807	53,349	341,544	17,387	54,150	302,204	493	1,037,504		6,942,522	14,410,951			
		Other investment	11,801	123,314	10,193	17,019	51,914	2436	70,956	120,676	5821	1,216,557		868,535	4,912,688			
	Others	Direct investment	31,438	1,483,349	252,158	0	0	57,731	145,628	449,143	17,049	485,179	1,619,105		11,917,188	35,627,689	0	35,627,689
		Portfolio investment	27,405	467,587	31,671	57,238	555,794	44,230	179,495	331,020	427	901,801	4,105,577		15,012,857			
		Other investment	101,165	282,483	50,161	68,266	343,240	11,890	109,971	141,094	18,065	479,661	1,571,560		8,697,644			
	Total asset of financial instruments	Direct investment	159,328	5,681,939	346,593	0	0	246,439	600,106	1,488,345	44,496	1,749,177	5,801,025	6,136,100				
		Portfolio investment	55,095	1,919,175	68,551	223,789	1,251,217	146,966	722,733	1,312,224	1143	3,131,812	11,282,286	19,213,209				
		Other investment	117,812	1,017,853	128,326	218,930	655,546	35,702	603,859	664,503	50,371	4,132,695	3,454,403	5,887,927				
	Total assets		332,235	8,618,966	543,470	442,719	1,906,763	429,107	1,926,698	3,465,072	96,010	9,013,684	20,537,714	31,237,236				
	Difference (L > A)		512,084	0	128,411	0	0	0	287,521	0	277,165	2,413,682	3,236,100	4,390,453				
	Total		844,319	8,618,966	671,882	442,719	1,906,763	429,107	2,214,219	3,465,072	373,175	11,427,365	23,773,814	35,627,689				

(continued)

Table 2.3 (continued)

Financial Instruments	Holder of claim (creditor) MX	NL	RU	SA	SG	ZA	ES	CH	TR	GB	US	Others	Total liability of Financial Instruments	Total liabilities	Difference (A > L)	Total
Financial net worth	**−512,084**	2,276,109	**−128,411**	260,808	167,744	73,421	**−287,521**	402,903	**−277,165**	**−2,413,682**	**−3,236,100**	**−4,390,453**				
Reserve assets	176,373	38,117	468,478	496,589	287,292	51,631	70,652	788,582	93,028	171,823	449,070					
Monetary gold	4935	25,275	86,903	433	0	5169	11,621	42,894	20,130	12,737	334,457					
Special drawing rights	3810	6537	6725	8089	1052	2078	3834	4526	1343	13,270	50,803					
Reserve position in the fund	2438	1921	3117	1651	1065	845	2368	1210	157	6434	22,016					
Other reserve assets	165,190	4384	371,733	486,417	285,181	43,540	52,828	739,952	71,398	139,381	41,794					
Adjustment item	**−248,048**	**−1,677,379**	34,335	**−99,771**	244,074	**−79,952**	**−875,120**	**−445,265**	**−185,491**	1,831,010	**−6,887,413**					
Net Financial Position	**−583,760**	636,847	374,402	657,626	699,109	45,100	**−1,091,988**	746,220	**−369,628**	**−410,849**	**−9,674,443**					

Data Sources
IMF's CDIS: Coordinated Direct Investment Survey - CDIS Home - IMF Data
IMF's CPIS: https://data.imf.org/regular.aspx?key=61227426
IMF's BOP/IIP: https://data.imf.org/regular.aspx?key=60587815
BIS international banking statistics: http://stats.bis.org/statx/toc/LBS.html

Notes **(1)** There is a clear criterion to distinguish direct and PIs (i.e., investment of 10% or more of the voting power in DI enterprises). The IMF's CPIS and CDIS strictly follow this criterion. Therefore, there is no overlapping between these two datasets. Moreover, the data on "Other Investment" in Table 2.2 are from LBS. Because the data of LBS are consistent in concept and scope with those of IIP, CDIS, and CPIS, LBS should be selected instead of CBL. The data on BIS's LBS overlaps with the CPIS data, so to prevent double counting, we selected data of LBS, which covers all instruments, **Out of which loans and deposits** are used to compile Table 2.3. **(2)** The data of financial derivatives are not included in Table 2.3 because of the absence of statistics on financial derivatives in many G20 countries

Table 2.3 shows the cross-border liabilities of debtors (rows) and cross-border claims of asset holders (columns). The GFF matrix reveals the following structural equilibrium relationships. First, we can determine both the distribution and scale of external assets and liabilities of a country and show the basic structure of its external investment position. By analyzing the rows of the matrix, we can determine the sources of inward financial investment to a country (the debtor). Moreover, by analyzing the columns of the matrix, we can also identify the destinations of outward financial investments from a country (the creditor).

We used data from CDIS, CPIS, IIP, and LBS instead of OIs to compile the GFF matrices for each country. Table 2.3 shows the cross-border liabilities of debtors (rows) and cross-border claims of asset holders (columns). The GFF matrix reveals the following structural equilibrium relationships. First, we determine both the distribution and scale of external assets and liabilities for a country and show the basic structure of its external investment position. By analyzing the rows of the matrix, we determine the sources of inward financial investment to a country (the debtor), and by analyzing the columns of the matrix, we also identify the destinations of outward financial investments from a country (the creditor).

Moreover, we know that the rows in the matrix will always sum up to the columns, i.e., total global assets = total global liabilities. Second, the point on a row, the total liabilities of financial instruments a country holds = the total liabilities of the country. Moreover, the point on the column, the total assets of financial instruments a country holds = the total assets of the country. Therefore, we derive the structure of the external assets and liabilities of a country. Third, from the balance of external financial assets and liabilities, we derive the relationship as total liabilities of a country − total assets of a country = the country's net financial assets, which reveals the balance between domestic and foreign financial assets and liabilities.

2.4 Using the GFF Matrix

Table 2.3 also indicates the scope of external financing conditions, such as (1) the proportion of and relationship with the international financial market, (2) the risk of imbalance in external financial assets and liabilities, and (3) transmission route of the impacts of an outbreak of a financial crisis in a country or region, to enable the implementation of an effective financial policy that considers the impacts of other countries. For brevity, we focus on CN, JP, and the US to trace the effects of external financing, such as DI, PIs, and bank credit funds.

2.4.1 *The Composition of Bilateral Investment Between CN, JP, the US*

Utilizing the information presented in Table 2.3, we have constructed a matrix, displayed in Table 2.4, which specifically centers on CN, JP, and US. This matrix delineates the composition and features of mutual financial investments among CN, JP, and the US based on a W-t-W benchmark. In Table 2.4, the rows mean funds raised, and the columns mean funds used. The table indicates that, as at the end of 2018, CN received $121 billion from JP and $109 billion from the US through DI. In terms of PI, JP's investment in CN was $23 billion, whereas the US' investment in CN was $159 billion. This implies that the US focuses on securities investment in CN, whereas JP focuses on DI and bank loans in CN.

The columns in Table 2.4 indicate that CN's investment in the US in terms of DI, PI, and OI exceeds the scale of its investment in JP. In 2018, CN's DI in the US was $67 billion, PI was $132 billion, and OI was $124 billion, both exceeding those in 2016 (Zhang & Zhan, 2019). CN's PI in the US accounts for 26.5% of its total PI, which is mainly holding the US treasury bonds. CN's investment in the US ranks first, accounting for 10% of its total foreign investment, whereas its investment in JP accounts for 1.15% of its total foreign investment. GB and AU are also large financial investment targets of CN, accounting for 3.19% and 1.6% of CN's total foreign investment, respectively (see Table 2.3).

The "row" of JP shows that the US investment in JP is much higher than that of CN, with DI of $113 billion, PI of $1,008 billion, and OI of $37.5 billion, accounting for 42% of total DI from the US to JP, 40.6% of the total PI, and 28.1% of the total OI. This indicates that JP and the US focus on DI, PI, and OI, and the investment scale is large. CN and the US focus on DI and PI, but the investment scale is small. Regarding JP's external investment, as shown in the columns in Table 2.3, the scale of JP's investment in the US is also much larger than that of CN. JP's DI in the US was $488 billion or 31% of its total FDI; PI accounts for 37%, and OI accounts for 31%. This implies that JP and the US focus on PI and OI, whereas JP and CN focus on DI (7.7%) and OI (1.6%). In addition to the US and CN, the UK, and FR are also larger recipients of JP's external investments.

Table 2.4 shows three characteristics of foreign investment between CN, JP, and the US First, the forms of mutual investment between CN, JP, and the US are different. The investment between the US and JP is mainly PI and OI, whereas the investment between CN and the US is mainly DI and PI. Second, the US occupies an absolute dominant share in the foreign financial investment market. Compared with that of the US and JP, the scale of CN's foreign investment is still relatively low. Third, as at the end of 2018, the net IIPs of both CN and the US were negative, at −$392 billion and −$3,236 billion, respectively, whereas that of JP was $5,227 billion. Moreover, from 2015 to 2018, CN, JP, and the US maintained the same positive and negative signs.

Table 2.4 The composition of bilateral investment by W-t-W (as of end-2018, USD bn.)

Creditor		China			Japan			United States			Others			Total liabilities
Debtor		DI	PI	OI	DI	PI	OI	DI	PI	OI	DI	PI	OI	
China	DI				121 (7.7%)			109 (1.9%)			1224			1455
	PI					23 (0.56%)			159 (1.4%)			995		1178
	OI						58 (1.6%)			34 (1%)			899	991
Japan	DI	3.16 (0.16%)						113 (2%)			152			269
	PI		10(2%)						1008 (9%)			1465		2482
	OI			24 (3.2%)						375 (11%)			937	1336
United States	DI	67 (3.8%)			488 (31)						3895			4450
	PI		132 (26.5%)			1516 (37.3%)						12,763		14,411
	OI			124 (16.5%)			1133 (30.8%)						3656	4913
Others	DI	1912 (96.5)			956 (61.2%)			5578 (96%)						11,917
	PI		350 (70.3%)			2530 (62.2%)			10,115 (89.7%)					15,013
	OI			60. (80.3%)			2487 (67.6%)			3045 (88%)				8698

(continued)

Table 2.4 (continued)

Creditor	China			Japan			United States			Others			Total liabilities
Debtor	DI	PI	OI	DI	PI	OI	DI	PI	OI	DI	PI	OI	
Total assets	1982	498	751	1569	4069	3677	5801	11,282	3454	6136	19,213	5888	124,409
Net worth	527	**−680**	**−240**	1300	1586	2341	1351	**−3129**	**−1458**	**−5781**	4200	**−2810**	

2.4.2 *The Matrix for a Financial Instrument*

Table 2.3 provides an overview of the distribution of DIs, PIs, and international bank credit funds in each country. From a row-based perspective, we can understand the countries that raised funds, how much they raised, and in what ways. From a column-based perspective, we can see the countries that used funds in each type of instrument and how much they used. This information can clarify the following relationships. First, it shows a country's overall external positions, the extent of its holdings of creditor's rights and its debt, and through which financial instruments and counterparties, i.e., from-whom-to-whom and by what. Second, it shows a country's influence on the GFF and the structure and scale of its financing. Third, structural changes and equilibrium conditions in the DI, global bond, and international banking credit markets are revealed. Fourth, the effect of a financial crisis that extends from a country or region to others is shown. Finally, it provides a way to monitor the stability of GFF and its equilibrium state.

A breakdown of cross-border exposures categorized by the instrument type and country offers major advantages than using BIS and IMF data on all instruments as developments in cross-border exposures vary across instruments and countries and such differences cannot be recognized using aggregate financial statistics. For example, whereas cross-border exposures in loans significantly decreased after the financial crisis, cross-border exposures in debt portfolios remained generally stable and increased with government debt. The central motivation for extending the analysis is to observe the interconnectedness of G20 economies and the relevance of bank cross-border exposures.

Based on the matrix model presented in Table 2.3, we construct the international DI, international PI, and cross-border banking credit matrices of a financial instrument. We have displayed these matrices in Tables A.3, A.4, A.5 and A.6 in the Appendix, and we will use these matrices to conduct a financial network analysis.

2.5 Interpreting Financial Networks in G20

Financial networks displaying the interconnections between countries as at the end of 2018 are constructed based on the type of instrument. The analysis focuses on FDI, PI, and cross-border bank credit because of the absence of statistics on financial derivatives in many G20 countries. The networks are constructed using stock data instead of flow data as the interest lies in the total exposures of a country with other countries, and flow data can be volatile.

A network is merely an alternative representation of a matrix, where the graphical representation allows for a faster interpretation of the interconnectedness among countries. A network consists of nodes and the links connecting them. The nodes in the financial network below represent different countries and a link from country i to j represents country i's claims (exposure) on country j. The positions of the nodes are arbitrary, but to facilitate the identification of systemically important countries, the sizes of the nodes are proportional to the countries' holdings of a given liabilities. For example, if the US is represented by a large node in the financial network that depicts exposures in debt securities, it means that the US is a large issuer of debt securities. Likewise, the width of the link is also proportional to the size of each country's exposure to another country. As networks are constructed to assess financial stability instead of drawing a link proportional to the absolute value of a bilateral claim, its width is based on the lending country's capacity to absorb the potential loss of this claim. A smaller country is less able to absorb the loss of a claim than a larger country; therefore, the links' widths are proportional to the ratio of a bilateral claim to the lending country's total consolidated assets. Representing claims relative to the size of a country is a novel contribution of this study to the literature because previous papers that conducted network analysis with national data used absolute claims.

In the previous section, we created the GFF matrix based on (W-t-W); here, we use GFF matrix data to conduct a financial network analysis. Then, we use financial network theory to analyze the mutual influence and shock of G20 members in FDI, PI, and cross-border bank credit markets. To identify the transmission of negative economic shocks at the national sector level and quantitatively analyze the mechanism of balance sheet contagion, which is the core concept of macro-financial risk analysis, we need a W-t-W financial network model that is based on each type of financial instrument. As the networks below show, the differences in interconnectedness as at the end of 2018 depend on the financial instrument and country under consideration. For a broader understanding, the networks below should be interpreted in conjunction with Tables A.3, A.4 and A.5 of the Appendix, showing the largest increases and decreases in financial exposures by FDI, PI, and international loans as at the end of 2018 by country.

2.5.1 Basic Concepts Related to Network Theory

A typical system stability assessment emphasizes the analysis, identification, and response of risk factors and weaknesses in the system. New research methods in this field regard the whole economy and financial system as an interrelated network involving the internal entities, clarify the links between the entities in the system to effectively identify the trajectory of a negative economic shock to the system, and quantitatively predict the severity of a secondary contagion.

The global financial crisis, which occurred from 2007 to 2008, also revealed the complexity of the macro-financial system and highlighted the importance of network analysis methods. Hendricks (2006, 2009) pointed out that new financial stability models depict the financial system as a network structure or population set, whereas more traditional research in this field pays more attention to inter-agency interaction and portfolio effects, such as credit risk contagion and collective selling, caused by a decline in the value of a particular class of asset. In a traditional model, market liquidity, deposit maturity, fundraising, and leverage ratios are important factors that affect the severity of the contagion and its feedback cycle. Empirical research of network theory mostly adopts the following methods: a relationship model between multiple nodes is constructed using balance sheet information. Then, the stability of the network is tested by simulating the impact of a shock.

The W-t-W data of the GFF can be seen as a network of interrelationships in which the nodes—the elements interlinked in the network—are countries, and the edges—the links between nodes—are asset/liability links. The edges in the network are "weighted" by the amounts involved in every asset/liability relationship. Limited by space, we focus on degree centrality in network analysis to illustrate the importance and influence of G20 countries in the GFF network.

Degree centrality uses the most direct metric to describe node centrality in network analysis (Girón et al., 2018; Zhang, 2020). The greater the degree of a node, the higher the degree centrality of the node and the more important the node is in the network. In an undirected graph, degree centrality measures the extent to which a node in the network is associated with other nodes. For an undirected graph with g nodes, the degree centrality of node i is the total number of direct connections between i and other g-1 nodes, expressed by the following matrix:

$$C_D(N_i) = \sum_{j=1}^{g} x_{ij}(i \neq j) \tag{2.14}$$

where $C_D(N_i)$ represents the centrality of node i, which is used to calculate the number of direct connections between node i and other $g - 1$ j nodes ($i \neq j$ excludes the connection between i and j, so the data in the main diagonal can be ignored). $C_D(N_i)$ is simply calculated as the sum of the values of the cells in which the corresponding row or column of node i in the network matrix is located. Because undirected relationships form a symmetric data matrix, cells with the same rows and columns have the same value.

The GFF and W-t-W provide a lot of analytical possibilities; we will emphasize its connection using the network theory. Using Eq. (2.14) and based on matrix C, which is in Tables A.3, A.4, A.5 and A.6 in the Appendix, we calculate the degree centrality of FDI, PI, and cross-border bank credit, and draw the network diagrams, which are shown in Figs. 2.2, 2.4, and 2.9, 2.10 and 2.11. This is because matrix C represents the network of interconnections better.

Tsujimura and Mizosita (2002) proposed the indicators for observing the influence coefficient (IC) and the sensitivity coefficient (SC), but according to the network theory (Kimmo & Samantha, 2016), IC and SC are also considered as network centrality measures of a network represented by the inverse of Leontief (degree centrality). IC and SC can be regarded as a certain network centrality measure, i.e., degree centrality (in-degree and out-degree) of the weighted network represented by $(I - C)^{-1}$. Here, we define in-degree as external claims and out-degree as external debts (see Appendix 1 for the method for calculating IC and SL). Therefore, we also use the matrix of Tables A.1, A.2 and A.3 in the Appendix to obtain the inverse of Leontief by $(I - C)^{-1}$ and then measure the degree centrality of FDI, PI, and cross-border bank credit of the G20 countries and make a network diagram showing degree centrality, as shown in Figs. 2.3, 2.5, 2.7, and 2.8.

2.5.2 *Degree Centrality in the Network of FDI and PIs*

2.5.2.1 FDI Network

FDI is an investment made by a firm or individual in a country in business interests located in another country. Generally, FDI occurs when an investor establishes foreign business operations or acquires foreign business assets, including establishing ownership or a controlling interest in a foreign company. FDI is different from PIs in which an investor merely purchases equity securities issued by foreign-based companies. The IMF's CPIS and CDIS strictly follow this criterion; therefore, there is no overlapping of these two datasets.[10]

To illustrate the scale of FDI and its directional relationships among G20 members, we construct Fig. 2.1 based on Eq. (2.14) with the network theory using data from Table A.3 in the Appendix. The sizes of the nodes in Fig. 2.1 indicate that the US, NL, LU, GB, and CN were the top five countries as at the end of 2018. FDI in the US was $4,450.2 billion, whereas FDI from the US was $5801 billion. FDI in the NL was $3204 billion, whereas FDI from the NL was $5681.9 billion. To understand the mutual relationships among the G20 countries' external financial investments in detail, we segregate them into G7, BRICS, and other countries.

First, we observe the relationship between CN and the G7 in terms of FDI. Regarding outward DI, as at the end of 2018, the US, NL, LU, GB, and CN had the five largest positions, whereas JP ranked last in the G7. The outward DI of the US mainly flows to NL (13.97%), GB (13.73%), LU (12.52%), CA (6.35%), JP (1.95%), and CN (1.88%), JP and CN ranked ninth and tenth. Most outward DIs from CN to the G7 countries are to the US (3.38%), GB (0.83%), DE (0.6), and CA (0.56), which receives a relatively low proportion of CN's outward DI to G7. However, CN's outward DI in other economies accounted for 87.2% of its total outward DI.

We also consider the relationship between CN and the G7 in terms of inward DI. The US remains the world's largest recipient of FDI among the G7 countries. As at the end of 2018, the total DI the US received from all countries is $4,025.5 billion, accounting for 12.27% of the world's total FDI, which ranks first in the world. However, the scale of inward DI to CN was ranked sixth in the world, as it received $ 1454.9 billion, accounting for 4.01% of the total FDI in the world. Among the G7, CN mainly attracted DI from JP (8.33%), the US (7.51%), and DE (6.26%). We also note that CN's share of inward DI in other economies is 53.61%, which is much higher than that of the G7.

[10] See data which from IMF, http://data.imf.org/?sk=40313609-F037-48C1-84B1-E1F1CE54D6D5&sId=1482334777935.

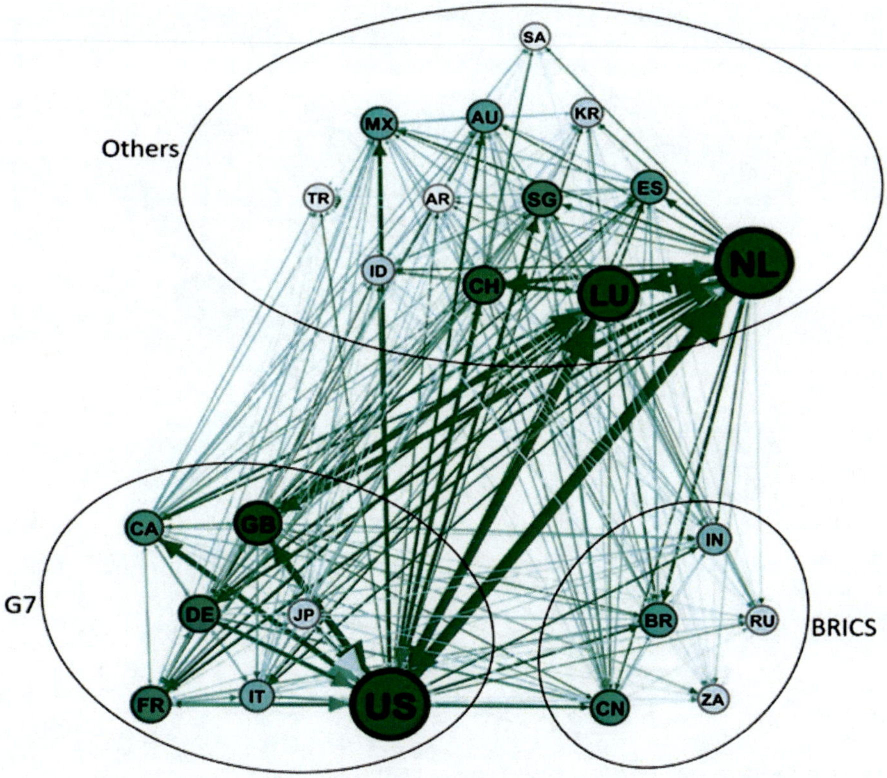

Fig. 2.1 Degree centrality in FDI network (as of end-2018)

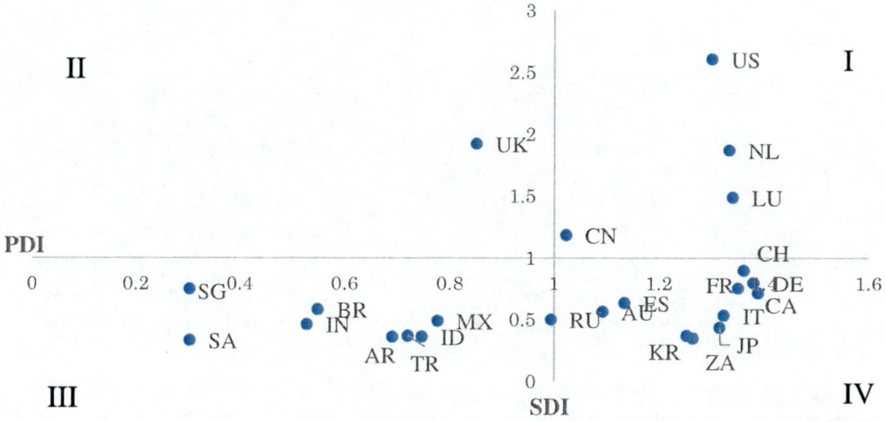

Fig. 2.2 The position of PDI and SDI by CDIS (as of end-2018)

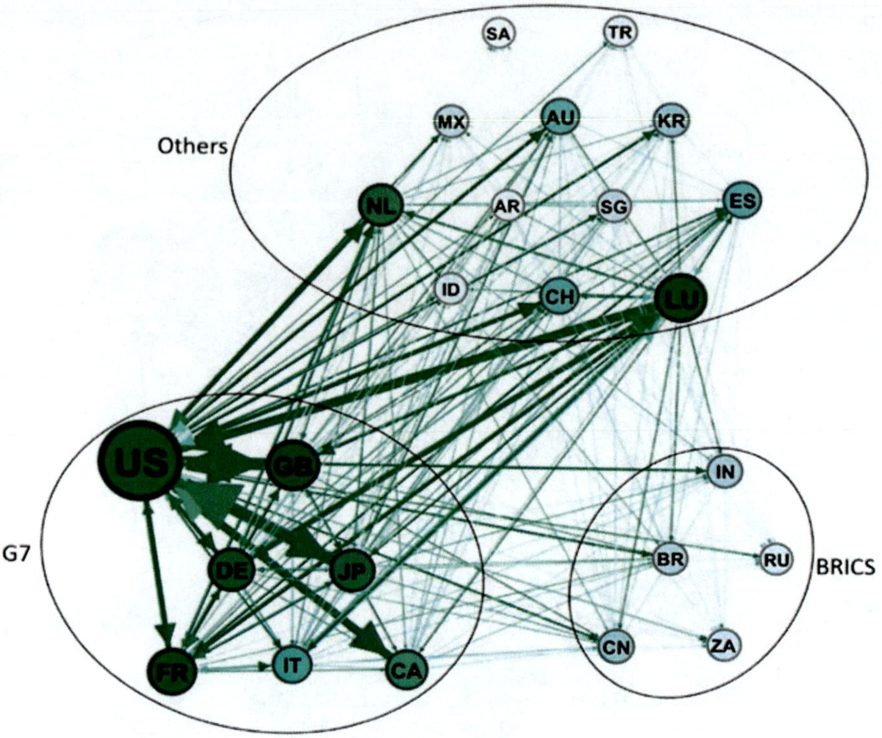

Fig. 2.3 Degree centrality in PI network (as of end-2018)

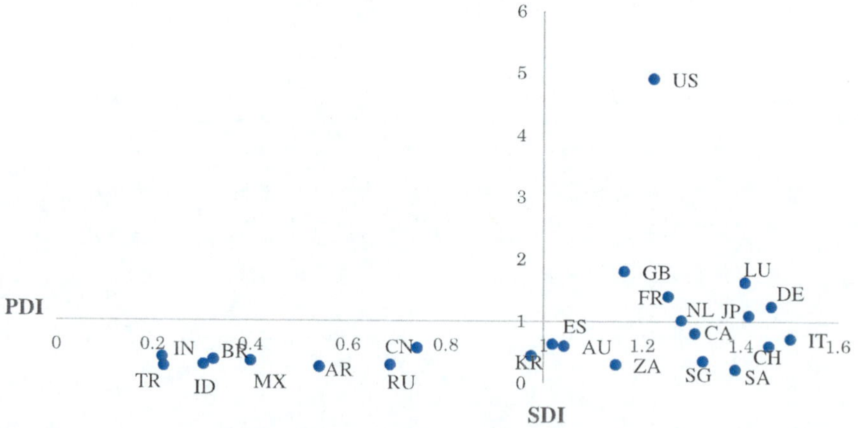

Fig. 2.4 Positions by PDI and SDI per the CPIS (as of end-2018)

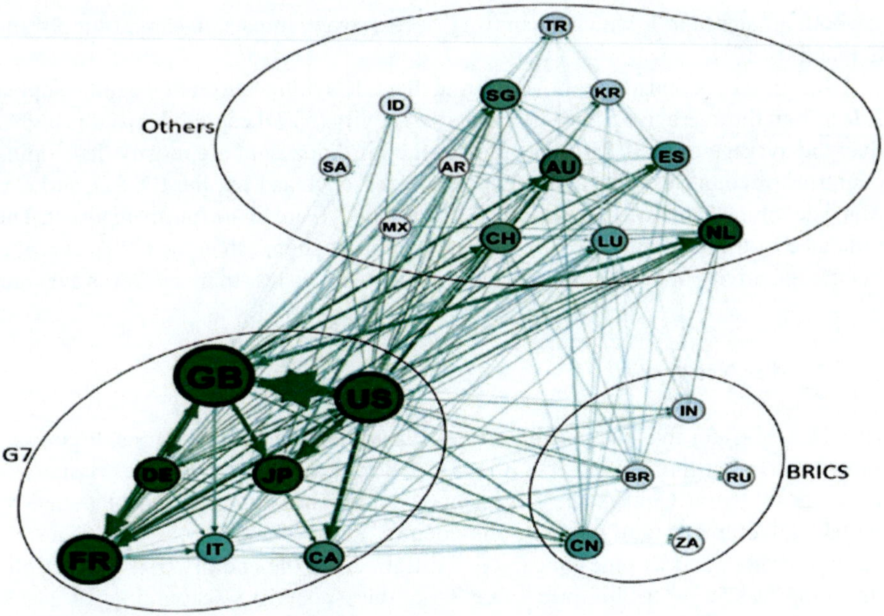

Fig. 2.5 Degree centrality in cross-border bank credit (as of end-2018)

FDI, as a form of transnational participation in each other's real economies, is conducive to expanding employment, technology transfer, and the introduction of advanced management methods. Moreover, the financial risk of DI is relatively low. Figure 2.2 was drawn using the method shown in Appendix, we plot the countries' positions with the PDI on the horizontal axis and the SDI on the vertical axis. This provides a visual representation of the degree centrality of countries in the FDI market. Figure 2.2 can be divided into four quadrants. Moving anticlockwise, the PDI and SDI in the upper right quadrant are higher than average (greater than 1). In the second quadrant, the PDI is less than 1, but the SDI is greater than 1. In the third quadrant, both the PDI and SDI are less than 1, i.e., below average. In the fourth quadrant, the PDI is greater than 1, but the SDI is less than 1. The quadrant in which a given country lies indicates its influence tendencies in global financial markets (Zhang & Zhao, 2019).

Figure 2.2 shows the differences in the G20 countries' status and influence in the FDI market. The US, NL, LU, and CN are in the first quadrant, indicating that these countries have a strong influence in the FDI market. In particular, the PDI of the US is 1.3, and its SDI is 2.6, which is the largest in the world. In terms of FDI, the NL ranked second in the world after the US, with the PDI of 1.33 and an SDI of 1.87. CN is also in the first quadrant, with the PDI of 1.02 and SDI of 1.18, indicating that CN has a strong ability to attract FDI, and its SDI is greater than 1, indicating that CN's sensitivity to the capital liability of FDI is above the G20's average. Except CN, the countries in the first quadrant are all advanced economies, their PDI and SDI are both greater than 1, which means they have a strong impact on capital supply and demand for FDI.

UK is in the second quadrant, indicating that UK's influence on FDI capital supply is less than the average of the G20, but its sensitivity to FDI capital demand is higher than the average of G20. Those located in the third quadrant are mostly developing countries or countries with small economic sizes, such as MX, IN, ID, SG, and ZA. Many developed countries, such as CH, JP, FR, and DE are in the fourth quadrant. The influence of these countries on FDI capital supply is higher than the G20's average, but the sensitivity of these countries to capital demand is less than the G20's average.

2.5.2.2 PIs Network

Here, we consider the degree centrality of PIs among the G20 countries. Figure 2.3 shows that, as at the end of 2018, the US, GB, and LU were the largest issuers of PI as these countries have the largest nodes. The high amount of debt issued by the US government partly reflects the country's fairly low savings rate, a current account deficit, and a domestic capital shortfall. As at the end of 2018, its portfolio financing was $1.44 trillion (see Table 2.4 in the Appendix), accounting for 24.5% of the global portfolio market. However, the US holds the largest share of assets in the global portfolio market, providing strong liquidity; its portfolio assets were $1.13 trillion in 2018, accounting for 19.1% of portfolio assets of the global market.

Here also, we divide the G20 countries into three groups based on their different stages of economic development—the G7, BRICS, and other economies. The financial networks show that within the other category, LU, the NL, CH, and ES have the largest nodes with thick edges. This indicates that securities trading between these developed countries is large. The network graph also shows that the nodes of the BRICS are still small, with thin edges, indicating that the BRICS are still developing in terms of securities investment.

Two additional points stand out. First, as at the end of 2018, JP was the largest foreign holder of debt issued by the US, holding $1516 billion, accounting for 10.5% of the US debt portfolio held by the G20 countries. LU ranked second, holding $1097.15 billion, whereas the United Kingdom was third, holding $1037 billion. Despite trade frictions with the US in recent years, CN holds $132 billion of the US debt portfolio, accounting for 1% of the total debt portfolio. This shows the trading relationships between major countries and the US with respect to PIs. Second, compared with other G20 countries, the US, GB, LU, FR, DE, and JP finance themselves more through debt portfolios, as shown by the sizes of their countries' nodes. In this regard, CN ranks 10th in the G20.

There are noteworthy differences in the G20 financial network of debt portfolios between 2015 and 2018. The size of most countries' nodes increased over the period, implying that in recent years, these countries have considerably increased their amount of outstanding debt. Over these three years, the US and GB increased their debt portfolios by a little over 20%; that of JP increased by 28.6%, and that of CN increased by 38%. Moreover, the net debt portfolio for the US decreased, but that of GB and CN increased. IN holds 0.96% of the G20's total PIs, and RU only holds 0.32% (Zhang, 2020, 319).

The distribution of the degree centrality of PIs in international portfolios is different from that of DIs, as shown in Fig. 2.4. In the figure, the US, GB, LU, FR, DE, JP, and NL are in the first quadrant, implying that the influence and sensitivity of these countries in the stock market are strong.

No country is in the second quadrant, whereas the third quadrant contains KR, CN, RU, AR, MX, BR, IN, ID, and TR. These countries are less influential in and responsive to PI than the other G20 countries. In addition, the more developed countries, such as IT, CH, SG, SA, NL, CA, ZA, ES, and AU, are distributed in the fourth quadrant. The influence of these countries on PI is less than the G20's average, but their sensitivity to the markets is higher than average.

2.5.3 *Changes in Degree Centrality in Cross-Border Bank Credit*

Cross-border bank credit is dominated by a small number of very sizeable links between banks in one country and borrowers in another. The largest-sized cross-border banking links are mainly between major advanced economies as shown by Fig. 2.5. Degree centrality increased up until the Great Financial Crisis (GFC) and has abated only slightly since. It is higher for interbank credit than for credit to the non-bank sector. Despite the substantial decline in interbank credit in the aftermath of the GFC, concentration in the interbank segment has remained high (Aldasoro & Ehlers, 2019).

2.5.3.1 Cross-Border Bank Credit Network

Cross-border links can be divided into two types. The first is as a result of receiving loans from other countries (debt), and the second arises from issuing credit to other countries (creditor rights). These links are measured by the in- and out-degree values, which are equal to claims and debts, respectively. The higher the in-degree value, the more the banking industry in a country is affected by the operations of the banking industry in other countries. The higher the out-degree value, the stronger the ability of a country's banking industry to spread its operations to other countries and the greater its influence on the operation of the banking industry in other countries.

To compare cross-border bank credit between 2018 and 2022, we employed the approach outlined in Sect. 2.2.2. The LBS data was utilized to compile the cross-border bank credit matrix for the years 2018 and 2022 (see Appendix Tables A.5 and A.6). Subsequently, applying formulas 2.3–2.7, we generated the cross-border bank credit network for 2018 and 2022, illustrated in Figs. 2.5 and 2.6.

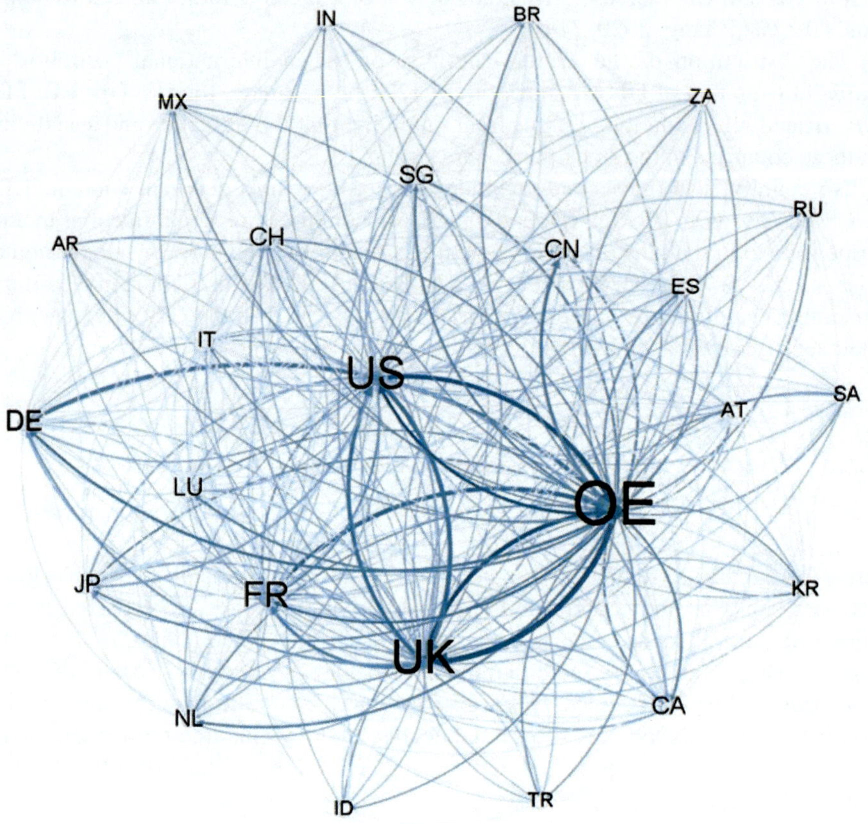

Fig. 2.6 Degree centrality in cross-border bank credit (as of end-2022)[11]

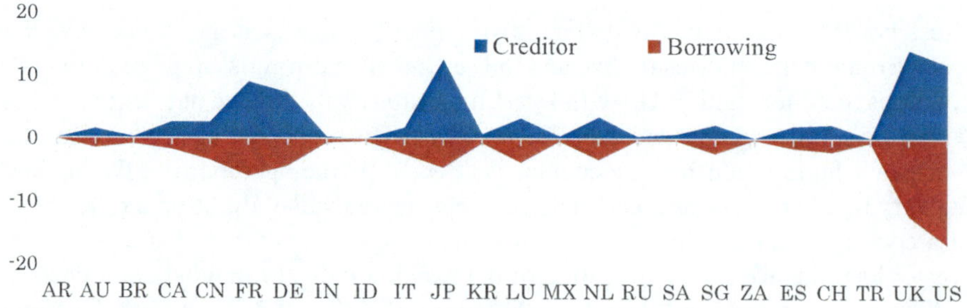

Fig. 2.7 Degree centrality by recipient and lender countries (%, as of end-2018)

[11] OE is others economies.

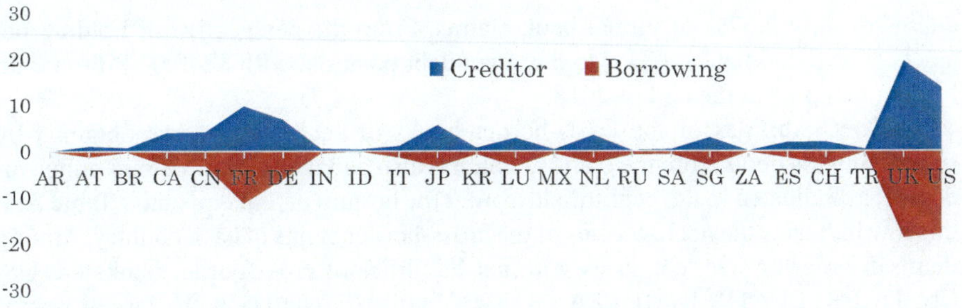

Fig. 2.8 Degree centrality by recipient and lender countries (%, as of end-2022)

Figure 2.5 is a network diagram that indicates the relationship between W-t-W and the scale of the credit position held by 24 countries and other economies in the G20. In Fig. 2.5, the lower left circle represents the G7 countries; the lower right circle represents the BRICS, and the upper circle represents the rest of the G20 countries. The node size is determined by degree centrality (Zhang, 2020, 384), which represents countries, and the thickness of the edge depends on the weight of the loans held by the G20 countries, which is based on the number of credit funds held by each other.

The nodes in Fig. 2.5 indicate that, from the perspective of the country's influence, the top eight countries are GB, US, FR, JP, DE, LU, NL, and CA. These countries have more influence than others, and this observation is the same as what is indicated by the fourth quadrant in Fig. 2.10. Additionally, the width of the edge of the network graph represents the amount of cross-border credit of these countries. The wider the edge, the larger the scale of cross-border bank credit between the country and other countries. In network terms, the weighted in-degree and weighted out-degree represent the relationship between the capital inflow and outflow of a country's cross-border credit.

In Fig. 2.5, we observe that the global bank credit is concentrated in the G7 (but the nodes of CA and IT are not very large), LU, NL, and CN. In the global bank credit market, the US still holds the largest share of loans, with 16.8% of global bank loans ($4.91 trillion). The US also held $3.45 trillion in foreign bank claims, accounting for 11.8% of global bank claims. Based on the share of loans, GB is in the second place; it held 12.24% of global bank loans ($3.58 trillion). GB held $4.13 trillion in foreign bank claims, accounting for 14.13% of global bank claims, which makes it easy to view GB as a "banking economy." FR was third and held a 5.92% share of loans, accounting for 9.01% of global bank claims.

JP's share of loans in the market is slightly lower than that of FR, GB, and the US, accounting for 4.57% ($1.34 trillion) of global bank loans. However, JP's proportion of financing through an international bank is larger than that of CN. JP also held $3.68 trillion in foreign bank claims, accounting for 12.57% of global bank claims, which is the world's second-largest holder of foreign bank assets.

CN is the seventh-largest holder of bank loans, accounting for 3.39% ($990.8 billion) of global bank loans. However, CN held $750.8 billion in foreign bank claims or only 2.57% of global bank claims. From the perspective of holding net assets of external banks, CN also has a net debt position, with $240.07 billion with foreign banks as at the end of 2018.

The net liabilities of the cross-border banks of each country are obtained by subtracting the debts from the claims, which is also equal to the out-degree minus in-degree, as indicated in the penultimate row at the bottom of 3 in Appendix Table A.5, this row indicates the net liabilities of the cross-border banks of each country. Among them, in order of size, countries with net liabilities of cross-border banks are US, CN, IT, BR, LU, TR, IN, ID, ZA, SG, MX, and AR. The US is the largest debtor of the cross-border banks, with $1.46 trillion, whereas CN ranks second at $240.07 billion. From the perspective of the net assets of cross-border banks, in order of size, they are JP, FR, DE, GB, CA, ES, SA, CH, NL, RU, KR, and AU. JP is the largest creditor of cross-border banks with $2.34 trillion, and FR is in second place with $903.9 billion.

Figure 2.6 shows the network diagram of cross-border bank credit as of end-2022. Compared with 2018, total cross-border bank credit in 2022 fell by $5.339 trillion after COVID-19 and the Russia-Ukraine war, and how degree centrality in cross-border bank connections responded highlights the structural nature of degree centrality.

As can be seen from the nodes in the network diagram in Fig. 2.6, the UK and the US are still the largest holders of cross-border bank credit. Among them, the UK holds cross-border bank credit assets and liabilities of $4.648 trillion and $4.521 trillion, respectively, while the United States is $3.293 trillion and $4.295 trillion, which is still the top 10 of G7 and LU and NL in the G20. Among them, a prominent change is that the balance of cross-border bank credit assets and liabilities held by China increased from 2018 to $988 billion and $761 billion, respectively, ranking sixth in assets and seventh in liabilities. Figures 2.7 and 2.8 show this change in degree centrality.

As shown in Figs. 2.7 and 2.8, a structural feature of cross-border banking from 2018 to 2022 is high concentration, which is high degree centrality, with a small number of very large bilateral links accounting for the majority share of total global cross-border bank credit. The largest linkages are almost exclusively between advanced economies, while links involving emerging market economies tend to be of smaller size. Cross-border banking links are highly concentrated. A small number of very large country-level links, mostly between advanced economies, dominate cross-border bank credit. Just five major creditor countries—UK, US, FR, DE, and JP–account for 50% of global cross-border credit (see Table A.6).

2.5.3.2 Changes in Degree Centrality from 2007 to 2022

The PDI and SDI are viewed as network centrality measures. The PDI and SDI, which are calculated from the inverse Leontief matrix, can be regarded as nodes in the W-t-W network. The PDI is a relative indicator of the amount of funds supplied to international markets, including indirect effects, when a country increases its use

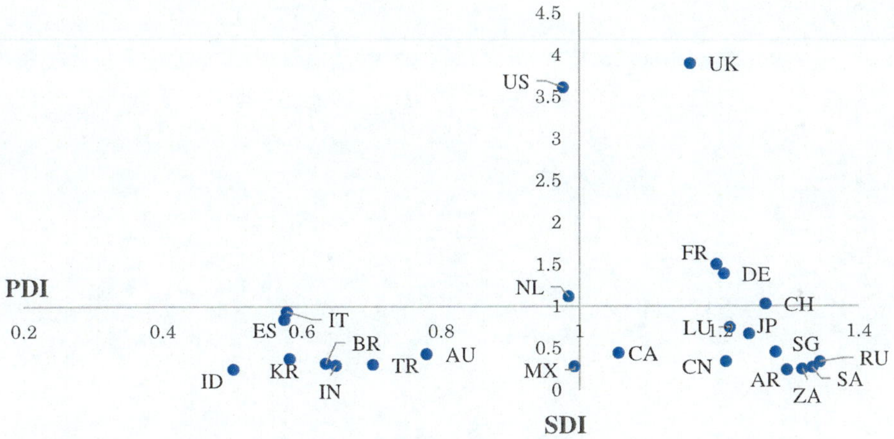

Fig. 2.9 Position of PDI and SDI by LBS (as at the end of 2007)

of funds. If direct funds are supplied to a country holding external net debt, the PDI will be small. In contrast, if countries with financing channels, including global and regional financial markets, supply funds, PDI will be larger. On the one hand, from the perspective of fund demand, when the global fund demand increases, the SDI of a country will be relatively lower when it obtains direct financing from other countries' banks. On the other hand, when the country obtains indirect financing from international markets or regional banks, its SDI will increase. Therefore, the size of a country's PDI largely depends on the asset portfolio of the country, whereas the size of the SDI largely depends on the liabilities portfolio of other countries. To facilitate the comparison of the impact of the 2007–2008 US financial crisis, COVID-19, and the Russia-Ukraine war on cross-border bank credit, we use the same method to draw the G20 network location map in 2007, 2018, and 2022 to reflect the changes in degree centrality, which are shown in Figs. 2.9, 2.10, and 2.11.

Figure 2.9 shows the position of the G20 countries in international credit markets as at the end of 2007 and after the 2007–2008 US financial crisis. In the figure, UK, FR, DE, and CH are in the first quadrant. Thus, in the international credit market, the asset influence and liability sensitivity of these four countries are higher than the average of the G20. Among them, the PDI and SDI of UK are 1.16 and 3.89, respectively, indicating that, as at the end of 2007, UK had the strongest sensitivity to bank lending. The US and NL are in the second quadrant; their SDI is higher than the G20's average, but their PDI is less than the G20's average. The US has an SDI value of 3.61, which is second only to that of UK. The capital needs of the international credit market have a strong ripple effect on the US and UK. When the capital needs of the international credit market doubled, the capital needs of UK and US banks increased by 3.89 and 3.61 times, respectively.

Countries in the third quadrant include ID, KR, ES, IT, BR, IN, TR, AU, and MX, whose PDI and SDI values are less than the G20's averages. The countries in the

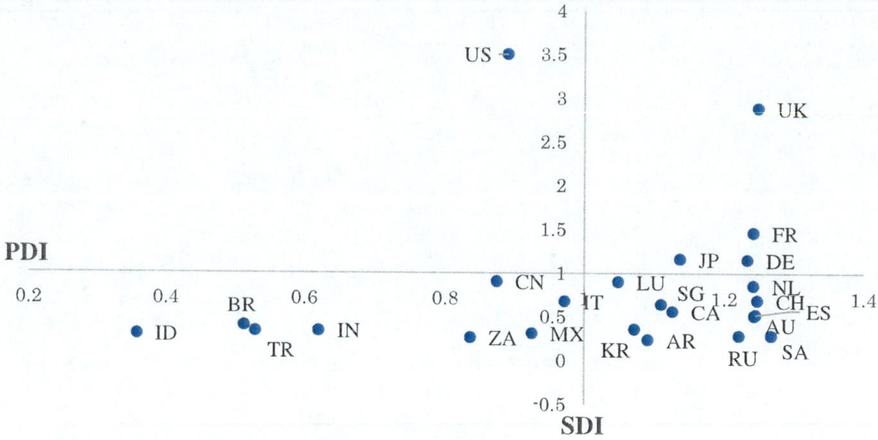

Fig. 2.10 Position of PDI and SDI by LBS (as of end-2018)

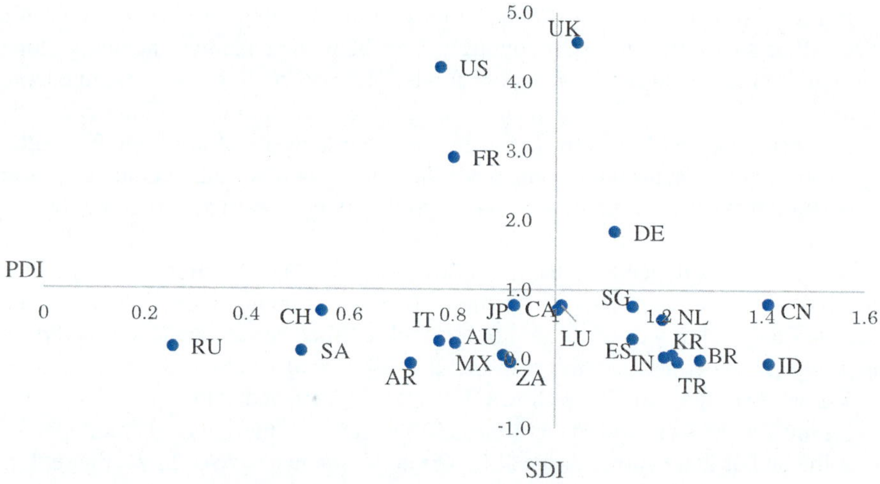

Fig. 2.11 Position of PDI and SDI by LBS (as of end-2022)

fourth quadrant are CA, CN, LU, JP, SG, AR, ZA, SA, and RU, which have more influence on bank assets than the G20's average but have weaker sensitivity to their liabilities. JP and CN are in the second quadrant; however, JP's PDI and SDI are slightly higher than CN's. Thus, as at the end of 2007, JP's influence and sensitivity in the international credit markets were greater than CN's.

To observe the changes in international bank funding and credit during the 2007 financial crisis in the US, we used the same data source (see Appendix Table A.5) to map the G20 network of cross-border loans as at the end of 2018, as shown in Fig. 2.10. The figure indicates that, relative to 2007, UK, FR, and DE continued to be in the first quadrant, whereas JP entered the first quadrant in 2018. The US was still in the second quadrant; although its SDI did not change much, its PDI declined slightly to 0.89. A significant change in the third quadrant is CN's placement in that quadrant. Thus, as at the end of 2018, CN's PDI and SDI were less than the G20's average. Moreover, KR, AU, and ES moved from the third to fourth quadrant. The changes in the international loans market from 2007 to 2018 show that the US has basically maintained its original position in this coordinate; JP has upgraded, but CN has declined.

Referring to Fig. 2.11, we can observe the current degree centrality status in cross-border bank credit networks for the year 2022 as compared to 2007 and 2018. The dominance of the UK, US, DE, and FR in cross-border bank credit remains consistent and has not undergone any significant changes. Specifically, throughout this period, the UK consistently occupied the first quadrant, with the US consistently positioned in the second quadrant. The asset influence of the UK and the financing capability of the US ultimately held the most favorable positions. On another hand, China has steadily enhanced its standing in the realm of cross-border bank credit, transitioning from the third quadrant to the fourth quadrant. In contrast, Japan has exhibited a declining trend, moving from the first quadrant in 2018 to the third quadrant in 2022. Furthermore, due to the economic sanctions imposed by Europe and the US in response to the Russia-Ukraine war, RU is in a most obvious decline position in 2022. Both its PDI index and SDI index are lower than those of any other country in the G20.

2.6 Conclusions

This study uses a new statistical approach to measure GFF and establishes a new statistical model based on the economic theory of the GFF. This model depicts the structure, influence, and sensitivity of the GFF at stock levels. The approach and data sources are elaborated. Moreover, the structure and equilibrium of the GFF matrix of the G20 countries are detailed to provide a meaningful case study using a GFF matrix of CN, JP, and the US. This study makes the following four main contributions.

First, Table 2.3, which builds on prior theoretical constructs in the literature, is an innovation due to its provision of an operational statistical system framework and is the core of this study. The data in Table 2.3 make GFF a reality, which serves as the basis for useful metrics contained in Tables A.3, A.4, A.5 and A.6 in the Appendix (the External Asset and Liabilities Matrix for 2018 and 2022). Therefore, based on the tables, other financial instrument matrices can be constructed to meet the needs of policy-making authorities.

Second, this is the first study to compare national financial exposures across G20 economies using the GFF analysis framework. We used CDIS, CPIS, and LBS data to estimate bilateral financial exposures between G20 economies and connected national financial networks through cross-border exposures by merging information from the CDIS and CPIS datasets. We calculated the PDI and SDI of the G20 countries for DI, PI, and cross-border banks and identified the advantages and disadvantages for each country.

Third, we introduced the financial network analysis method into the cross-border bank credit of the G20, and network correlation and EC analyses were conducted. Thus, the structural relationship between CN, JP, and the US is clarified.

Fourth, the analysis uses cross-border bank credit in the G20 countries to construct a cross-border financing matrix based on the W-t-W benchmark, and statistical description and analysis are conducted based on this matrix. Moreover, the assessment of changes in interconnectedness between countries following the 2007–2008 financial crisis and the COVID-19 pandemic in 2022 is contingent upon the specific financial instruments and countries being examined. Cross-border interbank links are highly concentrated. A small number of very large country-level links, mostly between advanced economies, dominate cross-border bank credit. CN has steadily enhanced its standing in the realm of cross-border bank credit, JP has exhibited a declining trend, and RU is in a most obvious decline position in 2022.

National exposures in cross-border bank credit among countries have witnessed an increase, notably in the US, CN, and JP. Generally, the exposures to cross-border bank credit and the expanding financial bubble in CN, particularly within the financial sector, have experienced a more pronounced increase compared to the exposures in the US and JP.

There are some limitations of this study, which can be addressed in future studies. First, the accuracy of the GFF table has to be improved, especially the processing of the reserve data. The data about reserves are not included in the current external asset and monetary matrix because of the mismatch of data sources. CPIS, CDIS, and LBS have their own information system; these information systems can be kept in accordance with the W-t-W-based matrix. However, the data about reserves are from IIP and cannot be kept in accordance with the W-t-W based matrix. Therefore, the integration and matching of IIP, CPIS, CDIS, and LBS data systems should be strengthened.

Second, the function of the GFF matrix should be enhanced. Based on the established stock table of GFF, the function should be extended. For example, the GFF matrix should be extended to flow categories, such as transactions and revaluations, counterparty country sectors, and domestic interactions.

Third, in the future, we will improve the financial network analysis method, explore new approaches, and expand the network theory. This will include the development of centrality measures of GFF, which directly represents the net interconnections, particularly eigenvector centrality, capturing direct and indirect links between financial instruments.

Appendix A: The Method for Constructing LBS Matrix

The process of converting BLS account data to a matrix is as follows.

Table A.1 Canadian example

Table A6.2-S	Banks' cross-border positions on residents of Canada Outstanding at end-December 2022, in millions of US dollars
Country	Canada
Dataset	Locational banking statistics (LBS_D_PUB)
Data updated	'23/10/2023 09:32
Data URL	http://stats.bis.org:8089/statx/srs/table/A6.2?c=CA&p=20224&f=xlsx

Selection and Download of Relevant Data

Relevant data can be selected from LBS and its data source can see the Composition of Locational Banking Statistics (LBS).[12] It consists of two parts, **Global tables** and **Country tables**. Select A6.2 By country (residence) of counterparty and location of reporting bank from Country tables which shows the location of reporting bank.

Select Database

Select and download the G-20 data, and the countries are listed in 24 columns[13] in order A, B, and C, etc., such as Canada in A6.2 (see Table A.1).

Setting of "Columns" and "Rows"

In the all Countries state, the columns of the matrix are set to assets (or liabilities), and the data in the columns are taken from the "**Of which: loans and deposits**" of all sectors in the liabilities side (or Assets side).[14] Some countries, such as Argentina, China, India, Indonesia, Russia, Saudi Arabia, Singapore, and Turkey are not listed in the list of countries on the left of Table 2.1. Therefore, the above countries need to be inserted into the columns of the matrix in order of A, B, and C.

Refer to Table 2.1 and set relevant countries in the order of A, B, and C that was set with the "rows". When some countries are not listed in Table 2.1, insert countries such as the above eight countries according to the order of the rows. Select the data on the asset side (or liabilities side) of A6.2 for these countries, and put them into the rows. Pay particular attention to the corresponding relationship between the column sequence and the row sequence of the object countries.

---

[12] https://stats.bis.org/statx/toc/lbs.html.

[13] In 2022, we mainly selected 24 countries from the G-20.

[14] Setting the "columns" of the matrix as assets or liabilities that depends on the purpose of the study. Please refer to the part of Data sources from BIS which written in the Sect. 2.2.2.

Handling of Row and Column Sums and the Items of "Others" and "Totals"

The data of other terms are calculated by subtracting the data of the observed country from the sum of columns or rows. Since the data of the above eight countries are inserted in the row, the items of "others" which is by total of countries in each column minus the observed country may have a negative number. Therefore, we may go to use three aggregate numbers to get the items of the "totals" such as the following ways:

(1) BLS'A5 Location of reporting bank (All reporting countries). Positions reported by banking offices located in the specified country regardless of the nationality of the controlling parent. By instrument "Of which: loans and deposits". Table A.2 shows the summary information from A5 and A6.2 of BLS, and the summary data of the asset side and the liability side are exactly the same in A5 and A6.2.Once the total items are determined, the other items are determined by subtracting the countries of observation from the totals.
(2) All countries (total) in A6.2 (See Table A.1), A6.2's data is the same as that of A5. But if there are no relevant country data in All countries (total) in A6.2, such as China, India, etc., then use the following method.
(3) If there are no relevant national data in **A6.2 all countries (total)**, such as the above 8 countries, select the total data of these countries in **A6.2_Select a country**. It needs to make sure that the data ranges of "Other" and "Total" for each country are consistent.

Appendix B: The Correspondence Between the Summarized Data in A5 and A6.2 of LBS

See Table A.2 in Appendix.

Table A.2 The correspondence between the summarized data in A5 and A6.2 of LBS

From A5 All reporting countries of BLS

	Outstanding					
	Cross-border positions		Local positions		Unallocated positions by residence	
Q4 2022	Claims	Liabilities	Claims	Liabilities	Claims	Liabilities
	Q4 22	Q4 22	Q4 22	Q4 22	Q4 22	Q4 22
Total	*36,499,974*	***33,382,494***	81,136,407	78,415,201	223,569	2,953,334
	…	…	…	…	…	…
By instrument						
Of which: loans and deposits	**22,619,197**	**23,909,589**	48,847,045	70,410,917	23,149	23,540

From A6.2 all countries (total) of BLS

	Claims				Liabilities			
Q4 2022	All sectors		Of which: non-banks		All sectors		Of which: non-banks	
	All instruments	Of which: loans and deposits	All instruments	Of which: loans and deposits	All instruments	Of which: loans and deposits	All instruments	Of which: loans and deposits
Cross-border positions	36,499,974	**22,619,197**	16,717,122	8,991,179	***33,382,494***	**23,909,589**	11,484,627	9,233,783

Appendix C: Calculation Method of PDI and SDI

The influence and sensitivity coefficients are defined as follows. We set the position of the two-way financial investment from country i (as a row) to country j (as a column) and set the number of observation objects as n. Then, Table A.3 in the Appendix is set by forms with matrix Y formed by n rows and n columns, as shown in Table A.3 in the Appendix.

Set $T_i = T_j = \max\left(\sum_{i=1}^{n} y_{ij}, \sum_{j=1}^{n} y_{ij}\right)$, $\varepsilon_j = T_j - \sum_{i=1}^{n} y_{ij}$, and $\rho_i = T_i - \sum_{j=1}^{n} y_{ij}$,

T is the total of rows or columns for the matrix Y of external assets/liabilities; the total of the rows equals the total of the columns of each country. Designating ε_i as the net liabilities of country i, and ρ_j as the net assets of country j, if the net assets of country i are nonnegative, $\varepsilon_i = 0$ and $\rho_j > 0$; and if the net assets of country i are negative, $\varepsilon_i > 0$, and $\rho_j = 0$. To illustrate the effect of the influence and sensitivity coefficients, we first need to define the input coefficient c_{ij}. The input coefficient c_{ij} is the ratio of funds raised from country i to the total external financing of country j, i.e.,

$$c_{ij} = \frac{y_{ij}}{T_j}$$

From the direction of the rows in Table A.3 in the Appendix, we arrive at the following equilibrium equation:

$$\sum_{j=1}^{n} y_{ij} + \varepsilon_i = \sum_{j=1}^{n} c_{ij} T_j + \varepsilon_i = T_i \tag{2.15}$$

where C is the $n \times n$ matrix composed of the elements of c_{ij}. Thus, the equilibrium equations can be rewritten as

$$CT + \varepsilon = T \tag{2.16}$$

Solving for T yields.

$$T = (I - C)^{-1} \tag{2.17}$$

where Eq. (2.17) is the Leontief inverse. Denoting the inverse matrix as $\mathbf{\Gamma} = (\mathbf{I} - \mathbf{C})^{-1}$, which has elements $\gamma_{i,j}$, we can denote country j's PDI as μ_j^y and its SDI as σ_i^y; then, they can be defined as follows. And the PDI and SDI of Tables A.3, A.4, A.5 and A.6 in the Appendix can be calculated separately using the same method.

$$\mu_j^y = \frac{\sum_{i=1}^{n} \gamma_{i,j}}{\frac{1}{n}\sum_{j=1}^{n}\sum_{i=1}^{n} \gamma_{i,j}} \tag{2.18}$$

$$\sigma_i^y = \frac{\sum_{j=1}^{n} \gamma_{i,j}}{\frac{1}{n}\sum_{i=1}^{n}\sum_{j=1}^{n} \gamma_{i,j}} \tag{2.19}$$

The numerator in Eq. (2.18) is the sum of the eigenvector of the column (asset side for a country) of the Leontief inverse, and its denominator is the average of the total of rows in Leontief inverse. We can derive the country j's IC using Eq. (2.18). The numerator in Eq. (2.19) is the sum of the eigenvector of the row (liability side for a country) of the Leontief inverse, and the denominator is its average column total. We can derive the country i's SDI using Eq. (2.19).

.

.

Table A.3 FDI matrix (as of end-2018, millions of USD)

	AR	AU	BR	CA	CN	FR	DE	IN	ID	IT	JP	KR	LU	MX
AR		0	4908	2010	1010	2156	2388	6	0	1594	0	378	4004	1631
AU	0		176	29,235	24,988	18,453	13,370	276	608	2694	65,321	11,764	18,231	0
BR	467	3100		9177	3052	27,089	13,986	260	136	13,250	21,266	6274	58,000	8158
CA	0	23,772	**−2503**		11,081	9176	16,733	555	35	3199	17,070	4135	102,405	1169
CN	0	9812	918	9963		24,017	91,014	831	17,602	11,720	121,165	77,643	11,581	**−46**
FR	1	2228	776	7190	6360		98,798	111	15,841	33,901	15,516	1065	143,383	0
DE	0	0	82	8345	11,988	81,426		395	1	43,767	22,467	4409	148,667	**−127**
IN	0	1145	**−13**	1788	2156	6333	26,329		88	6959	24,280	12,443	2756	180
ID	0	1611	2	2317	8835	318	2792	242		855	31,155	7735	981	0
IT	1	0	757	305	1857	104,196	38,717	85	0		4141	909	78,163	0
JP	0	927	221	6423	3160	28,803	10,225	57	**−8**	2135		7089	6634	0
KR		765	0	1812	6128	5619	10,225	463	**-285**	1717	38,372		1608	0
LU		0	8289	75,148	14,316	60,091	182,689	99	**−0**	44,692	12,698	6183		0
MX	1172	601	1270	19,505	744	5588	14,187	108	0	2446	11,914	3047	25,911	
NL	1571	5424	**−17,746**	26,762	18,129	177,372	173,772	13,015	**−1033**	64,234	116,814	8399	789,281	26,137
RU	0	179	6	748	6578	20,885	20,511	88	0	12,163	1523	3052	12,166	0

(continued)

Table A.3 (continued)

	AR	AU	BR	CA	CN	FR	DE	IN	ID	IT	JP	KR	LU	MX
SA		0	0	1	484	5795	1912	238	0	5300	5203	210	124	0
SG		17,852	231	6985	35,970	10,249	15,518	15,280	30,161	940	71,573	16,182	44,443	0
ZA		759	93	630	6000	2049	6430	389	107	2036	7104	223	8302	0
ES	342	34	7579	5864	829	57,169	67,328	191	7	41,468	7358	955	81,935	11,395
CH	80	258	**−21,644**	7090	4748	57,366	42,151	2945	28	9661	5909	444	418,379	465
TR		0	167	947	1602	3941	9624	41	3	7706	0	1195	5630	0
GB	0	83,217	6406	74,194	16,542	148,105	144,983	4721	368	25,292	154,551	10,352	730,851	13,372
US	4666	91,697	20,371	459,192	67,038	237,198	304,209	11,808	320	41,987	487,940	90,626	523,722	65,555
Others	33,928	253,632	198,084	230,148	1,728,676	414,532	343,595	30,689	8786	175,185	325,424	109,309	1,261,168	31,438
Total assets	42,228	497,011	208,431	985,781	1,982,270	1,507,926	1,651,488	82,893	72,765	554,900	1,568,766	384,021	4,478,326	159,328
Net liabilites	41,725	43,064	342,590	0	0	0	0	231,055	50,089	0	0	0	0	206,625
Total	83,953	540,075	551,021	985,781	1,982,270	1,507,926	1,651,488	313,948	122,854	554,900	1,568,766	384,021	4,478,326	365,953

	NL	RU	SA	SG	ZA	ES	CH	TR	GB	US	Others	Total liabilites	Net assets	Total
AR	13,804	11	0	0	57	24,869	3067	0	0	9522	12,538	83,953		83,953
AU	80,673	455	0	0	8343	1505	6592	**−12**	47,867	163,999	45,537	540,075		540,075

(continued)

Table A.3 (continued)

	NL	RU	SA	SG	ZA	ES	CH	TR	GB	US	Others	Total liabilites	Net assets	Total
BR	180,722	0	0	0	544	60,307	10,433	0	13,334	79,032	42,433	551,021		551,021
CA	257,820	1669	0	0	271	11,690	28,304	16	35,472	368,498	38,096	928,665	57,116	985,781
CN	25,615	254	0	0	120,425	3554	22,628	191	16,665	109,332	780,047	1,454,931	527,339	1,982,270
FR	185,172	2979	0	0	1056	22,763	63,888	99	98,704	68,035	92,240	860,105	647,821	1,507,926
DE	333,825	8125	0	0	3931	24,687	59,647	1727	38,573	137,148	130,349	1,059,432	592,056	1,651,488
IN	25,379	121	0	0	1211	2921	6743	201	18,457	42,444	132,028	313,948		313,948
ID	22,262	5	0	0	50	84	1528	**−49**	6782	10,240	25,108	122,854		122,854
IT	172,061	2775	0	0	820	12,262	18,221	140	20,228	33,080	34,600	523,317	31,584	554,900
JP	37,312	55	0	0	68	471	16,914	3	8934	113,254	26,147	268,824	1,299,941	1,568,766
KR	22,176	33	0	0	0	1881	3909	2	7154	39,021	12,145	152,746	231,275	384,021
LU	319,537	**−12,113**	0	0	2129	9838	193,568	535	149,011	726,121	648,112	2,440,945	2,037,381	4,478,326
MX	114,681	6	0	0	163	42,007	5536	**−8**	12,908	95,873	8296	365,953		365,953
NL		40,415	0	0	3859	**−4744**	154,576	17,573	189,099	810,238	590,920	3,204,069	2,477,870	5,681,939
RU	94,345		0	0	1389	1020	23,300	695	13,483	14,071	182,015	408,217		408,217
SA	11,751	0	0	0	**−1**	1089	1135	**−47**	5727	10,884	6328	56,136		56,136
SG	57,274	3471	0	0	358	673	25,363	32	11,713	254,670	127,353	746,290		746,290
ZA	24,181	35	0	0	0	1199	2506	8	13,250	7313	11,976	94,591	151,848	246,439
ES	168,339	6441	0	0	323		10,958	207	92,871	35,425	63,668	660,688		660,688
CH	483,950	17,760	0	0	4250	12,414		205	54,608	253,253	123,604	1,477,925	10,420	1,488,345
TR	22,773	8229	0	0	2	5760	2309		8189	3903	27,447	109,467		109,467

(continued)

Table A.3 (continued)

	NL	RU	SA	SG	ZA	ES	CH	TR	GB	US	Others	Total liabilites	Net assets	Total
GB	669,120	6378	0	0	23,556	118,636	81,713	4114		796,564	365,805	3,478,840		3,478,840
US	875,817	7332	0	0	15,903	99,592	296,364	1815	400,970		346,053	4,450,174	1,350,851	5,801,025
Others	1,483,349	252,158	0	0	57,731	145,628	449,143	17,049	485,179	1,619,105		11,917,188		11,917,188
Total assets	5,681,939	346,593			246,439	600,106	1,488,345	44,496	1,749,177	5,801,025	6,136,100			
Net liabilites	0	61,624	56,136	746,290	0	60,582	0	64,971	1,729,663	0	5,781,088			
Total	5,681,939	408,217	56,136	746,290	246,439	660,688	1,488,345	109,467	3,478,840	5,801,025	11,917,188			

Data Source IMF's CDIS, http://www.imf.org/external/data.htm, May 10, 2021

Table A.4 PI matrix (as of end-2018, millions of USD)

	AR	AU	BR	CA	CN	FR	DE	IN	ID	IT	JP	KR	LU	MX
AR		252	364	1452	234	753	2625	0	2	2951	1060	178	15,552	14
AU	0		11	25,525	9319	34,286	48,098	2	455	6788	146,000	13,018	45,831	
BR	471	3695		15,068	1638	8160	5047	52	0	1290	12,906	10,723	45,461	3100
CA	1	24,546	121		5126	31,076	60,762	1	3	4518	67,610	8304	72,932	45
CN	0	13,089	8	24,615		11,170	5829	536	1086	735	22,982	14,822	54,441	119
FR	0	26,766	220	39,837	6023		403,500	22	6	171,721	258,069	23,457	418,172	168
DE	60	30,047	221	31,757	10,583	186,620		1	15	77,313	121,766	7759	351,651	179
IN	0	6868	4	17,531	983	7091	4255		1857	314	19,352	3060	53,070	5
ID	0	0	0	5677	999	1432	7439	62		1320	10,134	1428	29,282	22
IT	0	4412	65	8282	1769	227,970	135,159	7	28		54,498	1853	169,275	19
JP	0	40,476	25	62,038	10,130	148,094	32,835	19	133	10,434		21,858	144,482	8
KR	0	12,229	1288	18,222	5652	10,511	8298	5	120	623	22,025		41,998	14
LU	298	11,792	1707	11,676	12,044	463,586	625,880	264	1963	650,383	110,059	24,929		1629
MX	11	2776	1148	10,011	872	11,314	17,000	0	6	4303	19,572	1274	40,655	
NL	0	22,044	377	24,881	3027	283,453	276,843	6	7036	60,339	112,967	8058	202,031	94
RU	3	1374	1	3082	1570	2547	4217	0		1058	3230	881	20,930	16
SA	0	127	0	250	52	1656	1085	5	67	126	1003	495	1921	5

(continued)

Table A.4 (continued)

	AR	AU	BR	CA	CN	FR	DE	IN	ID	IT	JP	KR	LU	MX
SG	0	11,375	22	5688	5986	2202	6715	53	664	686	16,481	5124	20,026	
ZA	0	2532	1	6343	579	1626	4314	2	0	1199	7507	1024	21,555	
ES	66	4802	1661	9519	1103	177,367	145,493	0	10	102,194	52,062	2814	109,685	464
CH	0	12,404	1461	29,534	3938	28,131	54,502	8	9	9401	31,243	5163	83,951	76
TR	0	497	5	2682	306	2662	4132	0	40	1897	5069	334	17,234	9
GB	39	67,516	518	87,458	15,927	239,266	194,598	35	9	60,938	170,246	31,099	362,506	239
US	28,484	333,372	16,422	988,562	132,022	298,069	387,734	4475	3669	130,317	1,515,981	207,054	1,097,152	21,461
Others	214	159,776	15,238	170,081	268,075	544,008	862,239	270	4918	276,874	1,286,953	70,274	1,019,581	27,405
Total assets	29,649	792,767	40,889	1,599,772	497,957	2,723,051	3,298,601	5825	22,094	1,577,724	4,068,775	464,979	4,439,373	55,095
Net liabilites	50,656	321,111	370,306	168,563	679,585	689,014		543,768	194,950			134,459		300,744
Total	80,305	1,113,879	411,195	1,768,335	1,177,542	3,412,065	3,298,601	549,593	217,044	1,577,724	4,068,775	599,438	4,439,373	355,839

	NL	RU	SA	SG	ZA	ES	CH	TR	GB	US	Others	Total liabilites	Net assets	Total
AR	2481	55	154	0	13	625	734	3	2433	34,430	13,940	80,305		80,305
AU	28,909	90	1473	27,729	986	2083	24,963	0	64,851	332,553	300,910	1,113,879		1,113,879
BR	12,732	75	3772	0	544	0	4601	16	26,443	168,692	86,709	411,195		411,195
CA	27,984	257	287	18,535	1525	2260	35,351	10	55,360	981,173	370,549	1,768,335		1,768,335
CN	12,516	28	12,357	111,724	1105	187	6860	9	57,967	159,127	666,232	1,177,542		1,177,542

(continued)

Table A.4 (continued)

	NL	RU	SA	SG	ZA	ES	CH	TR	GB	US	Others	Total liabilites	Net assets	Total
FR	192,320	426	7188	0	1041	69,381	74,126	8	176,736	556,593	986,285	3,412,065		3,412,065
DE	226,997	2490	5664	0	552	28,142	78,708	33	146,754	398,767	1,164,484	2,870,564	428,036	3,298,601
IN	9941	32	2198	50,851	631	0	3917	9	30,004	176,016	161,604	549,593		549,593
ID	7923	63	2457	23,866	182	141	2415	3	10,667	66,617	44,913	217,044		217,044
IT	35,168	63	2361	0	1274	129,386	9427	1	46,271	115,191	330,081	1,272,560	305,164	1,577,724
JP	45,039	40	31,930	0	442	4386	32,440	1	145,272	1,007,631	744,695	2,482,409	1,586,366	4,068,775
KR	14,622	19	6131	47,323	31	424	12,396	1	33,042	213,376	151,090	599,438		599,438
LU	108,943	18,582	1074	24,052	14,783	168,673	222,249	45	118,562	138,919	941,392	3,673,484	765,889	4,439,373
MX	9091	483	3117	0		5853	6825	2	19,661	146,286	55,580	355,839		355,839
NL		4841	1806	12,532	542	38,320	62,913	12	104,557	452,853	489,618	2,169,151		2,169,151
RU	9683		1603	0	106	0	2994	16	10,898	56,599	48,677	169,486		169,486
SA	101	10		0	0	0	268	1	1986	7584	16,218	32,961	190,828	223,789
SG	9411	389	666		535	128	4337	0	18,860	88,738	119,080	317,169	934,048	1,251,217
ZA	9420	56	2384	0	0	0	2700	0	17,438	91,279	32,065	202,024		202,024
ES	47,179	15	5766	0	91		12,084	15	39,075	138,743	280,519	1,130,725		1,130,725
CH	24,589	109	6154	0	5663	6105		7	58,207	458,052	172,386	991,095	321,129	1,312,224

(continued)

Table A.4 (continued)

	NL	RU	SA	SG	ZA	ES	CH	TR	GB	US	Others	Total liabilites	Net assets	Total
TR	4660	714	1112	0	58	333	1658		7463	27,911	20,585	99,362		99,362
GB	118,662	4235	13,549	37,267	55,246	32,660	77,034	32		1,359,579	1,440,965	4,369,623		4,369,623
US	493,217	3807	53,349	341,544	17,387	54,150	302,204	493	1,037,504		6,942,522	14,410,951		14,410,951
Others	467,587	31,671	57,238	555,794	44,230	179,495	331,020	427	901,801	4,105,577		15,012,857	4,200,352	19,213,209
Total assets	1,919,175	68,551	223,789	1,251,217	146,966	722,733	1,312,224	1143	3,131,812	11,282,286	19,213,209			
Net liabilites	249,976	100,935			55,058	407,992		98,219	1,237,811	3,128,665				
Total	2,169,151	169,486	223,789	1,251,217	202,024	1,130,725	1,312,224	99,362	4,369,623	14,410,951	19,213,209			

Data Source IMF CPIS, http://www.imf.org/external/data.htm, May 10, 2021

Table A.5 Cross-border banking credit matrix (as of end-2018, millions of USD)

	AR	AU	BR	CA	CN	FR	DE	IN	ID	IT	JP	KR	LU	MX
AR	0	1	513	3	0	683	468	0	0	71	579	57	61	132
AU	23	0	9	5516	17,500	12,173	20,021	4888	780	1010	91,087	4774	4957	0
BR	17	515	0	0	911	17,943	3329	7	0	502	11,416	2507	10,771	2618
CA	52	9216	42	0	11,285	17,913	31,249	249	131	958	48,331	2128	9122	127
CN	0	41,568	34	7017	0	34,012	22,713	0	0	2309	57,687	46,319	18,753	33
FR	188	19,798	5662	5559	8250	0	215,538	2308	193	84,719	184,912	1846	116,045	237
DE	524	9125	473	5589	18,925	102,405	0	1558	359	73,985	80,881	5061	93,509	217
IN	0	3072	0	1419	0	6584	10,263	0	0	322	22,315	4388	262	0
ID	0	1399	0	0	0	4601	3627	0	0	240	20,221	6495	259	0
IT	141	448	139	0	684	330,215	76,909	30	18	0	30,108	536	36,481	6
JP	21	15,833	528	34,966	23,890	187,625	34,083	2062	1396	3361	0	7620	6230	137
KR	13	3517	169	1101	22,764	14,014	7305	904	315	51	29,939	0	755	4
LU	79	1123	2825	31,524	11,271	191,802	186,655	107	37	30,695	152,636	752	0	7
MX	0	185	209	0	100	3738	3115	0	0	492	11,631	3157	237	0
NL	364	6535	1590	4257	0	120,658	176,064	0	0	10,866	61,991	1350	24,188	52
RU	0	10	0	4	0	9184	8108	0	0	6590	4185	1494	4316	5

(continued)

Table A.5 (continued)

	AR	AU	BR	CA	CN	FR	DE	IN	ID	IT	JP	KR	LU	MX
SA	0	56	0	176	0	7157	2300	0	0	200	3137	4203	2666	0
SG	0	32,060	1	6224	0	34,415	31,497	0	0	244	156,478	5757	7194	100
ZA	3	286	2	35	2238	3254	2588	602	1	359	5608	226	997	16
ES	959	269	8445	0	4914	118,213	68,704	15	8	44,390	27,178	318	9917	728
CH	3328	2220	23	783	1346	68,344	61,495	227	158	5080	26,530	660	42,052	30
TR	0	13	0	706	0	11,527	22,147	0	0	10,292	4314	1904	1796	0
GB	372	147,320	4502	65,088	70,742	516,543	332,537	22,394	706	78,085	273,357	5290	68,032	397
US	13,970	51,080	8614	399,548	124,076	287,595	254,255	9159	6267	50,335	1,132,671	35,098	40,798	11,801
Others	1518	102,095	51,653	184,701	431,866	535,303	551,050	28,236	7934	105,225	1,240,012	80,099	459,626	101,165
Total assets	21,572	447,744	85,433	754,216	750,762	2,635,901	2,126,020	72,746	18,303	510,381	3,677,204	222,039	959,024	117,812
Net liabilites	2555	0	170,283	0	240,072	0	0	106,265	98,648	214,068	0	0	130,737	4715
Total	24,127	447,744	255,716	754,216	990,834	2,635,901	2,126,020	179,011	116,951	724,449	3,677,204	222,039	1,089,761	122,527

	NL	RU	SA	SG	ZA	ES	CH	TR	GB	US	Others	Total liabilites	Net assets	Total
AR	2209	0	0	0	0	1464	739	0	4063	6384	6700	24,127	0	24,127

(continued)

Table A.5 (continued)

	NL	RU	SA	SG	ZA	ES	CH	TR	GB	US	Others	Total liabilites	Net assets	Total
AU	0	91	2206	24,559	474	719	4353	25	58,040	46,598	147,267	447,070	674	447,744
BR	7359	0	0	68	261	11,409	1966	1	16,949	61,973	105,194	255,716	0	255,716
CA	8732	74	241	2428	139	1885	7226	94	89,839	207,529	53,833	502,823	251,393	754,216
CN	4625	0	0	0	537	5391	3529	0	71,739	34,340	640,228	990,834	0	990,834
FR	77,774	19,491	14,576	18,251	1206	65,461	58,279	326	526,711	121,665	183,025	1,732,020	903,881	2,635,901
DE	86,634	8756	2817	13,293	1322	34,910	44,826	3726	347,916	75,670	252,177	1,264,658	861,362	2,126,020
IN	1064	0	0	0	475	426	1545	0	31,937	16,938	78,001	179,011	0	179,011
ID	0	0	0	0	0	121	493	0	4579	2662	72,254	116,951	0	116,951
IT	10,395	1850	3397	124	34	72,747	6225	812	76,032	4491	72,627	724,449	0	724,449
JP	2958	106	202	72,820	204	503	7623	68	344,607	375,144	214,060	1,336,047	2,341,157	3,677,204
KR	0	103	21	18,557	0	87	2902	2	16,337	17,730	65,891	202,481	19,558	222,039
LU	27,206	1560	315	7823	299	11,248	56,397	95	136,969	62,380	175,956	1,089,761	0	1,089,761
MX	1169	0	0	2	2	30,591	3001	0	4537	40,271	20,090	122,527	0	122,527
NL	0	0	0	14,672	1481	44,398	22,463	3161	251,442	41,244	182,862	969,638	48,215	1,017,853
RU	2591	0	0	0	9	983	2085	0	17,355	168	37,091	94,178	34,148	128,326

(continued)

Table A.5 (continued)

	NL	RU	SA	SG	ZA	ES	CH	TR	GB	US	Others	Total liabilites	Net assets	Total
SA	0	0	0	0	5	611	3389	0	30,927	3917	34,071	92,815	126,115	218,930
SG	46,285	0	0		260	1349	18,754	0	88,901	47,528	198,513	675,560	0	675,560
ZA	656	4	5	51	0	235	554	55	21,601	1780	17,915	59,071	0	59,071
ES	15,522	2363	11,809	676	3	0	4719	282	53,192	6915	43,267	422,806	181,053	603,859
CH	26,819	15,583	10,696	13,936	436	7232	0	7191	212,861	16,315	69,804	593,149	71,354	664,503
TR	10,937	0	0	0	55	7653	4853		29,943	1931	56,275	164,346	0	164,346
GB	279,121	17,991	87,360	73,132	14,174	123,509	146,812	10,647	0	689,270	551,521	3,578,902	553,793	4,132,695
US	123,314	10,193	17,019	51,914	2436	70,956	120,676	5821	1,216,557		868,535	4,912,688	0	4,912,688
Others	282,483	50,161	68,266	343,240	11,890	109,971	141,094	18,065	479,661	1,571,560		8,697,644	0	8,697,644
Total assets	1,017,853	128,326	218,930	655,546	35,702	603,859	664,503	50,371	4,132,695	3,454,403	5,887,927			
Net liabilites	0	0	0	20,014	23,369	0	0	113,975	0	1,458,285	2,809,717			
Total	1,017,853	128,326	218,930	675,560	59,071	603,859	664,503	164,346	4,132,695	4,912,688	8,697,644			

Data Source BIS international banking statistics, https://www.bis.org/statistics/about_banking_stats.htm, June 20, 2021

Table A.6 Cross-border banking credit matrix (as of end-2022, millions of USD)

	AR	AT	BR	CA	CN	FR	DE	IN	ID	IT	JP	KR	LU	MX
AR		21	2	0		622				135	0	10	29	0
AT	1		0	7038	20,494	18,897	2493	1709	474	561	16,832	1536	804	4
BR	165	68		257	1	521	327	0		154	38	17	1536	29
CA	1	1956	1571		15,615	5398	1136	0	82	179	47,346	577	1186	958
CN	0	19,296	925	22,081		32,277		0	0	1601		19,484	12,891	196
FR	539	11,771	3446	8956	20,764		203,683	4129	3060	98,604	203,311	4263	82,816	183
DE		262	1	370	0	52,001		0		3272	0	1254	29,376	0
IN	0	2672	9	855	0	3604		0		35		905	152	0
ID	0	747	1	87	0	206		0		9		788	36	0
IT	17	1379	427	217	328	138,945	33,305	63	42	0	9150	7	23,796	1
JP	0	11,698	856	5829		35,774	0	0		2194	0	8897	4307	488
KR	12	1331	418	259	26,211	1836	3391	3931	5406	19	9739		76	324
LU	41	1490	1239	4738	8135	252,691	128,577	680	71	24,477	5366	146		18
MX	280	20	5	11,309	12	4431	178	0	0	64	144	27	498	
NL	0	13,777	4748	6365		149,496	72,533	634	0	8144	2564	237	21,304	28
RU	0	52	0	58		22,763				2855		898	3935	0
SA	0	70	0	0		9334				2877	17	28	496	0

(continued)

Table A.6 (continued)

	AR	AT	BR	CA	CN	FR	DE	IN	ID	IT	JP	KR	LU	MX
SG	0	43,790	21	5758		26,831				184	561	23,312	4133	3
ZA		539	0	89	812	2913	2199			110	299	8	274	0
ES	1183	223	110	493	1018	85,594	41,246	604	155	32,278	561	15	15,347	1156
CH	454	3728	223	7916	2591	120,754	24,390	502	303	2674	11,464	288	59,911	22
TR	0	30	0	220		914				153		48	309	0
UK	257	32,760	1795	149,516	48,495	744,366	427,761	28,366	1753	23,904	340,959	7797	29,995	555
US	3394	81,951	43,992	424,598	20,947	313,303	83,889	18,371	3325	6893	447,702	25,845	9741	3978
Others	6688	32,384	132,187	293,258	822,868	307,659	619,343	68,731	72,086	61,876	250,418	91,384	387,133	82,284
Total assets	13,032	262,015	191,976	950,267	988,291	2,331,130	1,644,451	127,720	86,757	273,252	1,346,471	187,771	690,081	90,227
Net liabilites	6976	47,002				391,125				65,184			115,316	51,076
Total	20,008	309,017	191,976	950,267	988,291	2,722,255	1,644,451	127,720	86,757	338,436	1,346,471	187,771	805,397	141,303

	NL	RU	SA	SG	ZA	ES	CH	TR	UK	US	Others	Total liabilites	Net assets	Total
AR	62				2	1089	1572		239	13,056	3169	20,008		20,008
AT	5194	5	64	27,639	83	317	1101	6	80,176	59,218	64,371	309,017		309,017

(continued)

Table A.6 (continued)

	NL	RU	SA	SG	ZA	ES	CH	TR	UK	US	Others	Total liabilites	Net assets	Total
BR	345		0	1	5	593	1589	0	22,577	25,011	17,707	70,941	121,035	191,976
CA	0	15	65	15,861	71	363	2354	266	131,872	248,339	263,596	738,807	211,460	950,267
CN	0				1373	274	4820		72,469	134,937	437,980	760,604	227,687	988,291
FR	89,071	4755	5339	32,159	519	93,334	66,968	6778	248,444	225,397	1,303,966	2,722,255		2,722,255
DE	26,998		0	0	10	9847	13,381	0	25,488	12,915	1,283,736	1,458,911	185,540	1,644,451
IN	0		0		175	24	1290		38,455	22,870	35,403	106,449	21,271	127,720
ID	0				0	10	112		3621	9038	11,423	26,078	60,679	86,757
IT	8510	2287	143	340	30	20,252	18,293	3453	23,706	20,590	33,155	338,436		338,436
JP	0				88	107	1224	0	175,824	326,627	225,967	799,880	546,591	1,346,471
KR	0	630	2980	7468	6	545	849	2587	16,685	44,110	35,974	164,787	22,984	187,771
LU	13,547	2433	1399	2998	33	12,127	35,659	1375	80,630	71,928	155,599	805,397		805,397
MX	0	0		49	1	2848	7493	0	11,362	87,000	15,582	141,303		141,303
NL		657	154		1190	27,573	23,650	4472	157,017	45,808	105,049	645,400	180,586	825,986
RU	0			0	6	895	15,346		6888	5950	137,872	197,518		197,518
SA	0			0	8	3629	4301	0	76,904	17,949	62,182	177,795		177,795
SG	0	0		0	92	230	27,638		142,442	89,245	380,649	744,889		744,889

(continued)

Table A.6 (continued)

	NL	RU	SA	SG	ZA	ES	CH	TR	UK	US	Others	Total liabilites	Net assets	Total
ZA	681	0	3	600		50	1264	1	16,728	4819	8232	39,621		39,621
ES	0	511	1526	5551	20		7241	1200	44,201	31,379	87,432	359,044	97,755	456,799
CH	25,624	864	3245	9673	144	9185			226,246	49,690	174,230	734,121		734,121
TR	2322				7	108	2810	2759	10,434	6310	19,354	45,778	52,187	97,965
UK	88,210	6932	28,677	111,322	9358	33,389	174,823	22,097		1,062,079	1,145,408	4,520,574	126,955	4,647,529
US	87,821	730	5223	78,269	1777	17,243	22,158	1468	1,337,720		1,254,258	4,294,596		4,294,596
Others	477,601	31,545	37,195	409,505	15,005	222,767	31,169	51,503	1,697,401	678,966		3,687,380	381,338	4,068,718
Total assets	825,986	51,364	86,013	701,435	30,003	456,799	467,105	97,965	4,647,529	3,293,231	4,068,718	23,909,589		
Net liabilites		146,154	91,782	43,454	9618		267,016			1,001,365	0			
Total	825,986	197,518	177,795	744,889	39,621	456,799	734,121	97,965	4,647,529	4,294,596	4,068,718			

Data Source BIS international banking statistics, https://www.bis.org/statistics/about_banking_stats.htm, December 20, 2023

References

Aldasoro, I., & Ehlers, T. (2019). Concentration in cross-border banking. *BIS Quarterly Review*, June 2019. https://www.bis.org/publ/qtrpdf/r_qt1906b.pdf

Antoun de Almeida, L. (2015) A network analysis of sectoral accounts: Identifying sectoral interlinkages in G-4 economies. In *IMF Working Paper WP/15/111*.

Castrén, O., & Rancan, M. (2014). Macro-networks: An application to the Euro area financial accounts. *Journal of Banking & Finance, 46*, 43–58.

Copeland, M. A. (1952). *A study of money flows in the United States* (pp. 103–285). National Bureau of Economic Research.

ECB Website for Journalists. www.euro-area-statistics.org

European Communities, International Monetary Fund, Organisation for Economic Co-operation and Development, United Nations and World Bank. (2009). System of National Account 2008, Sales No. E.08.XVII.29, United Nations, New York.

Errico, L., Walton, R., Hierro, A., AbuShanab, H., & Amidžic, G. (2013). Global flow of funds: Mapping bilateral geographic flows. In: *Proceedings 59th ISI World Statistics Congress* (pp. 2825–2830), Hong Kong.

Errico, L., Harutyunyan, A., Loukoianova, E., Walton, R., Korniyenko, Y., Amidžic, G., AbuShanab, H., & Shin, H. S. (2014). Mapping the shadow banking system through a global flow of funds analysis. In: *IMF Working Paper WP/14/10*, Washington, DC.

Girón, C., Vives, M. R., & Matas, A. (2018). Propagation of quantity shocks in who-to-whom networks. In *The 35th IARIW General Conference*, Copenhagen, Denmark. http://www.iariw.org/copenhagen/giron.pdf

Hendricks, D., Kambhu, J., & Mosser, P. (2006). Systemic risk and the financial system. In *Back Ground Paper for Conference*, Federal Reserve Bank of New York.

Hendricks, D. (2009). Defining systemic risk. Financial Reform Project. https://www.issuelab.org/resources/8957/8957.pdf

IMF, BIS, & Financial Stability Board. (2009). The financial crisis and information gaps. In *Report to the G-20 Finance Ministers and Central Bank Governors*, October 28, 2009. https://www.imf.org/external/np/g20/pdf/100109.pdf

IMF. (2006). *Financial soundness indicators compilation guide* (pp. 17-63). International Monetary Fund, Publication Services, Washington, DC.

IMF. (2013). *Balance of payments and international investment position manual* (6th ed.) (BPM6).

IMF. (2016). *Monetary and financial statistics manual and compilation guide (MFSMCG)* (pp. 55–90). International Monetary Fund, Publication Services, Washington, DC.

IMF. (2022a). CDIS Table 6: Direct investment positions by all reporting economies cross-classified by counterpart economies—IMF Data. Accessed 25 December 2023.

IMF. (2022b). Coordinated portfolio investment survey—data tables—IMF Data. Accessed 22 May 2023.

Ishida, S. (1993) *Flow of funds in the Japanese economy* (in Japanese). Tokyo, Keizai Shimpo-Sha (pp. 169–190).

Kimmo, S., & Samantha, C. (2016). *Network theory and financial risk*. Risk Books, a Division of Incisive Media Investments Ltd.

Klein, L. R. (1983). *Lectures in econometrics* (pp. 35–44). North-Holland.

Li, Y., & Zhang, Y. J. (2020). *China's national balance sheet 2020*. China Social Sciences Press.

Robert, H. (2013). Why are the G20 data gaps initiative and the SDDS plus relevant for financial stability analysis? In *IMF Working Paper WP/13/6*.

Shrestha, M., Mink, R., & Fassler, S. (2012). An integrated framework for financial positions and flows on a from-Whom-to-Whom basis: Concepts, status, and prospects. In *IMF Working Paper WP/12/57*.

Stone, R. (1966). The social accounts from a consumer's point of view. *Review of Income and Wealth, 12*(1), 19–24.

The People's Bank of China. (2021). The People's Bank of China Quarterly Statistical Bulletin.

Tsujimura, K., & Mizosita, M. (2002). *Flow-of-funds analysis: Fundamental technique and policy evaluation* (pp. 30–49, 63–98). Keio University Press (in Japanese).
Tsujimura, K., & Tsujimura, M. (2018). A flow of funds analysis of the US quantitative easing. *Economic Systems Research, Taylor & Francis Journals, 30*(2), 137–177.
Zhang, N. (2016) Measuring global flow of funds: Theoretical framework, data sources and approaches. In *Contemporary works in economic sciences: Legal information, economics, OR and mathematics* (pp. 47–60). Kyushu University Press.
Zhang, N., & Zhao, X. (2019). Measuring global flow of funds: A case study on China, Japan and the United States. *Economic Systems Research, 31*(1); The International Scholarly Journal of the International Input-Output Association (IIOA).
Zhang, N. (2020). *Flow of funds analysis: The innovation and development* (pp. 283–368). Springer.
Zhang, N., & Zhu, L. (2021). Global flow of funds as a network: The case study of the G20. *Japanese Journal of Monetary and Financial Economics, 9*, 21–56.

Chapter 3
Structural Changes in China–US External Flow of Funds: Statistical Estimates Based on the VEC Model

Abstract This study constructs an analytical framework of the external flow of funds (EFF) to observe the process and obstacles of China and the United States (US) decoupling, examining the China–US structural relationship in savings and investment imbalance during 1980–2022. We observe the issues between China and the US in the external financial assets and liabilities by stock data, focusing on the external adjustment mode in 2008–2022. We construct a vector error correction model to calculate the quantitative relationship between short-term fluctuations and long-term trends of the EFF in China and the United States and explore the basic causes of economic conflicts between the two sides. This chapter discusses the risk of China–US economic decoupling and US debt, the strategic challenges both sides face, and the prospect of countermeasures.

Keywords Global flow of funds · Mirror image · Balance sheet · Vector error correction model

3.1 Introduction

The 1976 docudrama film "All the President's Men" popularized the catchphrase, "Follow the money," which became part of the American lexicon to cut through the lies and deceptions and find the truth during the Watergate scandal.[1] Interestingly, Mr. Nixon, the subject of the Watergate scandal, was also the first US President to visit China in 1972. Significant changes have occurred in the relationship between China and the United States (US) since the 1970s; however, after 50 years of diplomatic relations between the countries, widespread talks of decoupling are currently taking place. Since the beginning of the twenty-first century, political and economic cooperation and disputes have occurred between China and the US; in the last decade, disputes have exceeded cooperation. Therefore, this paper uses the phrase "follow

[1] NPR (2012), https://www.npr.org/2012/06/16/154997482/follow-the-money-on-the-trail-of-watergate-lore.

N. Zhang and Y. Zhang, *Global Flow of Funds Analysis*,
https://doi.org/10.1007/978-981-97-1029-4_3

the money" to focus on the global flow of funds (GFF) and explore the nature and strategic challenges of China–US decoupling from a statistical perspective.

The following studies focus on the balance of payments and international capital flows in the US since the twenty-first century. First, the thesis of Greenspan's Conundrum (2004) indicates that while the US maintained its current account deficit since the 1980s, a large amount of international capital flowed into the US because the short-term interest rate in the US was higher than the long-term interest rate in the early twenty-first century. This disparity triggered capital inflow from emerging market countries like China and India to the US. These factors contribute to the current US account deficit, which Federal Reserve (Fed) presidents later interpreted as a "global saving glut" (Bernanke, 2005). Almost simultaneously, Obstfeld and Rogoff (2005) argued that this view was overly optimistic and proposed a plan to devalue the real effective exchange rate of the US dollar (USD) by 33% to restore the current account balance and resolve the global imbalance. Cavallo and Tille (2006) and Gourinchas and Rey (2007a, 2007b) concluded that the increase in capital gains from the combination of foreign assets and liabilities denominated in different dollars and foreign currencies maintained the current account deficit. Gourinchas and Rey (2007a) constructed a well-defined measure of cyclical external imbalances, emphasizing the problem of observing the US economy's external imbalance, which must be considered through the traditional "trade channel" and a previously unexplored "valuation channel."

The US current account deficit has remained unchanged since then. Thus, this strategy has failed, and exchange rate depreciation may have a short-term effect on the persistence of the current account deficit; however, it is not the primary method for addressing unbalanced growth. Lu (2008) used the term "mirror relationship" in research on China–US foreign trade relations to observe the practical manifestations and causes of the unbalanced relationship between the Chinese and US economies. Thus, the study could interpret the structural characteristics of economic growth in recent years and the realistic adjustment the two countries face.

Iwamoto (2007, 2009, 2012, 2013, 2015) examined the mystery of the long-term current account deficit in the US through the lens of the outflow and inflow of foreign capital and net capital flow, indicating that the US earned enormous capital gains from the operation of foreign financial investment while maintaining its current account deficit. The study also separates one conjecture from the exchange rate fluctuation—a valuation channel in which the USD depreciation causes foreign exchange gain in US holdings of foreign assets denominated in local currency.

Blanchard and Milesi-Ferretti (2009, 2011) addressed two complex issues concerning current account deficits and surpluses. First, they examine why a country would seek to reduce its current account deficit or surplus. Second, they ask why the international community should demand more. They responded to the G20's request for the IMF to help develop "rules of the game" and reduce global current account imbalances by addressing these two issues and developing "indicative guidelines" for countries to follow.

Another viewpoint suggests that differences in the maturity of financial market systems could have created a "mirror image" between the US and China—the term "mirror image" is frequently used in the economics literature and financial statistics to describe and analyze opposite symmetric changes in economic phenomena and financial markets.[2] According to Willen (2004), if one country's financial market maturity is lower than another country's, the first country's savings rate will be higher, resulting in an external flow of funds (EFF) imbalance. Caballero (2008) investigated the relationship between global imbalances and low-interest rates using an equilibrium model, claiming that differences in the capabilities and assets each country provides to the world determine global imbalances. Mendoza et al. (2009) agreed that promoting financial market integration, reducing savings for financial development, and expanding loans from abroad could result in a global imbalance in the long run. Cauley (2015) questioned the asserted pecuniary benefits conferred by the dollar's international role. Furthermore, Gourinchas et al. (2019) investigated the implications of currency hegemony for the US external balance sheet, the international adjustment process, and the USD exchange rate predictability. Moreover, Gourinchas (2019) discussed the implications of living in a "dollar world" for policymakers and some potential challenges to the USD's hegemony.

However, the long-term time series of data relating to the US balance of payments and international investment positions since 2008 does not support this view. The US net external asset-to-gross domestic product (GDP) ratio fell from 7.49% in 1980 to −86.7% in 2021; conversely, the current account deficit increased from −6% of GDP in 2006 to −3.5% in 2021. Therefore, we believe that a financial operation, such as a currency mismatch, could postpone the change in the US current account deficit for some time; however, the fundamental reason for the change in the current account deficit can be found in the fundamental structure of a country's economic growth, namely, the structural relationship between savings and investment in the real economy.

Referring to the above research results, combined with the 2008 US financial crisis, the COVID-19 pandemic in 2020, and the changes in the international environment caused by the Russia–Ukraine war in early 2022, this paper attempts to analyze the causes, results, and prospects the mirror-image and decoupling relationship between China and the US from the perspective of the GFF. This study expands on the concept, research object, statistical description, and econometric analysis based on the vector error correction (VEC) model. GFF analysis connects the gap between domestic savings and investment with the surplus and shortage of funds, links the current balance with international capital flows, and extends the previous analysis of domestic capital flows to the analysis of GFF. We statistically describe and estimate the mirror-image relationship and exposure risks of China–US external fund flows since the 1990s, potential debt risks associated with economic decoupling, and the relationship's prospects and challenges.

The remainder of this paper proceeds as follows. Section 3.2 establishes a theoretical framework for the GFF analysis based on the equilibrium relationship between savings and investment and foreign trade and capital flows. The savings–investment

[2] IMF (2021), Coordinated Direct Investment Survey (CDIS), https://data.imf.org/regular.aspx?key=60564262.

equation is used to observe the real economies of China and the US. Section 3.3 discusses the potential reasons for the mirror-image relationship between China and the US in the GFF, including the countries' external adjustment, foreign investment returns and risks, and the sustainability of this relationship. Section 3.4 refers to the previous research of the VEC model, tests the unit root and co-integration relationship of eight variables related to the US GFF, and establishes the econometric model for measuring American foreign capital outflow and inflow. Section 3.5 statistically estimates the quantitative relationship of short-term fluctuations and long-term trends for US external capital flow and examines the structural problems of the China–US mirror-image relationship. Finally, Sect. 3.6 summarizes the results of statistical description and quantitative estimation, indicating that the root of the conflict between China and the US in the GFF over the past 40 years is the imbalance of domestic economic development. The final Section also addresses the main challenges and prospects currently facing China and the US.

3.2 Structural Issues in Economic Growth Between China and the United States

GFF is the international capital flow caused by financing and current account imbalances caused by the savings–investment gap; the GFF is divided into three convergent components: savings–investment flows, trade flows, and external capital flows. From the fund flow mechanism perspective, domestic capital surpluses and deficits in the flow of funds table correspond to the current account of the balance of payments. Moreover, the net financial investment in the rest of the world (ROW) sector in the flow of funds table corresponds to the financial account of the payment balances.

3.2.1 A New Framework for GFF Analysis

The GFF analysis relates the domestic savings–investment gap to the external financial surplus or deficit and observes international capital flows caused by current account adjustments. Furthermore, it investigates the relationship between the real and financial economies and the mutual influence of domestic and international capital flows through savings–investment, trade, and foreign capital flows. The GFF analysis is a broader extension of the flow of funds analysis and an expansion from domestic to international capital flows.

From the perspective of expenditures, GDP can be divided into final consumption, investment, and net exports of goods and services. When a country's domestic savings cannot meet its investment needs, there is a shortage of funds, and it is necessary to raise capital from overseas, producing an inflow of international capital. In contrast, when savings exceed the country's domestic capital needs, excess funds will be used to provide capital to other economies, such as through the purchase of bonds issued by other countries and so on, generating an outflow of domestic funds. According to the definition of the flow of funds statistics system, the real economy and the financial economy have the following ex-post equilibrium relationship with the balance of payments.

$$(S_p - I_p) + (T - G) = (\Delta A - \Delta L) = EX - IM \tag{3.1}$$

where, ΔA is the change in assets, ΔL is the change in liabilities, EX is the output and IM is the input. Namely, the difference between Private Savings (S_p) and Private Investment (I_p) + the excess of Government Revenues (T, for Taxes) over Government Expenditures (G) = Financial Surplus or Deficit = Current Account. Government expenditures consist of government consumption expenditures (GC) and government investment (GI), that is, $T-G=T-GC-GI$. Therefore, the equilibrium for savings and investment for the government sector can be expressed as:

$$(T - G) = (T - GC - GI) = S_g - I_g \tag{3.2}$$

In other words, Government Revenues–Government Expenditures = Government Savings–Government Investment. As can be seen in Eqs. (3.1) and (3.2), the sum of the differences between savings and investment for the private sector and for the government sector is equal to the net financial investment in the rest of the world sector, that is, the external current account.

If $\Delta A-\Delta L$ in (3.1) is further decomposed into domestic and foreign financial transactions ($\Delta A_d-\Delta L_d$) and ($\Delta A_f-\Delta L_f$), the following formula is valid:

$$(S_p - I_p) + (S_g - I_g) = (\Delta A_d - \Delta L_d) + (\Delta A_f - \Delta L_f) = EX - IM \tag{3.3}$$

Since ($\Delta A_d-\Delta L_d$) cancels out across domestic sectors, the following relationship can be obtained from the balance of payments statement.

$$EX - IM = \Delta A_f - \Delta L_f \text{ (Current Account = Fiancial Account)} \tag{3.4}$$

Equations[3] (3.3) and (3.4) indicate the theoretical equilibrium relationship between savings and investment, financial surplus, or deficit and balance of payments. According to this equilibrium relationship, the financial surplus or deficit with the rest of the world sector is consistent with the deficit or surplus of the current account and has a post-event equilibrium relationship corresponding to the balance of domestic savings and investment.

Let $r_{t-1} \vartriangle L_{t-1}$ be the interest payments on external debt, and we set $CRA = FRA_t - FRA_{t-1}$, where CRA = the change in reserve assets and FRA = the stock of

[3] According to the Balance of Payments and International Investment Position Manual (BPM6), the Current Account + Capital account − Financial Account + error or omission = 0. To highlight the main relationship between external physical transactions and financial transactions, capital account, and error terms are omitted here.

foreign reserves assets. This allows us to transform Formula (3.3) into

$$(EX_t - IM_t) - (\triangle A_t - \triangle L_t - r_{t-1} \triangle L_{t-1}) - (FRA_t - FRA_{t-1}) = 0 \quad (3.5)$$

Formula (3.5) presents an equilibrium, highlighting several areas where a crisis can occur in the international flow of funds. The first is when the current account deficit is too large *(IM > EM)* for pre-foreign exchange reserves to handle. The second source of short-term capital outflows includes changes in stock market returns, market interest rates, and foreign exchange rates, which cause short-term capital outflows to significantly exceed international capital inflows. In this case, a lack of foreign exchange reserves to meet the needs of domestic capital may precipitate a currency crisis. The third is an external debt payment crisis caused by current and capital account deficits. The fourth case is when exchange rates fluctuate rapidly, causing a currency to appreciate or depreciate significantly, eventually leading to frequent crises in the current account, capital account, external debt payments, and so on.

Equations (3.1) and (3.5) show the theoretical equilibrium relationship between savings and investment balance, capital surplus or shortage, and international balance of payments. This equilibrium relationship is essentially what a balance in GFF represents: the extension of the domestic flow of funds, which connects domestic economies with the rest of the world. The domestic capital surplus and deficit are consistent with the current account deficit (surplus), which corresponds with the financial account balance, according to this equilibrium relationship. We need a new analytical framework that corresponds to the GFF's operational mechanism and can serve as the foundation of a statistical monitoring system to test external financial stability and observe systemic financial risk using the GFF. The analytical framework, which links the domestic savings–investment balance, current account balance, and international capital flows, reflects the interdependence of the domestic flow of funds and international capital movements.

According to Equations (3.1)–(3.5), two key points determine whether China and the US can maintain their "external sustainability." The first is structural changes in savings and investment in the respective domestic real economies, and the second is the ability of the US to repay its massive and growing external debt. We examine each of these two significant issues in Section 3.3.

3.2.2 Construct an Investment–Savings Equation

To assess the relationship between savings and investment rates and the demand for foreign capital inflows represented by domestic savings deficits, we develop the investment–saving equation, shown in formula (3.6).

$$(I/Y)_i = \alpha + \beta_1 (S/Y)_i + \beta_2 (D * S/Y)_i \quad (3.6)$$

where $(I/Y)_i$ is the ratio of gross domestic investment to gross domestic product (GDP) in country i, and $(S/Y)_i$ is the corresponding ratio of gross domestic savings to GDP. Because we use cross-sectional data from China and the US from 1980 to 2022, and because the 2007 financial crisis in the US had a structural impact on the economic growth of both countries, we added a time dummy variable to Eq. (3.1) using 2007 as the transition point of the trend change. Before we present the results, let us discuss how to interpret this basic equation. We include a time dummy variable $(D * S/Y)_i$ in Eq. (3.6) because it is a long-term time series, and we consider the structural impact of the global financial crisis in 2007–2008 on the conversion of savings into investment in China and the US. First, we set D to 0 from 1980 to 2007, and then D is set to 1. Although China and the US have different political systems, economic models, and market maturity, this model demonstrates the heterogeneity of each country. The same elements in economic operations can be extracted from the common System National Accounts (SNA) dataset, and the ratio relationship between savings and investment can be compared internationally. Table 3.1 displays the statistical estimation results.

Although Eq. (3.6) analyzes domestic savings and investment to determine the extent of global capital mobility, the equation can also be interpreted in terms of foreign investment flows. Because the excess of gross domestic investment over gross domestic saving equals the net inflow of foreign investment, we can run a regression of the net foreign investment inflow to GDP ratio on the domestic saving ratio to generate a coefficient of β. If β = 1, domestic savings are fully used for domestic investment. If β < 1, it means that not all of the funds used for investment are from domestic sources and that foreign capital must be brought in to meet the domestic investment needs. Testing the hypothesis that β equals 1 is equivalent to testing the hypothesis that international capital flows do not depend on domestic savings rates.

In this case, sample data from China, and the US are used to perform a statistical F-test, assuming that all explanatory variable parameter estimates are 1.

$$H_0 : \beta_1 = \beta_2 = 1$$

The sample size is 42, the degree of freedom is (k − 1, n − k) = (2, 39), and the significance level is set at 1%, resulting in a critical value of 5.19. The sample parameters' estimated F-statistics are all greater than this critical value (Table 3.1), so the above hypothesis testing can be rejected.

Table 3.1 The statistical estimation of savings in investment (1980–2022)

	α		β_1		β_2		R-squared	F-statistic
	Coefficient	t-Statistic	Coefficient	t-Statistic	Coefficient	t-Statistic		
China	19.77	6.56	0.44	5.76	0.08	4.25	0.81	86.30
The U.S	13.01	9.01	0.49	6.58	−0.08	−5.04	0.68	41.86

Data Source IMF, World Economic Outlook Databases, April 2022

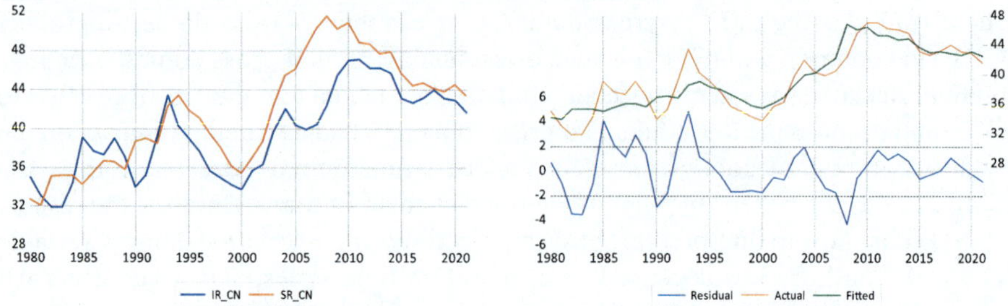

(a) (b)

Fig. 3.1 **a** China's savings and investment rate and **b** Estimates of China's investment rate

Estimates of β that are close to 1 indicate that most incremental savings in a given country remain in that country. Note that a finding that β is close to 1 could reflect that a high rate of return stimulates both domestic savings and domestic investment; however, this interpretation is inconsistent with the hypothesis of perfect world capital mobility,[4] in which the domestic savings rate is unaffected by domestic investment opportunities. Nonetheless, there is a positive correlation between savings and investment in China and the US, according to the change trajectory of the saving rate and investment rate in Figs. 3.1, 3.2 and 3.3. The correlation coefficient in China was 0.85 and 0.69 in the US. A country with a higher degree of capital market openness has a lower correlation between its savings rate and investment rate, and vice versa. Different domestic market systems and tax systems for foreign direct investment (FDI), the risk of exchange rate fluctuations, asymmetric information, and other factors all contribute to capital flow risks, resulting in the Feldstein–Horioka puzzle (1980).

The assumption of perfect capital mobility contradicts the traditional Keynesian interpretation, which holds that exogenous changes in the level of investment cause income to fluctuate until the resulting savings level equals investment. Whatever the merits of this argument for a closed economy, it is inapplicable if domestic savings are added to the global capital pool. A high observed value of β could reflect other common causes of variations in both savings and investment; however, finding high values of β would be strong evidence against the hypothesis of perfect world capital mobility, putting the burden of identifying such common causal factors squarely on those who defend that hypothesis.

With perfect global capital mobility, an increase in the savings rate in the country *i* leads to an increase in investments in other countries; the distribution of incremental capital among countries varies positively with each country's initial capital stock and inversely with the elasticity of the country's marginal product of the capital schedule. In the extreme case where country *i* is infinitesimally small compared to the global economy, the β value implied by perfect global capital mobility is 0; however, with perfect global capital mobility, even a relatively large country's β value would be

[4] Feldstein-Horioka (1980).

the order of magnitude of its share of total global capital. Table 3.1 shows that the coefficient of β_1 is greater than that of β_2 for China and the US. β_2 is a time dummy variable that indicates that the impact of the savings rate on the investment rate in China and the US has decreased since 2007. The negative value of β_2 in the US indicates the extent to which the country uses foreign capital to compensate for lacking domestic investment and scales of international capital inflows, which have increased significantly since 2007.

3.2.3 Unbalance of Savings and Investment in China and the United States

Table 3.1 indicates that the estimated values of β1 and β2 in China and the US are all between 0 and 1, β1 is greater than β2, and β2 is a time dummy variable. China's β1 and β2 maintain the same sign, but the estimated value of β1 and β2 in the US shifts from positive to negative. Table 3.1 shows that China's β1 is 0.44, while β2 is still 0.08, indicating a positive value despite the decline. In the US, the value of β1 is 0.49, less than the parameter estimates for China, where the value of β2 is −0.08. T statistics show the significance of each estimate.

Table 3.1 illustrates two important points. First, domestic savings rates, and foreign capital flows affect investment in China and the US. Second, the impact of the financial crisis in the US has reduced the impact of savings rates on investment rates in various countries since 2008; however, the scale and influence of international capital flows have increased, as has the proportion of foreign capital used to supplement the lack of domestic investment. As a result of domestic and foreign capital portfolio management, China's investment rate tends to decline; while the investment rates in the US have increased, they remain lower than China's. Furthermore, the coefficient of determination, R, which represents the degree of fitting between the predicted results and the actual investment equation, is close to 1. This result indicates that the fitting degree of China and the US investment equations with the savings rate as the explanatory variable has achieved a relatively ideal effect. These parameter estimates correspond to changes in actual data (Figs. 3.1 and 3.2).

We can roughly understand the changes in savings and investment and overseas capital inflows in China and the US during the analysis period based on the speculative results in Table 3.1 and the trend shown in Figs. 3.1 and 3.2. Investment and savings rates in the US fell significantly following the 2007 financial crisis and began to recover in 2010, but the savings rate was lower than the investment rate throughout the period. Figure 3.2 and the estimated results in Table 3.1 indicate that after 2007, the savings rate in the US fell faster than the investment rate, which hurt investing; the US required a large volume of foreign funds to compensate for lacking investment funds. During this period, the massive imbalance in domestic savings and investment in the US could only be filled by increased imports and capital inflows from abroad; however, it is suitable for savers in other countries who believe the US offers

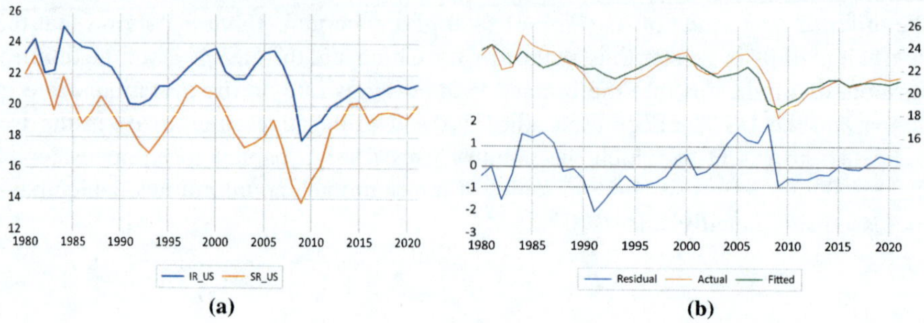

Fig. 3.2 **a** The US savings and investment rate and **b** Estimates of the US investment rate

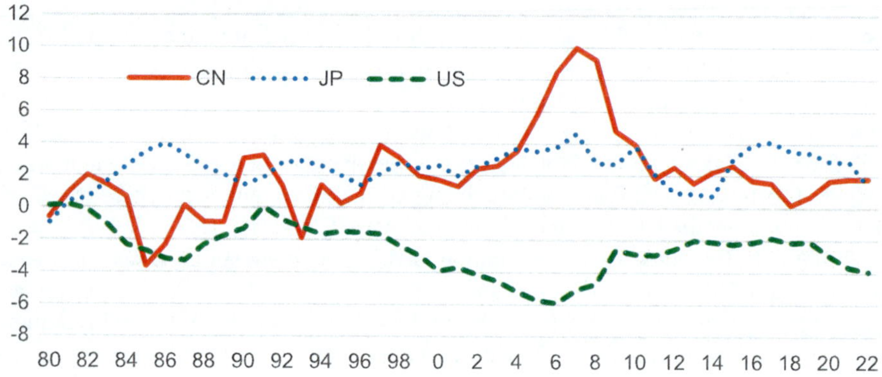

Fig. 3.3 Current account balance (% of GDP). *Source* IMF, World Economic Outlook Database: October 2022

better investment opportunities than their home countries. Since 1990, China's rapid economic growth, and sustained high savings have necessitated increased foreign exports. Regarding the external environment, China's accession to the World Trade Organization at the end of 2001 and the improvement in the international trade environment are favorable to the expansion of foreign trade; high-quality, low-cost Chinese goods are the best choice for US domestic demand.

Equations (3.1)–(3.5) of the equilibrium relationship in Sect. 3.2 shows that the proportion of domestic savings over investment is the primary source of capital for foreign investment. Therefore, we must demonstrate the change in net savings between China and the US during the analysis period and the impact on the current balance and foreign financial investment. From 1980 to 1989, China's net savings were low, but it has always been higher than investment since 1993. Net saving rates averaged 4% from 2004 to 2010, peaking at 9.9% in 2007–2008; however, from 2011 to 2022, China's net saving rate fell sharply, reaching only 0.17% in 2018. Meanwhile, the US has consistently had negative net savings, falling to around −5.3% in 2002–2005 and plummeting to −6.1% in 2007–2008 during the financial crisis, demonstrating a funding shortage. After 2012, the US deficit recovered and remained around −1.5%; therefore, net savings holdings in the US and China have shifted significantly since 2011. Although the US deficit has improved, China's previously high net saving rate has tended to fall, which is the fundamental reason for the resemblance between China and the US, which stems from the real economy.

However, the domestic economic structures of China and the US have changed, gradually forming a double surplus in China's current account and financial accounts and a double deficit in the US. The mirror image formed by the US trade imbalance over the last 20 years also depends on the structural basis of China–US economic growth.

3.3 Mirror Image Between China and the US in the EFF

Using the EFF analysis framework proposed in Section, we extend the analysis horizon from domestic savings and investment balance to foreign trade balance and from the real domestic economy to international financial investment.

3.3.1 Changes in the Current Account of China–US

The gap between domestic investment and savings must reflect changes in the current account. Figure 3.3 depicts the trends in current accounts from China, the US, and Japan from 1980 to 2022. During this time, the US ran a current account deficit and experienced two significant ups and downs; the first period lasted from 1980 to 1991. The US had a current account surplus of 2.317 billion USD in 1980; however, in 1987, it resulted in a deficit of 160,661 billion USD or 3.1% of the US GDP. Frequent trade tensions occurred between the US and Japan during this period.

Figure 3.3 also depicts the other major cycle in the US current account, which runs from 1992 to 2022. The US had a current account surplus of 2.895 billion USD, or 0.047% of GDP, in 1991 but a deficit of 50.614 billion USD the following year. In 2006, the US deficit reached 816.647 billion USD, or 5.9% of the US GDP, the highest level in 40 years. Since the 2008 financial crisis, the current account deficit in the US had been shrinking until 2017. However, it has risen since then, reaching a 40-year high of 985.254 billion USD (3.47% of GDP) in 2022, the highest since 1980.

According to China's balance of payments statistics, China's current account surplus accumulated 150 billion USD from 1980 to 2001, and from December 2001 to 2022, it accumulated to 4.2 trillion USD, 28.1 times that of the first two decades. In 2008, the current account surplus peaked at 420.6 billion USD (9.2% of GDP, see Fig. 3.3). In terms of China–US trade, China surpassed Japan in 2003 to assume the largest trade surplus with the US. From 2003 to 2010, China experienced a consistent economic growth rate of over 10%, indicating an overheating of the economy. During

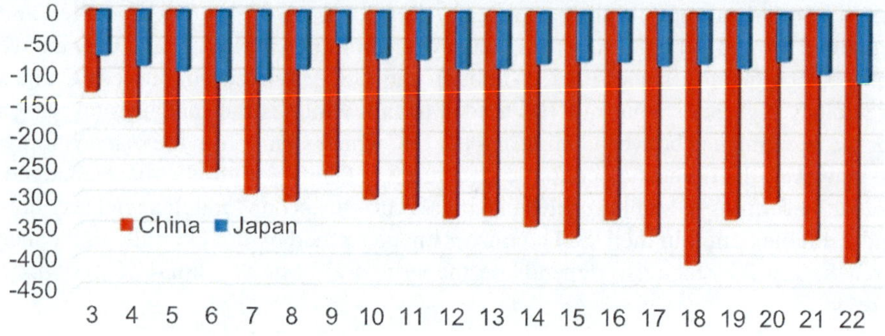

Fig. 3.4 The current US deficit with China and Japan (USD billions). *Source* Bureau of Economic Analysis (BEA), https://apps.bea.gov/iTable/iTable.cfm?ReqID=62&step=1

the period from 2007 to 2008, the growth in China's economy was accompanied by a notable increase in its current account surplus, which peaked at over 9% of GDP. Notably, the United States accounted for the majority of this current surplus during the period.

Figure 3.4 details the change in the US trade deficit with China. According to the graph, China's trade surplus with the US far exceeded Japan's trade surplus with the US between 2003 and 2022. Figure 3.4 indicates that the current deficit between the US and Japan is decreasing, while the current deficit between the US and China is increasing. The current US deficit with China increased from 2003 to 2022, while China's current surplus with the US increased from 131.792 billion USD in 2003 to a high of 407.447 billion USD in 2018. Even in 2022, when China–US relations were at their lowest, China's current surplus with the US was 402.7 billion USD. During this period, China's current surplus with the US accounted for 56% of the cumulative current account deficit in the US (Japan accounted for 16%), resulting in trade frictions between China and the US, becoming a significant issue affecting each country's economic development.

Thus, from the economic growth structure and EFF standpoint, we must ask why China and the US have had such a large and persistent trade deficit for so long. Moreover, we must determine how external adjustment benefits economic friction between the two countries and global economic growth.

3.3.2 External Adjustment of China and the United States

According to the Treasury International Capital System, China's holdings of US debt can be divided into three categories: long-term treasury debt (50%), long-term financial bonds (35%), stock shares, corporate bonds, and short-term bonds (15%). China surpassed Japan as the largest foreign creditor as of the end of 2008, owning 727.4 billion USD in American bonds. By July 2011, the figure had risen to 1,314.9

billion USD. Even at the end of 2021, when trade, and political tensions increased between China and the US, the figure remained at 1,069 billion USD, just below Japan; however, by the end of 2022, China's holdings of American treasuries had fallen by 18.9% to 867.1 billion USD (see Fig. 3.5).[5]

Equilibrium Formulas (3.1)–(3.5) show that a current account surplus (deficit) caused by excessive (insufficient) savings will inevitably result in an outflow (inflow) of overseas funds to achieve international payments balance. Simultaneously, because the US has the most mature financial market system and the function of global financial resource allocation, a large amount of international capital flows into the country, causing the American economy to evolve into a debt economy. Figure 3.5 depicts the evolution of financial investment in the US by the country's three largest global creditors—China, Japan, and the United Kingdom (UK)—from 2000 to 2022, using the foreign exchange earned from trade surpluses to purchase American bonds.

Figure 3.5 shows that China held the most US debt worldwide from 2009 to 2018, while Japan was the largest holder of US debt in other periods. Despite a slight decline in recent years, China remains the second largest holder of US debt, far more than the UK. The significant volume of foreign capital inflows was used for overseas investment and compensated for the fund shortage caused by the US current account deficit. Until 2018, this structure essentially played a role in stabilizing the trade balance between China and the US.

Examining inclusive trade flows and fund flows reveals an unstable symmetrical mirror image between China and the US. (1) High consumption in the US reflects high savings in China. (2) Massive Chinese exports of low-cost goods to the US

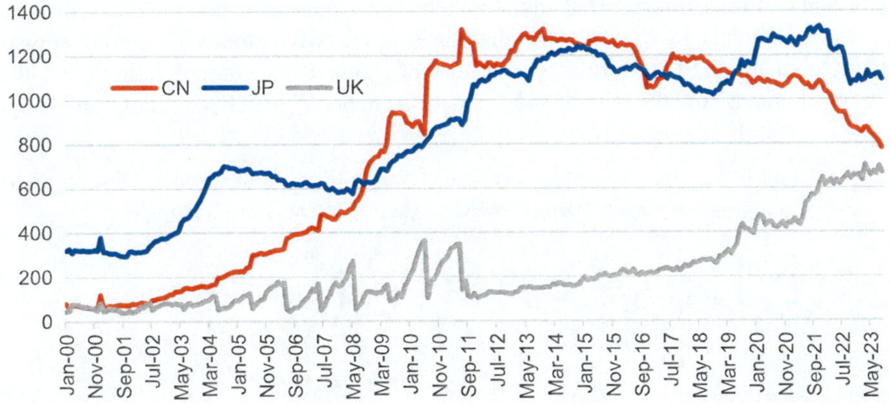

Fig. 3.5 Foreign holdings of US Treasury securities (billions of USD).[6] *Source* US Department of the Treasury, Sep. 2023

[5] The US Department of the Treasury, Treasury International Capital System http://www.ustreas.gov/tic/.

[6] The data in this table include foreign holdings of US Treasury marketable and non-marketable bills, bonds, and notes reported monthly under the Treasury International Capital reporting system. The data for 2022 is in August, while the data for other years are in December.

enable excessive consumption by US consumers. (3) The massive US trade deficit coexists with the Chinese trade surplus. (4) The rapid increase in Chinese foreign exchange reserves reflects the significant increase in US debt. (5) China has used trillions of USD to buy US Treasuries, indicating an urgent need to reinvest China's soaring current account surplus, which could provide financing for the US capital shortage. (6) Since 2016, China has consistently demonstrated a trend of decreasing its holdings of U.S. bonds. In September 2023, China's investment in U.S. Treasuries dropped to $778.1 billion, marking a significant decline of $474 billion from the peak observed in February 2016. Despite escalating trade and political tensions between the two countries, China has continued to buy US bonds. As a result, we necessary to examine the risks of massive foreign exchange reserves and US bond holdings for the development of the Chinese economy and and to find strategies to deal with the decoupling of the Chinese and U.S. economies.

From 1992, when China began publishing capital flow statistics, to 2014, China's EFF showed a trend of capital outflow exceeding capital inflow. For over two decades, the world's largest developing country has been exporting capital to the world's largest advanced country (Fig. 3.6). For a long time, China's external adjustment has maintained the unbalanced growth pattern of foreign trade between China and the US. Figure 3.6 shows China's foreign capital flow and current account surplus peaked from 2002 to 2007; however, from 2011 to 2022, a decline in China's economic growth rate and a decrease in net savings (Fig. 3.1a) caused China's external capital flow and current account surplus to contract significantly (Fig. 3.6). This reduction can be shown by the structural contraction trend in China's current balance and external fund flows. Since 2020, with the deepening of mutual distrust in China–US political relations, this external adjustment has been appearing no longer sustainable.

Many scholars have pointed out that the US uses two methods of external adjustment to ensure the long-term sustainability of its current account deficit. One is the "trade channel," and the other is the "valuation channel," which maintains the USD's

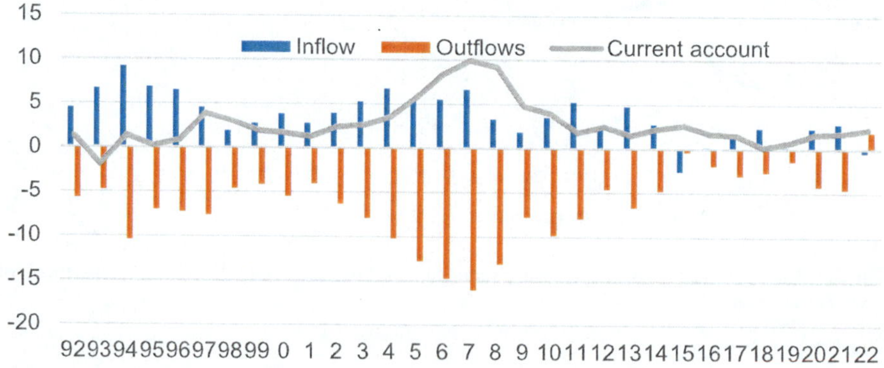

Fig. 3.6 China's EFF (% of GDP). *Source* The People's Bank of China, Quarterly Statistical Bulletin

"exorbitant privileges" status[7] by increasing capital gains earned from exchange rate changes on debts and claims denominated in various combinations of USD and foreign currencies.[8] Figure 3.7 depicts the changes in long and short interest rates in the US and the real effective exchange rate from January 1990 to January 2022. Since the beginning of the twenty-first century and 2010, the continuous depreciation of the USD's real effective exchange rate against CNY, JPY, and the EUR, among others, has significantly impacted US investment returns. Because of the depreciation of the USD, the use of foreign assets denominated in foreign currency has resulted in high investment returns for the US.

Figure 3.7 also shows that the US 10-year Treasury yield (UST) and interbank lending rate (FFR) have both fallen significantly from 5% in June 2007 to around 0.05% and 0.56% in May 2020, but in August 2023, it recovered to 5.12% and 4.17%, respectively. The financial market mechanism determines these changes. To examine macroeconomic vulnerabilities and prevent financial risks, it is necessary to observe structural changes in China and the US's international investment positions, returns from external investment, and financial shocks and transmission mechanisms caused by external financing of Chinese and US inter-sectors from the perspective of the balance sheet approach.

The FFR curve in Fig. 3.7 shows that, in response to the 2008 American financial crisis, the Federal Reserve Board (FRB) implemented the quantitative easing

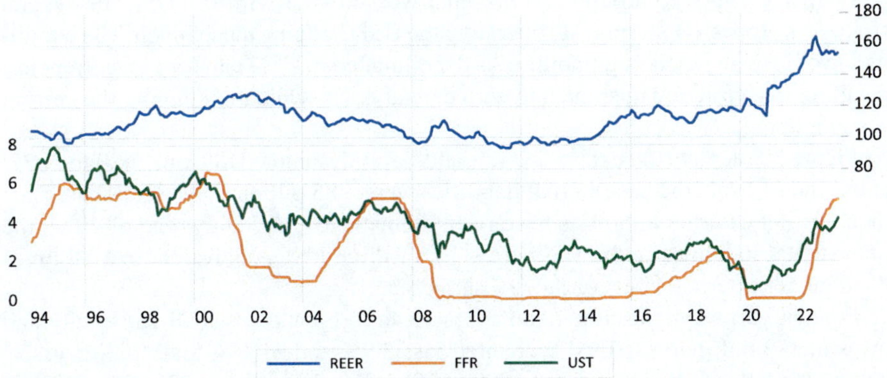

Fig. 3.7 Changes in interest rates and exchange rates of the US. *Source* FRB, https://www.federalreserve.g.,ov/data.htm. *Notes* REER (Real effective exchange rate), FFR (Federal funds rate),[9] UST (US Treasury 10 yields), and the right axis is REER. A rising REER means that the USD has appreciated in real terms relative to its trading partners, while a falling REER means depreciated in real terms

[7] Gourinchas et al. (2017) and Mc Cauley (2015).

[8] Iwamoto (2013) and Gourinchas et al. (2019).

[9] The federal funds rate refers to the short-term interest rate in the US interbank lending market. The change in such interest rates can sensitively reflect the surplus and shortage of funds between banks. The surplus and shortage of funds in the interbank lending market will spread to the market industry and commerce, affecting consumption, investment, and balance of payments.

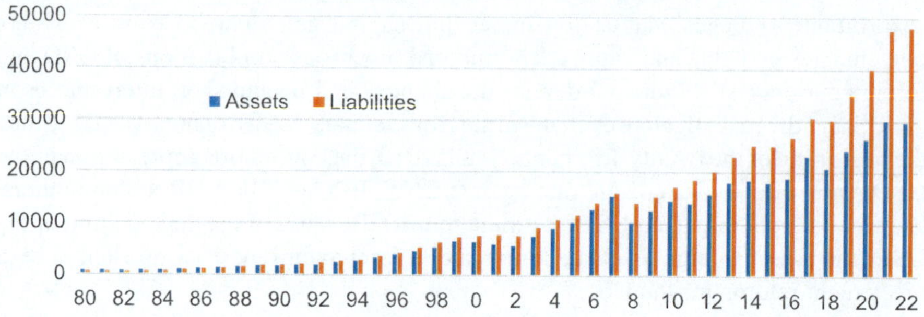

Fig. 3.8 External assets and liabilities of the US (billions of USD, By Stock data). *Source* FRB, Financial accounts

policy for more than 10 years, lowering market interest rates, and increasing capital supply, but also causing financial turbulence and increasing crisis risks. As a result, transferring the adjustment pressure of monetary policy to the outside to maintain the sustainability of external debt becomes the core interest of the US to maintain the USD's "exorbitant privilege" status—a higher return on US external assets than on US external liabilities. To that end, the US has generally taken two approaches: One strategy is to entice foreign money to return to the US at the right time, allowing other countries to pay for US monetary policy and its consequences.

From the 1990s to around 2015, China worked well with the US. The second strategy is properly applying and operating the USD hedging mechanism. The combination of foreign assets and liabilities is denominated in USD and foreign currencies, resulting in a currency mismatch to earn capital gains while maintaining the current account deficit.

Figure 3.8 depicts the external assets and liabilities in the US from 1980 to 2022; at the end of 1990, the net external liabilities were −85 billion USD. However, since then, the net external liabilities have risen dramatically, and at the end of 2022, the net external liabilities in the US were −15.3 trillion USD, equivalent to 1.88 times the total net external assets of the G7 plus China.[10]

Foreign currency is used to denote American foreign assets. FDI and equity and investment fund shares (EIFS) accounted for 33.9% and 21% of total foreign assets held in the US, respectively, from 1980 to 2007. From 2008 to 2022, FDI and EIFS accounted for 28% and 27% of the total foreign assets held, respectively, for 55%. In terms of US external debt, the total FDI, and EIFS accounted for 39.8% of US external debt during 1980–2007 and 40.5% of US external debt from 2008 to 2022.[11] Most of the remaining financing is in debt instruments or bank financing, with American external liabilities denominated in USD. In other words, the country's external claims are primarily in risky stocks denominated in foreign currencies, whereas its external debts are risk-free debt instruments denominated in USD. As a result, the US can

[10] At the end of 2022, data from IMF's IIP indicates that France and the UK had net external assets of −9.08 billion USD and −9.96 billion USD.

[11] Data from the US Bureau of Economic Analysis, BOP.

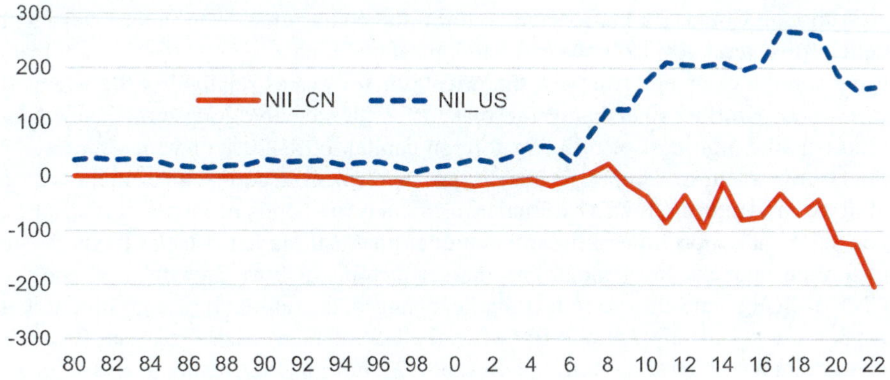

Fig. 3.9 Comparison of investment income (billions of USD). *Source* IMF, Balance of Payments Standard Presentation by Country

raise funds from emerging markets, such as China, and India, by issuing bonds at lower interest rates and reinvesting the proceeds in higher-yielding EIFS.

3.3.3 Comparison of External Investment Returns

With such a large amount of foreign investment, security, liquidity, and stable appreciation must all be considered, as well as investment returns.[12] Based on the above basic structure of foreign financial investment in China and the US, this paper discusses the benefits and risks of foreign financial investment. There are two methods for calculating the return on foreign investment. One method is to calculate net foreign investment income (the income from foreign investment less the payment from foreign investment), i.e., investment income, credit minus investment income, and debit. Based on international statistical standards, the figures come from the current account, the trade channel, and the real return on foreign investment. Because it is flow data, the data may sometimes contain negative values. The other is to define the difference between the net financial balance of the balance of payments (BOP) statistics and the change in the net international investment position in the international investment position (IIP) as capital gains, also called the valuation channel.[13] The first approach is used to focus the discussion here. Figure 3.9 shows the changes in the net investment income of China and the US from 1980 to 2022.

[12] According to the definition of BOP, investment income refers to the income generated by holding foreign financial assets. They include interest income, dividends, returns from overseas subsidiaries to the home company, and reinvestment returns from foreign direct investors. Specific items can be divided into direct investment income, securities investment income, and other investment income.

[13] For example, Obstfeld and Rogodd (2005) and Iwamoto (2007, 2012).

Although China is a net foreign creditor, the country has been in an investment deficit for several years. On the one hand, most of China's foreign assets result from using foreign exchange reserves; the private sector makes relatively little overseas investment. On the other hand, the reason for the negative long-term investment returns is the high cost of utilizing foreign capital, 60%–70% of which represents direct investment in China. Because direct investment is equity investment with a high required return, China's external investment is frequently negative. Furthermore, due to China's short time in the international financial market, a lack of experience in foreign financial investment, and market factors, such as interest rates, and the CNY exchange rate slow to reflect market changes, the return on foreign investment is poor. Except for 2007 and 2008, when the US financial crisis broke out (China's investment income was 3.7 billion USD and 22.2 billion USD, respectively), China's net external investment income was negative; it decreased significantly to −203.14 billion USD in 2022, with a total loss of outbound investment of −1033.2 billion USD from 2009 to 2022. There is a 2.4 percentage point (pp) difference exists between China's return on external investment and the cost of utilizing foreign capital.

Figure 3.9 also shows that perennial investment returns in the US were positive during the observation period. One important reason is that foreign capital in the US is primarily used for low-cost debt and equity investment; in contrast, foreign investment is primarily high-return FDI, so the relative cost is low. The smooth upward slope of the income curve of American foreign investment demonstrates the benefits of American financial investment. Even though the US achieved an investment income of 196.4 billion USD in 2020 during COVID-19, investment income continued to rise, rising from 30.05 billion USD in 1980 to a peak of 268.5 billion USD in 2017. From 2009 to 2022, the US received 2.84 trillion USD in foreign investment, with an average annual growth rate of 2.4%. Between 2009 and 2022, the total foreign investment income in China (the largest creditor to the US) and the US (the largest debtor to China) differed by a factor of four. As a result, China must gradually expand the capital market in an orderly manner to attract foreign capital inflows, use it to hedge capital outflows, optimize the structure of its foreign investment and financing, and improve the return on foreign investment in the long run.

Following the US financial crisis in 2007, the US current account deficit and foreign financial investment changed dramatically, with the current account deficit narrowing and the net foreign financial investment position declining significantly (see Fig. 3.10). These changes resulted in significant structural changes in the US external flow of funds.

Figure 3.10 depicts the evolution of the current account deficit (CA) and the net position in international financial investment (NIIP) from 1980 to 2022. As shown, after 2008, the CA, and NIIP of the US experienced a significant trend shift. Since 2008, the current account deficit in the US has narrowed significantly and become unsustainable, recovering from −5.9% in 2006 to −3.9% in 2022; however, the valuation implications of currency mismatches between external assets and liabilities have also demonstrated their limits. The net IIP in the US has declined significantly,

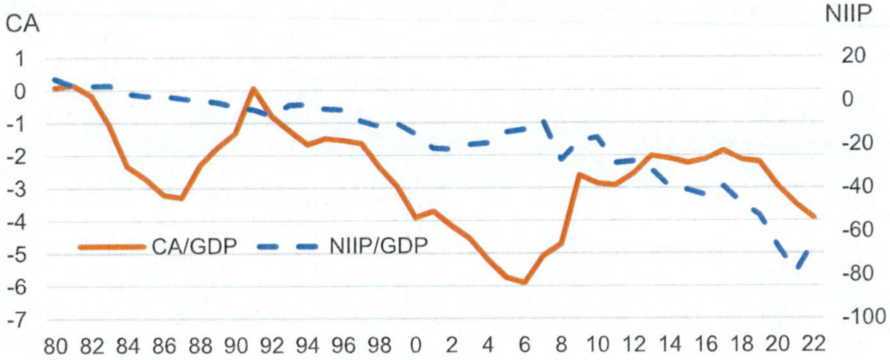

Fig. 3.10 The current US deficit and net international investment position (% of GDP). *Source* U.S. Bureau of Economic Analysis (BEA) Table 1.2. US Net International Investment Position at the End of the Period (March 29, 2023)

from 296.9 billion USD in 1980 to −18.1 trillion USD in 2021, with the NIIP-to-GDP ratio falling from 10.39% in 1980 to −86.6% (based on stock data) in 2021; however, NIIP in the fourth quarter of 2022 was negative at 16.1 trillion USD, and the downward trend increased slightly. The risk of a debt crisis caused by foreign financial investments has increased, and the period in which the US reaped enormous benefits from the USD's "exorbitant privilege" may end.[14] The US earns higher yields on its external assets than it pays on its external liabilities, but this undeniable advantage stems from direct investment, which is unrelated to the USD's international role.

3.3.4 Shock in External Adjustment to the Balance Sheet

The adjustment of the EFF in the US will inevitably affect the changes in the stock of foreign assets and liabilities; thus, we conduct a detailed analysis of the security and risks of foreign assets and liabilities in China and the US. Figure 3.11 depicts both countries' net positions in risky assets and safe liabilities to demonstrate the asymmetries of external balance sheets. Lane and Milesi-Ferretti (2018) provide data from 1982 to 2020 with annual frequency.

We calculate the net risky position by taking portfolio equity plus direct investment on the asset side and subtracting portfolio equity and direct investment on the liability side. We calculate the net safe position by taking debt assets (portfolio debt and other investments) plus reserve assets and subtracting debt liabilities.

China has been long safe and short risky. China has been rapidly accumulating safety (particularly in US Treasury bonds) to prevent the crisis and step-by-step financial liberalization from following direct investment and securities investment in the US and other developed economies. However, since 2007, China's net safe

[14] Gourinchas and Rey (2007a, 2007b) and Mc Cauley (2015).

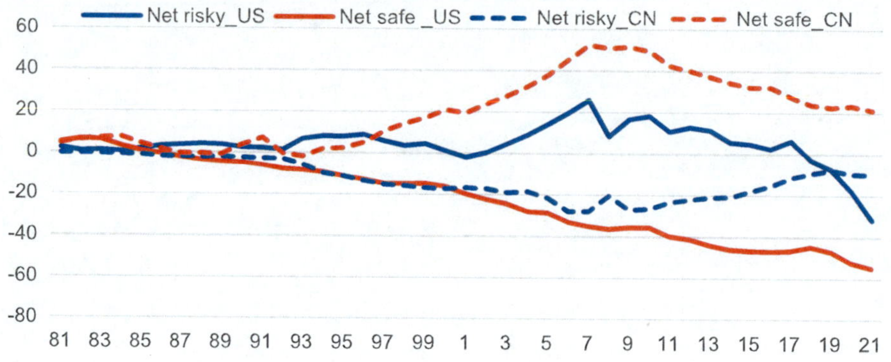

Fig. 3.11 Net risky and net safe holdings for the US and China (% of GDP). *Source* Lane and Milesi-Ferretti (2018) and IIP, published by IMF. *Notes* Net risky position = (portfolio equity assets + FDI assets)–(portfolio equity liabilities + FDI liabilities). Net safe position = reserve assets + debt assets–debt liabilities. Debt includes portfolio debt and other investments

asset holdings have continued to fall, from 52.39% in 2007 to 21.24% in 2021, while the country's net risky ratio has narrowed, from −27.95% in 2007 to only −9.91% in 2021. The reduced ratio of net safe assets to net risky assets indicates that China's external financial investment is in a state of structural contraction due to the international environment of deteriorating Sino-US political relations and rising financial risks. Figure 3.6 shows the structural contraction trend in China's current balance and external fund flows.

At the same time, the US has become increasingly long risky and, in particular, short safe. Furthermore, with 2008 as a turning point, the net safety to net risk investment ratio in the US showed a sharp downward trend, falling to −55.13% and −31.71%, respectively, in 2021 (a 40-year low), with the ratio especially beginning to decline sharply in 2018. That is, America's net safety position has fallen in lockstep with its net risky position since 2008. The US's commonly used external adjustment channel, namely, manipulating gains, or losses on future valuations through currency mismatches, is no longer functional. Figure 3.11 indicates a trend that the mirror image between China and the US from 1994 to 2021 is no longer sustainable, implying that China, and the US face significant structural risks in the EFF.

We provide a statistical description of various variables in the foreign capital circulation between China and the US; however, changes in foreign capital circulation are influenced by the real economy, domestic and foreign financial markets, and political factors, and the stability of the data itself must also be considered. Therefore, a comprehensive examination is still required, involving a systematic quantitative analysis framework that comprehensively examines the interrelationships between the inflow and outflow of foreign capital between China and the US, as well as their benefits and risks, and the "exorbitant privilege" of the USD. Furthermore, as of the end of 2021, China accounts for about 3% of the total market size of foreign direct investment, securities investment, and other investments, while the US accounts for

21.7%.[15] Therefore, the quantitative model focuses more on the factors in the US. The next section introduces a vector error correction (VEC) model that measures long-term trends and short-term fluctuations. We aim to examine the co-integration relationship between China and the US in the EFF and explore whether the "exorbitant privilege" status will continue. Specifically, we analyze the cross-inflow, and outflow of funds in the US to gain insights from multiple perspectives.

3.4 Co-integration Analysis and the VEC Model

The complexity of relations between variables in a single sequence cannot reflect changes in EFF; therefore, we introduced co-integration analysis and established a VEC model to observe interactions among multiple variables and to measure the structural impact on the formation of financial stress. Appropriate variables must be selected, and the stationarity of all variables must be tested to confirm their co-integration to observe changes in EFF with the VEC model. Then, the VEC model can be constructed with information about those relationships to enable analysis.

3.4.1 Data Sources and Selection of Variables

To monitor changes in the inflow and outflow of funds in the US according to the three dimensions of EFF in Eqs. (3.2–3.4), we selected the following explanatory variables from the considerations of real economic growth, US bond yield, China's CA surplus, FX market risk, bank credit market rate, and possibilities for data collection. That is, the economic growth rate of the US (*GDP_US*), US Treasury 10 yields (*UST*), current account surplus of China (*CA_CN*), the real effective exchange rate of the US (*REER*), and federal funds rate (*FFR*) are taken as explanatory variables for fund inflows (*FI*), fund outflows (*FO*), and net investment income (*NII*). We also take first-order lag variables of *FI, FO*, and *NII* as explanatory variables of *FI* and *FO*. According to the economic attributes of variables and market mechanism, *GDP_US*, *UST*, and *CA_CN* are set as long-term variables, while *REER*, and *FFR* are set as short-term variables.

We selected year data for these eight explanatory variables from 1980 to 2021 (see Fig. 3.12); the influence of these variables on the US's external capital flows was described and analyzed in Sect. 3.3. Data stationarity must be examined to establish the VEC model, and conducting a unit root test for each variable and whether there is a co-integration relationship between each variable is necessary. The selected variables are all expressed in logarithmic form because the difference in the variable's logarithm is approximately equal to its rate of change. Figure 3.1

---

[15] IMF (2022), CDIS, CPIS; BIS's LBS, http://stats.bis.org/statx/toc/LBS.html on March 3 22, 2023.

illustrates this relationship, showing trends in their variation and fluctuations in these variables during the sampled period differ, but all exhibit noticeable changes before and after 2008. Except for REER, all other variables showed a trend change during the observation and analysis period. *FI, FO, NII,* and *GDP_US* showed an upward trend, *REER, FFR,* and *UST* showed a downward trend, while *CA_CN* showed an upward trend at the turning point in 2008 and then a downward trend. Unit root tests of each variable are needed to verify a stationary time series.

3.4.2 Testing of Data Stationarity

A unit root test is a statistical test used to determine whether a time series data has a unit root, which indicates non-stationarity in the data. To determine whether *FI*, *FO*, and the relevant variables are co-integrated, we performed an Augmented Dickey–Fuller (ADF) test. Results for unit root tests appear in Table 3.2. Except for the absence of a unit root for *GDPR_US* and *REER*, all remaining time series variables are non-stationary but become integrated into order 1, i.e., I(1) after first-order differences.

Table 3.2 displays the results for unit root tests on *FI* and *FO* with all explanatory variables for non-stationary sequences. ADF is the Augmented Dickey–Fuller (1979) test, and DF-GLS is a GLS de-trending based on the Dickey–Fuller test proposed by Elliott, Rothenberg, and Stock (1996). Both tests contained constant terms. The lag for each variable is selected per the Schwarz Bayesian criterion, and the selected lag (*k*) is in parentheses. The lower half of Table 3.2 reveals critical values attain 10%, 5%, and 1% significance for ADF and DF-GLS.

Two cases warrant discussion. First, see the default test results for variables A. No differential processing of the original sequence of variables is conducted as a class test to maintain the default level. Figure 3.12 reveals that the time series of *REER* exhibits no apparent changes in trend; thus, only the test of the constant appears in the test equation. The time series for *FI, FO, NII*, *GDP_US, REER, FFR, UST*, and *CA_CN* trends upward or downward; therefore, we include the constant, and drift terms in the test equation. Tests of default variable A show that t-statistics for *FI* and *FO* and all explanatory variables exceed critical values at 1% significance; hence, the null hypothesis cannot be rejected. The sequences of variables have unit roots and are non-stationary.

Second, when a first-order difference treatment is applied to all variables, the test values of the difference variables B show that the t values for all variables are less than the critical value and statistically significant. The null hypothesis of a unit root is rejected, and the time series is considered stationary, suggesting that all non-stationary sequences become first-order single integer stationary after the treatment for first-order differences, confirming the criterion of I(1).

The time series of *FI* and *FO,* along with all explanatory variables treated with first-order differences, are co-integrated, satisfying the preconditions for the co-integration test. This result means that although the time series of *FI* and *FO* other

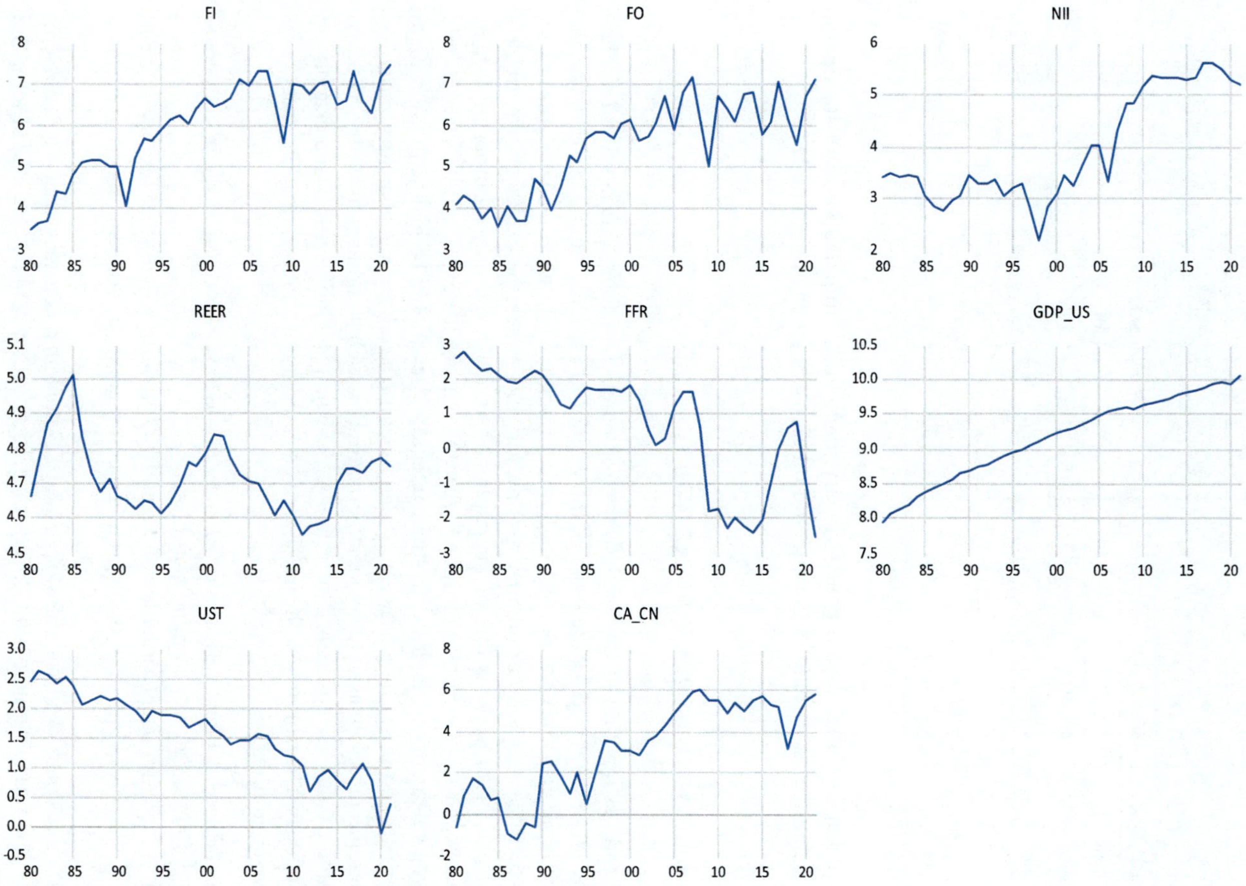

Fig. 3.12 *FI, FO,* and variations in explanatory variables. *Sources* FRB, http://www.federalreserve.g.,ov/econresdata/statisticsdata.htm, BEA, https://apps.bea.gov/iTable/iTable.cfm?ReqID=62&step=1. China Quarterly Statistical Bulletin, National Bureau of Statistics. *Note* Each variable is expressed in the logarithmic form

Table 3.2 Results for Unit Root Tests

Variable	A. Level variable		B. Difference variable	
FI	−0.95		−9.456	
FO	−0.964		−12.692	
NII	−0.473		−4.697	
GDP_US	3.46		−5.648	
REER	−4.207		−4.18	
FFR	−1.969		−5.875	
UST	−1.229		−7.88	
CA_CN	−2.337		−5.824	
	10%(*)	5%(**)	1%(***)	
ADF	−2.6079	−2.939	−3.6105	

explanatory variables are non-stationary, their linear combinations are stationary, indicating a stable long-term structural relationship among the variables.

3.4.3 Analysis of Co-integration

Before testing a set of time series for co-integration or long-term equilibrium, they should be inspected for the order of integration. If variables number more than two, i.e., more than one variable, the single-order number of the explained variable cannot exceed the single-order number for any variable. When the explanatory variable's integral order exceeds that of the explained variable, the integral order of at least two explanatory variables must be higher than that of the explained variable. If only two explanatory variables exist, their integral order should be identical. We seek to verify the co-integration between *FI* and *FO* with the other six explanatory variables; therefore, the Johansen (1991) test is required, which simultaneously observes, and captures multiple co-integration relations. The first and most critical step in applying the Johansen test is to test the number of co-integration relationships. The test covers five conditions.

i. The component variables of *FI* and *FO* and the co-integration vector contain no constant terms and trend variables.
ii. The component variables of *FI* and *FO* include no constant terms and trend variables, but the co-integration vector contains constant terms (restricted constant).
iii. The component variables of *FI* and *FO* contain a constant term—a variable in the form of time t—but the co-integration vector contains a constant term and no trend variable.
iv. *FI*, *FO*, and the co-integration equations contain a linear deterministic trend (restricted).

v. *FI* and *FO* contain a quadratic trend term, and the equation for the co-integration relationship contains a linear trend term.

Table 3.3 presents the results of the Johansen test. Although the results for the two tests are inconsistent, the conclusion of the max eigen statistics should be selected when its conclusion differs from the trace statistics. Table 3.3 displays these results, including those from testing the constant and trend terms. There were at least two simultaneous co-integrations, i.e., the co-integration vector has at least an order of two.

Testing for trace and max eigen statistics reveals at least two co-integration relations between *FI* and *FO* with each explanatory variable. This outcome aligns perfectly with our research objective to theoretically obtain the foundation for co-integration between *FI* and *FO* with each explanatory variable; the finding indicates a long-term equilibrium between *FI* and *FO* with the selected explanatory variable.

Table 3.3 shows that the co-integration vector order is two, indicating at least two co-integrations between *FI* and *FO* with the composite variables ($r = 2$). Therefore, we selected Case 3, which includes *FI*, and *FO* combination variables with a cointegrating relationship that includes a constant. Short-run dynamics include a constant. The associated two-dimensional vector autoregressive model (VAR) has both a constant and trend. Co-integration refers to the long-run equilibrium relationship between the variables in the model, and the short-run coefficients capture the immediate effects of changes in the variables on each other. The lag is set to 1 because our primary variables–long-run variables as *GDP_US*, *UST*, and *CA_CN*, and short-run variables as *REER* and *FFR*, undergo annual changes. Based on the conditions outlined above, we have made a conjecture on the co-integration equation (CE), presented in Table 3.4.

The top portion of Table 3.4 displays parameter estimates of the standardized *CE*, showing standardized results for the number of each possible co-integration. In row CE1, the *FI* coefficient (b_{11}) is normalized to 1, and *FO* is deducted from the co-integration vector ($b_{12} = 0$) of *CE1*, expressed as *FI* variables normalized to 1

Table 3.3 Length of lag and number of co-integrations

Sample: 1980 2021					
Included observations: 40					
Series: LOGFI_US LOGFO_US LOGUII LOGREER LOGFFR LOGGDP_US LOGUST LOGCA_CN					
Lags interval: 1 to 1					
Selected (0.05 level*) Number of Cointegrating Relations by Model					
Data trend:	None	None	Linear	Linear	Quadratic
Test type	No Intercept	Intercept	Intercept	Intercept	Intercept
	No Trend	No Trend	No Trend	Trend	Trend
Trace	3	4	3	4	5
Max-Eig	1	2	2	1	1

Table 3.4 Johansen Co-integration Test for Case 3

	Normalized cointegrating coefficients (standard error in parentheses)						UST	CA_CN
	FI	FO	NII	REER	FFR	GDP_US		
CE1	1	0	−0.6605	2.4479	0.4064	−9.7547	−6.0867	0.9104
			(0.2076)	(1.1881)	(0.1550)	(0.9216)	(0.9814)	(0.1340)
CE2	0	1	−1.7933	11.7144	0.1693	−15.8168	−9.1049	1.8982
			(0.4548)	(2.6027)	(0.3396)	(2.0189)	(2.1500)	(0.2936)
	Adjustment coefficients (standard error in parentheses)							
	D(FI)	D(FO)	D(NII)	D(REER)	D(FFR)	D(GDP_US)	D(UST)	D(CA_CN)
CE1	−0.5143	0.0143	−0.0864	0.0476	0.1096	0.0167	0.4018	0.6970
	(0.2458)	(0.3565)	(0.1804)	(0.0264)	(0.3560)	(0.0110)	(0.0717)	(0.4394)
CE2	0.2298	−0.0395	0.0690	−0.0261	0.0090	−0.0028	−0.1210	−0.5088
	(0.1112)	(0.1613)	(0.0816)	(0.0120)	(0.1611)	(0.0050)	(0.0325)	(0.1988)

with the quantitative structural relationship of other explanatory variables. In row CE2, the *FO* coefficient (b_{22}) is normalized to 1, and column *FI* deducts ($b_{21} = 0$) from the co-integration vector, expressed as *FO* variables normalized to 1 with the quantitative structure relationship of the other explanatory variables.

The bottom portion of Table 3.4 displays estimates of the adjustment coefficient for *CE1* and *CE2* (brackets indicate standard deviations). Table 3.4 displays the results for standardizing the equations for two co-integration relations, expressed as Eqs. (3.7) and (3.8).

$$\begin{aligned} \boldsymbol{FI} =& \mathbf{0.66}\boldsymbol{NII}-\mathbf{2.45}\boldsymbol{REER}-\mathbf{0.41}\boldsymbol{FFR} \\ &+\mathbf{9.75}\boldsymbol{GDP\_US}+\mathbf{6.09}\boldsymbol{UST}-\mathbf{0.91}\boldsymbol{CA\_CN} \end{aligned} \tag{3.7}$$

$$\begin{aligned} \boldsymbol{FO} =& \mathbf{1.79}\boldsymbol{NII}-\mathbf{11.71}\boldsymbol{REER}-\mathbf{0.17}\boldsymbol{FFR} \\ &+\mathbf{15.82}\boldsymbol{GDP\_US}+\mathbf{9.11}\boldsymbol{UST}-\mathbf{1.9}\boldsymbol{CA\_CN} \end{aligned} \tag{3.8}$$

Equation (3.7–3.8) shows the co-integration between *FI* and FO with other explanatory variables and represents a long-term equilibrium formed by the co-integration vector. Nonetheless, Table 3.4 reveals relatively large standard deviations for *GDP_US, FFR, UST*, and *CA_CN*, and parameter estimates exhibit no statistically significant influence. This finding can be explained as the influence of short-term fluctuations in economic variables on long-term equilibrium relationships when subjected to external shocks during a financial crisis. Concerning short-term market fluctuations, factors, and possibilities exist for deviating from co-integration at each moment *t*, i.e., economic variables often are unbalanced during short-term observations. Therefore, to consider disequilibrium in the model, the degree to which the variable deviates from its long-term equilibrium during short-term fluctuations can be measured by increasing or decreasing the order difference (ΔFI and ΔFO). Then the deviation can be corrected to approximate theorized long-run equilibrium so that parameter estimation of the co-integration equation can move toward the long-term mean. For this reason, we introduce the VEC model to simulate the change in this long-run equilibrium when it deviates from short-run equilibrium.

3.4.4 A VEC Model to Measure EFF of the United States

The VEC model is derived from Sargan's (1964) study of wage growth and prices. The representation theorem subsequently proposed by Engle–Granger (1987) combines Hendry and Mizon (1978) research into the VEC model with the co-integration concept. Equation (3.9) holds when only two variables are used to show this theorem (Minotani, 2007).

$$\begin{bmatrix} Y_t \\ X_t \end{bmatrix} = \begin{bmatrix} \alpha_{10} \\ \alpha_{20} \end{bmatrix} + \begin{bmatrix} \beta_{11} & \beta_{12} \\ \beta_{21} & \beta_{22} \end{bmatrix} \begin{bmatrix} Y_{t-1} \\ X_{t-1} \end{bmatrix} + \begin{bmatrix} \varepsilon_{1t} \\ \varepsilon_{2t} \end{bmatrix}. \tag{3.9}$$

Equation (3.9) represents a two-dimensional vector autoregressive model (VAR). When Y and X are first-order integrals, i.e., I(1), there is a co-integration between Y and X, which can be expressed as $Y_t = \alpha + \beta X_t + \mu_t$. Per Engle–Granger, when there is co-integration between X and Y by *I (1)*, the VAR model can be expressed as an error correction model (EC model). Conversely, if *ECM* can represent X and Y, then Y, *and* X are co-integrated. If Y and X are co-integrated, the error between Y_t and $E(Y_t) = \alpha + \beta X_t$ X cannot be large. Thus, there should be a mechanism for adjusting from $\mu_t = Y_t - E(Y_t) = Y_t - \alpha - \beta X_t$ to 0 to long-term equilibrium. In this manner, if the stable long-term relation is set to $Y_t = \alpha + \beta X_t + \mu_t$, the principle governing a typical EC model can be expressed as follows. In this way, if the long-term stationary relation is set to $Y_t = \alpha + \beta X_t + \mu_t$, the principle governing the EC model can be expressed as Eq. (3.10), a first-order error correction model.

$$\Delta Y_t = \gamma_i \Delta X_t - \gamma (Y_{t-1} - \alpha - \beta X_{t-1}) + \varepsilon_t \;\; (0 < \gamma < 1.) \tag{3.10}$$

When X and Y move at the same level, which puts $\Delta X_t = 0$, $\Delta Y_t = 0$ in a long-term situation; thus, Eq. (3.10) can be expressed as $Y_t = \alpha + \beta X_t$ concerning the long-term average. In the short run, however, $Y_{t-1} - \alpha - \beta X_{t-1} > 0$, i.e., $Y_{t-1} > \alpha + \beta X_t$. In turn, that means when Y_{t-1} exceeds the long-term expectation, with $\alpha + \beta X_{t-1}$, then for Y_t in the next period, t will be smaller than $Y_{t-1}(\Delta Y_t < 0)$. In contrast, when $Y_{t-1} < \alpha + \beta X_{t-1}$, since Y_{t-1} does not reach the expected level $\alpha + \beta X_{t-1}$ in the long run, Y_t in the period from t_1 to t exceeds Y_{t-1} ($\Delta Y_t > 0$). The short-term correction mechanism accelerates movement to the long-term mean, which is the principle underlying the error -correction mechanism in Eq. (3.10).

$Y_{t-1} - \alpha - \beta X_{t-1}$ in Eq. (3.10) is the error correction term. γ is the adjustment coefficient, representing the speed of adjustment toward long-term equilibrium. The closer γ is to 1, the faster the adjustment; thus $1/\gamma$ is the adjustment period. According to Eq. (3.10), some linear combination of these variables is stationary if there is a co-integration between variables. Such long-term equilibrium can be achieved through constant adjustment of short-term fluctuations. The EC model is a short-term unbalanced model. If the error correction is carried out, each variable returns to long-term equilibrium relationships.

Furthermore, let the estimated value of β_i be $\hat{\beta}_i (i = 1, 2, \ldots, n)$, and take the short-term imbalance model as the definition of the EC model. Then, the typical EC model is formalized as Eq. (3.11).

$$\Delta Y_t = \gamma_1 \Delta Y_{t-1} + \sum_{i=1}^{n} \hat{\gamma}_i \Delta X_{i,t-1} - \gamma (Y_{t-1} - \hat{\beta}_1 - \sum_{i=2}^{n} \hat{\beta}_i X_{i,t-1}) + \varepsilon_i \tag{3.11}$$

Per the principle governing EC models, the two-dimensional vector is extended to multiple variables to specify the model (Zhang, 2020), which can be set up in two ways. One is to estimate CE when the co-integration vector is known; another is to specify a theoretical EC model and then deduce the practical operational EC model. Since we have selected the co-integration vector and performed the co-integration test, we use the co-integration vector to specify the co-integration VEC model that describes *FI* and *FO*. Influenced by the Johansen co-integration test in Case 3, we selected it as the deterministic trend specification; CE includes an intercept but no trend and exhibits no deterministic trend and intercept term in VAR. Moreover, when we extend the discussion to the VAR(1) model with eight variables, we can directly use the relevant information of co-integration vectors from Tables 3.3 and 3.4 and specify the VEC model as Eq. (3.12).

$$\begin{aligned} \Delta FI_t =& \gamma_1 \Delta FI_{t-1} + \gamma_2 \Delta FO_{t-1} + \gamma_3 \Delta UII_{t-1} + \gamma_4 \Delta REER_{t-1} \\ &+ \gamma_5 \Delta FFR_{t-1} + C - \beta_1 \big(FI_{t-1} - \hat{\beta}_1 - \hat{\beta}_2 UII_{t-1} \\ &- \hat{\beta}_3 GDPR\_US_{t-1} - \hat{\beta}_4 UST_{t-1} - \hat{\beta}_5 CA\_CN_{t-1}\big) + \varepsilon_t \end{aligned} \tag{3.12}$$

$$\begin{aligned} \Delta FO_t =& \gamma_1 \Delta FI_{t-1} + \gamma_2 \Delta FO_{t-1} + \gamma_3 \Delta UII_{t-1} \\ &+ \gamma_4 \Delta REER_{t-1} + \gamma_5 \Delta FFR_{t-1} + C \\ &- \beta_1 \big(FO_{t-1} - \hat{\beta}_1 - \hat{\beta}_2 UII_{t-1} - \hat{\beta}_3 GDPR\_US_{t-1} \\ &- \hat{\beta}_4 UST_{t-1} - \hat{\beta}_5 CA\_CN_{t-1}\big) + \varepsilon_t \end{aligned} \tag{3.13}$$

where parameter $\gamma_i (i = 1, 2, \ldots, 5)$ is called the parameter of long-run influence and is intended to reflect the effect of short-term changes due to explanatory variables on the formation of capital inflow and outflows. β is the adjustment coefficient $(0 < \beta < 1)$, also known as the feedback effect. The bracketed $\hat{\beta}_i$ (i = 1, 2, ..., 5) is the short-run influence coefficient, and the difference between bracketed FI_{t-1} and

FO_{t-1} with variables (*t–1*) indicates a deviation adjustment to short-run equilibrium. It also captures the variables' short-run dynamics through the error correction term (ECT), which indicates how quickly the variables adjust to deviations from the long-run equilibrium. For example, ECT > *0* indicates that variables in the preceding period exceed equilibrium and negative adjustments are needed. ECT < *0* indicates that the value of each variable in the preceding period is below equilibrium, and positive adjustments are needed.

$\hat{\beta}_1$ is a constant estimate, and ε is a random error term. Since all variables in Eqs. (3.12–3.13) are I(1), the t-test can be used to evaluate the predicted results. After substituting 40 data points for each variable, Table 3.5 presents the CE with a value of 2 and estimated results for the VEC model with error correction. Table 3.6 displays the results of the model's correlation tests.

The conjecture results include four parts. The first represents the long-run parameter estimates of the CE (Cointegrating Eq: CointEq1 and CointEq2) in Table 3.5 to estimate *FI(–1)*, *FO(–1)*, etc. The second represents the short-run parameter estimates of the error correction term—the column headed *D(FI(–1))* and *D(FO(–1))*, etc., in Table 3.5—where the corresponding value of COINTEQ1 is the adjustment coefficient estimate of the error estimate term, with $\beta_1 = -0.5599$ and $\beta_2 = -0.0978$ (lower part of Table 3.5). The adjustment coefficient of statistical estimation is negative, consistent with the theoretical setting, and reflects the dynamic mechanism of long-term equilibrium deviation from self-correction to short-term fluctuations. When the imbalance occurs, the time series converges, and returns to long-term equilibrium. The statistical estimate of β_1 is greater than that of β_2, indicating that *FI* has a more substantial EC effect than *FO* in the current period. The third part is the correlation test results for a single equation in the model (Table 3.6). The fourth is the overall correlation for the model (lower portion of Table 3.6). Since we focus on *FI and FO*, the EC related to *D(FI(-1))* and *D(FO)-1))* in the second portion of Table 3.5 is embedded in the first part of CE in Table 3.5, and the EC model of the co-integration vector about *FS* and *FO* is obtained after sorting, as shown in Eqs. (3.14) and (3.15).

The VEC Model—Substituted Coefficients:

$$\begin{aligned} D(FI) =& -0.56(FI(-1) + 0.34NII(-1) - 1.69 + 0.24GDP\_US(-1) \\ & -0.001 * UST(-1) + 7.81CA\_CN(-1)) - 0.31(FO(-1) \\ & + 0.49NII(-1) - 1.64 + 0.32GDP\_US(-1) \\ & -0.13UST(-1) + 7.53CA\_CN(-1)) + 0.06D(FI(-1)) + 0.17D(FO(-1)) \\ & -0.13D(NII(-1)) + 0.08 + 1.41D(REER(-1)) - 0.04D(FFR(-1)) \end{aligned} \tag{3.14}$$

$$\begin{aligned} D(FO) =& -0.098(FI(-1) + 0.34NII(-1) - 1.69 + 0.24GDP\_US(-1) \\ & -0.001UST(-1) + 7.81CA\_CN(-1)) \\ & -1.05(FO(-1) + 0.49NII(-1) - 1.64 + 0.32GDP\_US(-1) \end{aligned}$$

Table 3.5 Estimated Results for the VEC Model

Sample (adjusted): 1982 2021 Included observations: 40 after adjustments Standard errors in () & t-statistics in [] Lags "interval (in first differences): 1 to 1 Endogenous variables: FI FO NII Exogenous variables (short-run only): D(REER(-1)) D(FFR(-1)) Exogenous variables (long-run only): GDP_US(-1) UST(-1) CA_CN(-1: Deterministic assumptions: Case 3 (Johansen-Hendry-Juselius): Cointegrating relationship includes a constant. Short-run dynamics include a constant

Cointegrating Eq:	CointEq1	CointEq2	
FI(−1)	1.000000	0.000000	
FO(−1)	0.000000	1.000000	
NII(−1)	0.335892 (0.17651) [1.90294]	0.490300 (0.12679) [3.86712]	
C	−1.691584 (0.67837) [−2.49361]	−1.640795 (0.48727) [−3.36735]	
GDP_US(−1)	0.237553 (0.56288) [0.42203]	0.316977 (0.40431) [0.78399]	
UST(−1)	−0.000861 (0.09922) [−0.00868]	−0.131254 (0.07127) [−1.84165]	
CA_CN(−1)	7.807983	7.528446	
Error Correction:	D(FI)	D(FO)	D(NII)
COINTEQ1	−0.559936 (0.23011) [−2.43332]	−0.097809 (0.27417) [−0.35675]	0.140762 (0.19460) [0.72335]
COINTEQ2	−0.309687 (0.19154) [−1.61686]	−1.048272 (0.22821) [−4.59349]	−0.230372 (0.16197) [−1.42228]
D(FI(−1))	0.056078 (0.21204) [0.26446]	0.227653 (0.25264) [0.90110]	−0.329901 (0.17932) [−1.83978]
D(FO(−1))	0.172796 (0.17162) [1.00685]	0.166333 (0.20448) [0.81345]	0.380353 (0.14513) [2.62074]
D(NII(−1))	−0.129377 (0.22201) [−0.58274]	−0.024901 (0.26452) [−0.09414]	0.029043 (0.18775) [0.15469]
C	0.077443 (0.06385) [1.21298]	0.036917 (0.07607) [0.48531]	0.042305 (0.05399) [0.78356]
D(REER(−1))	1.414893 (1.33817) [1.05734]	2.141252 (1.59437) [1.34301]	0.281154 (1.13163) [0.24845]
D(FFR(−1))	−0.038650 (0.10166) [−0.38019]	0.038215 (0.12112) [0.31550]	−0.042739 (0.08597) [−0.49713]

$$\begin{aligned}&-0.13UST(-1)+7.53CA\_CN(-1))\\&+0.23D(FI(-1))+0.17D(FO(-1))\\&-0.02D(NII(-1))+0.04+2.14D(REER(-1))\\&+0.04D(FFR(-1))\end{aligned} \tag{3.15}$$

A common problem exists in the short-term parameter estimation of VEC models; the parameter estimation of short-term change is less statistically significant than

Table 3.6 Tests of the Model's Overall Correlations

R-squared	0.515382	0.60047	0.19397
Adi. R-squared	0.409372	0.513073	0.017651
Sum sq. resids	4.661876	6.617898	3.333876
S.E. equation	0.381685	0.454763	0.322775
F-statistic	4.861629	6.870596	1.100111
Log likelihood	−13.7683	−20.77551	−7.062664
Akaike AIC	1.088416	1.438775	0.753133
Schwarz SC	1.426191 plePara>	1.776551	1.090909
Mean dependent	0.095181	0.069945	0.043005
S.D. dependent	0.496648	0.651708	0.325662
Determinant resid covariance (dof adj.)		0.00161	
Determinant resid covariance		0.000824	
Log likelihood		−28.25379	
Akaike information criterion		3.21269	
Schwarz criterion		4.732681	
Number of coefficients		36	

that for long-term equilibrium. There are two reasons for this problem. First, the model included three endogenous variables and five exogenous variables based on the concept of *FI* and *FO*, which affects passing the t-test when making statistical estimates. Second, taking a first-order difference for each variable makes it easy to reduce statistical significance; however, whether a single coefficient is statistically significant is not the focus of the VEC model. Its primary considerations are the stationarity and significance of the whole system. Table 3.6 displays the overall results for the VEC model.

In contrast, although some of the estimated coefficients in Table 3.5 are not statistically significant, the results of the F-test in Table 3.6 exceed the critical value of the F statistic, indicating statistical significance. Hence, the overall inference of each variable is effective. We tested the maximum lag interval at 2, but based on the Akaike information criterion (AIC) values, we chose the order as 1 instead. The lower section of Table 3.6 presents the AIC values for both CE equations, indicating that the CE we inferred provides the best balance between fit and complexity.

3.5 Empirical Analysis of Co-Integration for the EFF of the United States

CEs (3.14) and (3.15) are the calculated results of *FI* and *FO* estimated statistically by CE (3.12) and (3.13). The two CEs' coefficient symbols of long-term variables and short-term fluctuations differ, and the symbols for *FI* and *FO* estimation also differ. An estimate of the adjustment coefficient β_1 of CE (3.14) is –0.5599, β_2 and of CE (3.15) is –0.0978, which means that, with other difference variables unchanged, the first-order difference of *FI* and *FO* change in the period *t*, eliminating the 56% and 9.8% disequilibrium in the earlier period. Their co-integration restricts changes between *FI* and *FO* with other variables, and short-term fluctuations gradually approximate long-term stability. Figure 3.13 shows that the co-integration equation of *FI* and *FO* shows the dynamic change in the return to long-term equilibrium through short-term adjustment correction from 1980 to 2021.

3.5.1 *Analysis of Long-Run Relationship of CE*

We observe four aspects of the impact on *FI* and *FO*, i.e., external investment income, changes in the real domestic economy, fluctuations in financial markets, and external shocks. We first discuss the estimated value of long-term parameters in the first part of CE in parentheses in CE (3.14–3.15), reflecting the long-run impart of *FI* and *FO* with each variable on long-term equilibrium at time *t*.

First, based on statistical estimates, *NII(–1)* represents the long-term sustained returns of American foreign investment, with a 0.34% impact on capital inflow and a 0.49% impact on capital outflow. The inflow of foreign capital improves the investment return in the US, thus expanding foreign investment. Therefore, the investment return of the US positively correlates with *FI* and *FO*. The results of statistical estimation also show that the impact of *NII(–1)* on capital outflow is higher than that of foreign capital inflow to the US, which attracted money from around the world

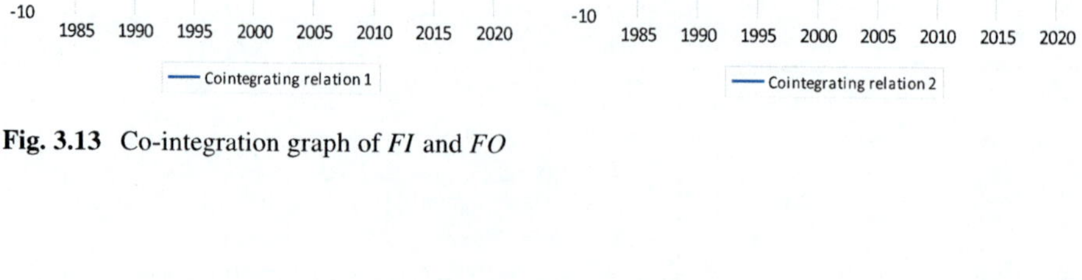

Fig. 3.13 Co-integration graph of *FI* and *FO*

and invested overseas, contributing to the economic growth of the US. This result is consistent with our statistical analysis in Sect. 3.3.

Second, during the observation period, the GDP growth of the US was relatively stable, with the highest rate of 7.23% in 1984 and the lowest rate of –3.4% in 2020. *GDP_US(–1)* represents the elastic impact of the real economic changes on *FI* and *FO*. We speculate that the correlation between *GDP_US(–1)* and *FI* and *FO* is positive. When GDP increases by 1%, the elasticity values of *FI* and *FO* are 0.24 and 0.32, respectively, suggesting that the change of *GDP_US(–1)* has a particular elastic influence on the capital inflow and outflow during the analysis period.

Third, Fig. 3.12 shows that *UST(–1)* negatively correlates with *FI* and *FO*, and the results estimated by VEC model statistics are also negative. The UST yields declined from 11.43% in 1980 to 1.45% in 2021, while UST_{t-1} decreased by 1 percentage point, *FI* increased by 0.001%, and *FO* increased by 0.13%. The change in *UST(–1)* in this period did not significantly impact the increase in capital inflow and outflow in the US. Furthermore, to maintain the continuous trade relationship between China and the US, the investment returns obtained by China through the purchase of US Treasury bonds also declined, which is one reason why the mirror image between China and the US in the EFF is not sustainable.

Fourth, the continuous increase in China's CA surplus positively impacts the capital inflow and outflow of the US. The effect of *CA_CN(–1)* on *FI* is 7.81%, while that on *FO* is 7.53%; the results of this statistical estimate have another implication. When *CA_CN* loses its state of continuous growth, *FI*, and *FO* in the US also lose their power of continuous growth to a certain extent, which the data from 2022 onwards should indicate.

3.5.2 Analysis of Short-Run Relationship on EC

Like the long-run parameter, short-run fluctuations in various variables affect capital inflows and capital outflows in four ways: first-order lagged variables of capital inflows and outflows, first-order lagged variables of external investment income, external shocks to the real effective exchange rate of the US exchange rate, and short-term fluctuations in financial market interest rates. In the short-run estimates of volatility (Eqs. 3.14–3.15), we used the four variables: *D(FI(1))* and *D(FO(1))*, *D(NII(1))*, *D(REER(1))*, and *D(FFR(1))*. Their parameter estimation results differ from those of long-term parameter estimation. The effects on *FI* and *FO* of short-run shocks originating in these variables differ from their long-run effects. Through these error corrections, both *FI*, and *FO* return to a stable long-term state and maintain a normal range. Notably, the parameter estimates for a given variable in the CE model are usually higher than their corresponding estimates in the error correction term.

The estimated short-run change parameters of the error correction term in Column *D(FI)* and *D(FO)* in Table 3.5 show that the effect of *D(FI(−1)* on *D(FI)* is 0.06%, and the effect of *D(FO(−1))* on D(FI) is 0.17%. Correspondingly, the effect of *D(FI(–1))* on *D(FO)* is 0.22%, while the effect of *D(FO(–1))* on *D(FO)* is 0.17%. For *D(FI)*, the short-term adjustment effect of capital outflow from the US is higher than that of overseas capital inflow. For *D(FO)*, the short-term adjustment effect of capital inflows is slightly greater than that of outflows. During the analysis period, the capital inflow, and outflow of the US showed a cycle of mutual coordination and improvement, thus maintaining the long-term net capital inflow of the US, i.e., the continuation of the huge foreign debt. The results of this statistical estimate are consistent with the characteristics of the statistical description developed in Sect. 3.3, which points out that the US's external debt (capital outflow) is mainly risky stocks denominated in foreign currencies. In contrast, the US's external debt (capital inflow) is risk-free bonds denominated in dollars. In turn, it also confirmed the path of investment income obtained by the US using the capital inflow to invest overseas.

Looking at the short-term impact of *D(NII–1))* on *D(FI)* and *D(FO)*, we see that the short-term adjustment effect of *D(NII(–1))* on *D(FI)* is –0.13%, while the short-term adjustment effect on *D(FO)* is –0.02%. The symbols of these estimated values are opposite to the long-term estimated values, which reflects the structural short-term dynamic adjustment of the long-term trend of *FI* and *FO*.

We next examine the short-term consequences of *D(REER(–1))* on *D(FI)* and *D(FO)*. Since the *REER* reflects the exchange rate and price changes, it reflects the change in the USD's external competitive strength better than the nominal exchange rate. During the observed analysis period, *REER* gradually rose to around 150 in the mid-1980s, appreciating since the beginning of the twenty-first century, reaching a peak of 95 in 2011 (see Fig. 3.12); however, there is no significant trend change in *REER*, so no unit root is detected in the time series (see Table 3.2). The results of the statistical estimation indicate that the short-term adjustment effect of *D(REER(-1))* on *D(FI)* and *D(FO)* are both positive, the effect of *D(REER(–1))* on *D(FI)* is 1.41%, and the effect of *D(REER–))* on *D(FO)* is 2.14%. The appreciation of *D(REER(–1)* has been a boon for capital inflows and outflows, and the short-term effect on capital outflows is slightly higher than capital inflows.

Finally, we examine the impact of short-term market interest rates (*FFR*) on *FI* and *FO*. The *FFR* refers to the short-term interest rate in the US interbank lending market, showing a long-term downward trend during the analysis period, from 16.4% in 1981 to 0.08% in 2021. However, *D(FFR(–1))* has different short-term effects on *D(FI)* and *D(FO)*. The effects of *D(FFR(–1)* on *D(FI)* and *D(FO)* were –0.04% and 0.04%, respectively. Compared with *D(FI(–1))*, *D(FO(–11))*, *D(NII(–1))*, and *D(REER(–1))*, *D(FFR(–1))* has a weaker short-term adjustment effect.

3.5.3 Analysis of Impulse Responses on FI and FO

In time series analysis, the impulse response is a concept that describes the dynamic relationship between a shock to a variable and its subsequent response over time. Specifically, it refers to the pattern of how a shock to one variable affects the values of other variables in a system over a certain period. Here in our VEC model, we attempt to use impulse response analysis to examine how a change in *FI(–1)*, *FO(–1)*, and

NII(–1) affects economic variables such as *FI*, *FO*, and *NII* over the next 10 years. Applying impulse shock to the *FI(–1)*, *FO(–1)*, and *NII(–1)* variables shows how the other variables in the system respond in the short- and long-term.

The results for the VEC model explain the influence of variables on long-term equilibrium and short-term fluctuations in *FI* and *FO*; however, changes in the coefficients in the model manifest only a local dynamic relationship. To systematize the model, we must determine the dynamic changes in variables within the VEC model after it is affected by a unit of random disturbance, the total influence on other variables, and the impact duration. For further impulse response analysis, we used a VEC model to observe standard deviations in disturbances caused by random impact on current and future values of endogenous variables. We attempt to reveal the impact of changes in US real economic growth, US bond yields, foreign exchange market risk, interest rates of the bank credit market, China's CA surplus on *FI* and *FO*, and the sustainability of large US external debt.

Each variable's impact duration on *FI* and FO differs; Fig. 3.14 displays the change in the impact of shocks on *FI* by each variable over 10 years. We first note the impact of the *FI* index itself. The change in the response of *FI* to *FI* was 0.3817 in the first year. Table 3.5 displays data derived from inference results for the model (0.3817 in the standard error equation). *FI* exhibits a strong response to new information about its standard deviation. In the first five years of the examined period, its random disturbance term exhibits significant influence, reaching its maximum the first year after impact and stabilizing after falling from 0.015 in the sixth year.

The impact of *FI* on *FO* presents repeated turnaround, and the initial impact is small, declining from –0.053 at the end of the second year to –0.115 in the third year; however, it starts to rise from the fourth year and has a slight relapse. In the fifth year, it approached 0.04 and remained stable at 0.03. The shocks eventually tend to dissipate after five years, and capital inflow to the US becomes a favorable factor for the outflow of American foreign capital and the quantitative relationship.

The initial impact of *FI* on *NII* (response of *FI* to *NII*) was significant, with a decrease from –0.15 in the second year to –0.154 in the third year, followed by a slight increase to –0.103 year four; however, from that point until the tenth year, the effect of *FI* on *NII* remained stable at around –0.1. As outlined in Sect. 3.3, despite being an external liability for the US, the cost of capital inflow is relatively low, which benefits the country's investment returns.

With one standard deviation of information, the impact of *FO* on *FI* and *NII* is generally lower than that of *FI* on *FO* and *NII*. Initially, the impact of *FO* on *FI* was 0.317; however, it decreased in the second year, reaching its lowest point of 0.04 in the third year before rising slowly and stabilizing in the fifth year, eventually disappearing. The influence of *FO* on *FO* is slightly greater than that of *FO* on *FI*, decreasing from 0.326 in the first year to 0.09 in the third year before recovering to 0.08 in the fifth year and then stabilizing. The impact of *FO* on *NII* is less significant than that of *FI* on *NII* and remains primarily negative throughout the observation period.

Lastly, we examine the effect of *NII* on *FI* and *FO*. Generally, the impulse effect of *NII* on both *FI* and *FO* is low, although the impact on *FI* is slightly more substantial

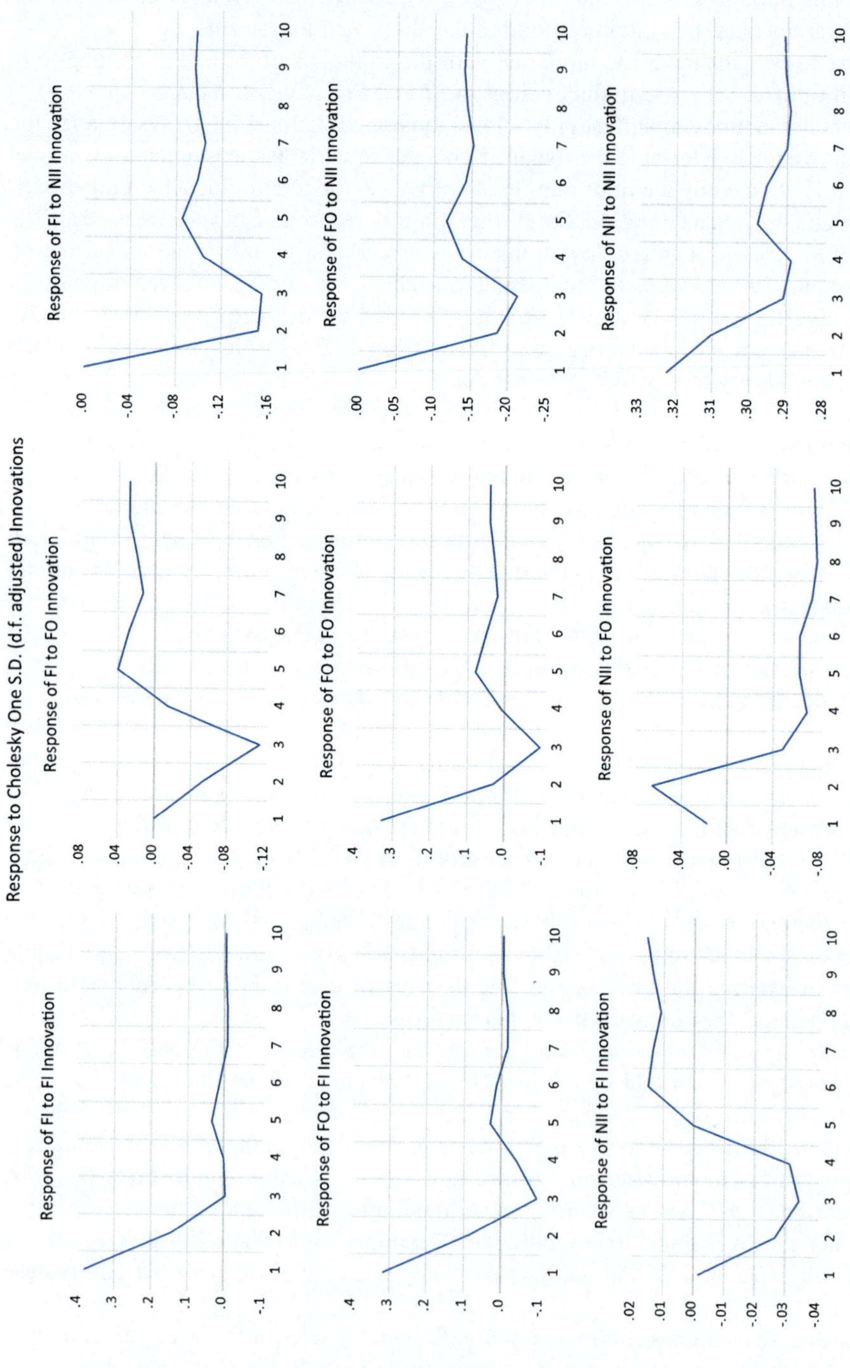

Fig. 3.14 Diagram of impulse influence. *Note* The horizontal axis represents years

than on *FO*. The impact of *NII* on *FI* initially displayed a downward trend before trending upwards; the *NII* ratio increased from its lowest value of −0.034 in the third year to its highest value of 0.015 in the sixth year before stabilizing. In contrast, the impact of *NII* on *FO* showed a long-term downward trend, declining from 0.065 in the second year to −0.077 in the eighth year. Despite a long-term downward trend in the influence of *NII* on *NII*, it displayed a relatively stable state with a high impulse effect, with a statistically estimated value remaining around 0.3 in the current decade. This finding aligns with the statistical description analysis in Sect. 3.3. Throughout the observation period, the US consistently achieved positive investment returns due to its reliance on low-cost debt and equity investment from foreign sources; foreign investments in the US were primarily high-return FDI.

After analyzing the impulse responses, we can see that the direction of the impulse response is not the most crucial factor. Instead, the intensity and duration of the response and the time required for the system to reach a new equilibrium state are crucial and have significant economic implications. If the response time is lengthy, policymakers must have excellent forecasting skills. For instance, if market information is delayed and policymakers fail to anticipate it promptly, inadequate policy regulation, or overcorrection can result.

3.6 Concluding Remarks

Based on our results, this paper analyzes the trend of the mirror-image relationship and decoupling between China and US.

Using the relevant data of more than 40 years to sort the results of statistical description and quantitative estimation, we determine that imbalances in the EFF between China and the US create an interdependence, allowing for forming a mirror-image relationship in the EFF. The US has a trade deficit with China, which China uses to increase its foreign exchange reserves, which it then uses to buy US treasuries, allowing the US to buy imports from China again. Since 2003, China, and the US have had an unbalanced trade relationship; however, this model's sustainability has reached a turning point as the economic relationship between the two countries faces four significant risks. Due to political distrust, a gradual decoupling of the two economies is now inevitable, which has caused a shift in the overall pattern of the GFF and created a global strategic challenge. Currently, four main issues must be addressed in the current economic situation.

First, the uneven economic growth, and development patterns have led to a structural imbalance. Second, the trade imbalance is not sustainable in the long run. Third, the mirror-image relationship between the US and China that has formed over the past two decades, as seen in the GFF, is unsustainable. Finally, there is a risk associated with China holding US bonds and the US holding foreign debt, and there is no longer a relationship of relative political trust between the US and China. The economic and political implications of these issues are summarized below.

3.6.1 Structural Imbalance in China–US Trade

The analysis in Sect. 3.2 suggests that the main reason for the China–US trade imbalance is a disparity in domestic industrial structures, specifically the structural imbalance of savings and investment between the countries. In 2021, the US GDP reached 23.3 trillion USD, 1.32 times larger than China's and 4.66 times larger than Japan's.[16] Services contributed more than 8 trillion USD to that total, ranking first in GDP share at more than 40%. The financial sector, which includes the consumer credit market, stock market, and insurance market, is the most crucial pillar of this economic structure and its ultimate beneficiary, particularly the mortgage market, the largest segment of the financial sector; however, American manufacturing has been shifting overseas to reduce costs, with manufacturing accounting for only 12% of GDP in 2021. To close the production gap, the US had to increase imports, which is one of the primary reasons why the US trade deficit has gradually increased. The economic structure in China is also not entirely reasonable, as the total proportion of China's services and finance in GDP is only 9.5%, whereas the proportion of real estate in GDP was 6.7% until 2021.[17] China's industrial value-added remains low, with revenue significantly lower than that in the US. Therefore, to reduce the China–US trade deficit, both countries must adjust their industrial structures to allow for the long-term balanced development of their respective economies.

EFF comprises two central causal relationships concerning China–US trade. (1) An increase in the CA deficit increases the inflow of funds, and (2) an increase in the outflow of funds increases the CA surplus. In other words, the US has been increasing its inflow of funds to compensate for its deficit caused by a lack of private and public savings. Meanwhile, China has seen a CA surplus versus the US due to its purchase of US bonds; the average Chinese household suffers from a consumption shortage due to the income gap, which contrasts with the excess savings of the industrial and public sectors.

3.6.2 The Unsustainable Mirror Image Between China and the United States

The fundamental cause of the China–US external imbalance is a mismatch between savings and investment in the real economy. As a result, China, and the US must adjust their economic growth rates to balance savings and investment and achieve BOP.

The US can run a current long-term account deficit due to (1) maturity mismatches between short-term liabilities and long-term assets, (2) currency mismatches between liabilities and assets, and (3) capital structure issues caused by an overreliance on

[16] IMF, World Economic Outlook Database October 2023.

[17] The national Bureau of Statistics of China, China Statistical Yearbook (2021).

debt. However, these three factors have also contributed to the severe fragility of the American balance sheet, increased financial risks, and the potential for solvency if a borrower cannot cover its debt.

Because of long-standing trends in economic growth and industrial structure imbalances, the US, and China have formed a mirror-image relationship in the EFF. The US trade deficit with China depends on China's CA surplus used to purchase US treasuries. Although the US can use capital gains to maintain its CA deficit for a limited time, the mirror relationship between China and the US is also approaching an unsustainable critical point due to unsustainable industrial and growth structures in both China and the US. Thus, China, and the US must alter their industrial structures and economic development patterns.

3.6.3 On the US Debt Risk

The US's net external liabilities reached 15.3 trillion USD at the end of 2022 (see Fig. 3.8), with annual interest payments of more than 500 billion USD at the end of 2022, which appears unsustainable. To sustain the increasing debt scale of the US and maintain GDP growth, the US must expand capital inflow to repay old debts with new debts and expand the capital outflow to earn investment returns; the models suggest that this is hardly sustainable. Our conclusions are summarized as follows.

(1) The statistical estimation of the model shows that *NII* has a higher impact on capital outflow than foreign capital inflow, indicating that only by maintaining capital inflow can it continue to create high returns on foreign investment.
(2) The so-called "valuation channel" is limited. The model's statistical estimation shows that the long-term Treasury bond interest rate and the short-term bank interest rate do not have strong growth elasticity on the US's capital inflow and outflow. The short-term adjustment effect of *D (FFR (–1))* is especially weak.
(3) The waning of the China effect is apparent. The statistical estimate shows that the elasticity of *CA_CN*'s capital inflow and outflow to the US is 7.81% and 7.53%, respectively; however, the strained political and economic relations between China and the US have reduced the *CA_CN* surplus in recent years. Naturally, the scale of China's purchase of US bonds will also decline, thus affecting the expansion of US foreign investment.

3.6.4 Strategic Challenge to China

From China's perspective, the economic crisis caused by the COVID-19 pandemic will undoubtedly affect the entire world, particularly vulnerable emerging markets. Foreign currency supports most of the premiums on emerging market assets. If the Federal Reserve lowers interest rates, the US bond yields fall dramatically. With more than 1 trillion USD of US debt held since 2011, China is naturally under pressure to avoid risk; however, Fig. 3.5 shows that from 2016 to 2022, China has gradually reduced its holdings of US bonds, reflecting China's strategic preparation for their decoupling from the US. Furthermore, the Federal Reserve has moved to raise interest rates since the Russia–Ukraine war. As the political relationship between China and the US has deteriorated, foreign trade relations have also changed, and economic decoupling between the two countries is also apparent in the decline of the CA_CN; thus, China, and the US face greater unknown risks. However, because a mutually beneficial relationship has been maintained since the 1990s, China should continue to hold US Treasuries even if it reduces the amount held, as it is in China's interest to do so.

Since the Russia–Ukraine war in 2022, Russia, Brazil, Saudi Arabia, and other countries, including China, have tried to reduce their holdings of USD assets and their demand for USD through currency swaps, a phenomenon known as "de-dollarization." However, the total amount of global external assets and liabilities suggests that by the end of 2021, the total amount of US external assets and liabilities will be 26.04 trillion USD and 31.67 trillion USD, accounting for the highest proportion of global financial assets and liabilities, 17.89% and 21.76%, respectively. In contrast, China only accounts for 3.35% and 3.1%[18]; thus, de-dollarization makes no economic sense in the short-term. See Chap. 4 for more details analysis about 2022.

As a result, China is attempting to modify its imbalanced EFF structure to reduce its foreign reserve balance through international market transactions; however, these policy changes have been ineffective in halting the increase in foreign reserves. As a result, China's economic structure should be adjusted by broadening domestic demand, diversifying its external financing, and internationalization of the CNY. These actions would alter the China–US relationship and create a new global economic structure. International cooperation with the G7 and emerging economies will become even more critical.

3.6.5 *Future of China–US Economic Relations*

Changes in relative power have resulted in political and economic tension between the US and China, with growing China exerting its influence and the US unwilling to give up its world dominance, the root of which can be traced back to the "survival of the fittest" in human historical relations. However, with China and the US accounting for 40% of the global GDP, economic friction between the two nations poses a significant risk to the global economy.

Since the 1990s, the US has been China's largest customer, trading partner, and investment partner. In turn, China is an essential supplier to the US and was once a close trading and investment partner; however, there is currently a lack of mutual political trust. Even if China offers more money for high-end American goods and

[18] IMF (the end of 2022), CPIS, and CDIS; BIS (the end of 2022) Locational banking statistics.

technology, the US will unlikely sell. However, the US and China are vying for technological superiority, which does not necessarily imply that the trade gap between the two countries will widen. Opportunities remain for beneficial interaction in other sectors related to people's livelihoods and financial commodities; there is still plenty of room for each to take what is required, and peace, and development should remain the goal of human society.

The US and China should avoid the zero-sum game of who conquers who and instead strive for peaceful coexistence, which would be a first in human history. In light of recent changes in global political and economic patterns, America's foreign financing mode based on superpower privilege is no longer viable. To begin dealing with the massive US debt risk, China, and the US must strictly adhere to and faithfully fulfill contract laws and regulations.

If "decoupling" is unavoidable for the time being, managing the separation in an orderly manner, minimizing the impact on the economy, avoiding new areas of conflict, and leaving room for the next historic cooperation will be a severe test of each country's political wisdom. As Alexander Hamilton, the first US Treasury Secretary, said at Congress in 1790, "It is essential that a nation's credit be well established."[19]

References

BEA. (2022). *U. S. Net International Investment Position at the End of the Period*, Table 1.2.

Bernanke, B. (2005). The global saving glut and the U.S. current account deficit, Speech at the Sandridge Lecture. https://www.federalreserve.g.,ov/boarddocs/speeches/2005/200503102/default.htm. Richmond, VA: Virginia Association of Economics.

BIS. (2009). Global imbalances: In Midstream? In *Reconstructing the World Economy, IMF Stuff Discussion Note*, Dec. 22. Washington: International Monetary Fund.

BIS. (2022). Locational Banking Statistics, http://stats.bis.org/statx/toc/LBS.html.

Blanchard, and Milesi-Ferretti. (2011). *(Why) Should Current Account Balances Be Reduced? IMF Stuff Discussion Note*, Mar. 1. Washington: International Monetary Fund.

Caballero, R. J., Farhi, E., Gourinchas, P., & Pierre-Olivier, G. (2008). An equilibrium model of 'Global Imbalances' and low interest rates. *American Economic Review, 98*(1), 358–393. https://doi.org/10.1257/aer.98.1.358

Cauley, R. N. Mc. (2015). Does the US dollar confer an exorbitant privilege? *Journal of International Money & Finance* 57:C, 1–14.

Cavallo, M., & Tille, C. (2006). *Could capital gains smooth a current account rebalancing? FRB New York Staff Report* vol. 237.

Engle, R. F., & Granger, C. W. J. (1987). Co-integration and error correction: Representation. *Estimation, and Testing, Econometrica, 55*, 251–276.

Feldstein, M., & Horioka, C. (1980). Domestic saving and international capital flows. *Economic Journal, 90*(358), 314–329. https://doi.org/10.2307/2231790

[19] Hamilton (1790).

Gourinchas, P. O., & Rey, H. (2007b). *From World Banker to World Venture Capitalist: US External Adjustment and the Exorbitant Privilege. G7 Current Account Imbalances: Sustainability and Adjustment*. University of Chicago Press, 11–66.

Gourinchas, P. O., Rey, H., & Sauzet, M. (2019). The International Monetary and Financial System, NBER WORKING PAPER SERIES, Working Paper 25782. http://www.nber.org/papers/w25782.

Gourinchas, P. O., & Rey, H. (2007a). International financial adjustment. *Journal of Political Economy, 115*(4), 665–703. https://doi.org/10.1086/521966

Gourinchas, P. O. (2019). The Dollar Hegemon? Evidence and Implications for Policymakers, Prepared for the 6th Asian Monetary Policy Forum. https://berkeley.app.box.com/s/oqtzizo8wch5snic1nys7ig01n6f65la. Singapore.

Greenspan, A. (2004). The Evolving U.S. payments imbalance and its impact on Europe and the rest of the world. *Cato Journal, 24*, 1–2.

Hamilton, A. (1790). *First Report on the Public Credit [To the Speaker of the House of Representatives]*. https://www.norton.com/college/history/archive/resources/documents/ch08_02.htm.

Hendry, D. F., & Mizon, G. E. (1978). Serial correlation as a convenient simplification, not a nuisance: A comment on a study of the demand for money by the bank of England. *Economic Journal, 88*, 549–563.

IMF. (2021a). Coordinated Direct Investment Survey (CDIS) https://data.imf.org/regular.aspx?key=60564262.

IMF. (2021b). Coordinated Portfolio Investment Survey (CPIS) https://data.imf.org/regular.aspx?key=60587815.

IMF. (2021c). International Investment Position. https://data.imf.org/regular.aspx?key=62805744.

IMF. (2022). *World Economic Outlook Database April 2022*.

Iwamoto, T. (2007). Sustainability of the US current account deficit. *Sekaikeizai Houron, 51*(9), 31–40.

Iwamoto, T. (2009). Global imbalances after the financial crisis. *Journal of JBIC International Research Office, 3*, 17–30.

Iwamoto, T. (2015). International investment positions, gross capital flows, and global liquidity. *International Economy, 18*, 1–19. https://doi.org/10.5652/internationaleconomy.ie2015.01.ti

Iwamoto, T. (2012). *External Imbalances and the Transfer of Wealth: Asymmetry of the Evaluation Effect in Japan and the United States, A Study of International Monetary Flows*. Chuokeizai Inc., 127–151.

Iwamoto, T. (2013). Structural changes of global economy based on gross capital flows and international investment positions. In *Proceedings Economic and Social Research Institute (ESRI)*. https://www.esri.go.jp/jp/prj/int_prj/2013/prj2013_01.html.

Johansen, S. (1991). Estimation and hypothesis testing of co-integrated vector autoregressive models. *Econometrica, 59*(6), 1551–1580.

Lane, P. R., & Milesi-Ferretti, G. M. (2018). The external wealth of nations revisited: International financial integration in the aftermath of the global financial crisis. *IMF Economic Review, 66*(1), 189–222. https://doi.org/10.1057/s41308-017-0048-y

Lu, F. (2008). A mirror image of external imbalance between China and the U.S.: For an understanding of the feature of the china recent-years economic growth and economic adjustment. *International Economic Review, 11–12*, 19–27.

Mendoza, E., Quadrini, V., & Ríos-Rull, J. (2009). Financial integration, financial development, and global imbalances. *Journal of Political Economy, 117*(3), 371–416. https://doi.org/10.1086/599706

Minotani, C. (2007). *Complete econometrics. Toyokeizai*, 645–673.

National Bureau of Statistics of China. (2021). *China Statistical Yearbook 2021,* China Statistics Press.

Obstfeld, M., & Rogoff, K. S. (2005). Global current account imbalances and exchange rate adjustments. *Brookings Papers on Economic Activity, 2005*(1), 67–146. https://doi.org/10.1353/eca.2005.0020

Sargan, J. D. Wages and prices in the United Kingdom. (1964). A study in econometric methodology (with discussion). In P. E. Hart, G. Mills, & J. K. Whitaker (eds.), *Econometric Analysis for National Economic Planning*, Vol. 16 of Colston papers, 25–63.

Willen, P. (2004). *Incomplete markets and trade* [Working paper]. Federal Reserve Bank of Boston.

Zhang, N. (2020). Flow of funds analysis. *Innovation & Development*. Springer, 137–169.

Chapter 4
A Global Flow of Funds Perspective on Debt, Assets, and Imbalances

Abstract This chapter establishes an analytical framework for examining the global flow of funds (GFF); expanding on the concept, research object, and analytical method for comprehending GFF. The structural changes of the G20, especially China–United States (US) decoupling, are examined alongside the possibility of a debt crisis using stock data to analyze the GFF matrix (GFFM) from 2018 to 2022. The financial network is used to analyze the basic characteristics and risks in the debt market between China (CN) and the US. Finally, CN's and the US' debt securities (DS) market positions and mutual financing relationships are analyzed using financial network technology. It also statistically estimates the impact of debt risk transmission. The issues of China–US are also observed in external financial assets and liabilities by stock data. By compiling the GFFM and using the financial network, we measure the risk exposure changes between CN and US external assets and liabilities, centrality, asset influence and liability sensitivity, and debt risk.

Keywords Balance sheet · Who-to-whom matrix · Financial network · Debt crisis · Shock dynamics

4.1 Introduction

The period spanning 2018–2022 represents a significant era of global transformation. Primarily, China–US relations have evolved from the 1998–2008 harmonious era of mutual exchanges to escalating political differences and conflicts since 2018. The economic relationship has transitioned from a reflective state of foreign trade, financial investment, and mutual closeness to the current scenario characterized by an increasing economic decoupling. Subsequently, the global spread of COVID-19 in 2021 and the outbreak of the Russia–Ukraine war in 2022 appeared to revert the global village to a situation reminiscent of the Cold War tensions. Furthermore, the economic globalization that emerged in the 1990s appears to be experiencing a trend toward anti-globalization.

N. Zhang and Y. Zhang, *Global Flow of Funds Analysis*,
https://doi.org/10.1007/978-981-97-1029-4_4

Since 2018, China (CN) and the United States (US) have experienced strained relations, but so too has CN's interactions with the European Union (EU), Japan (JP), Canada (CA), Australia (AU), India (IN), and other nations. China's financial transactions with most of them, notably Germany (DE), JP, US, and the United Kingdom (UK) have experienced a significant decline, as indicated in Tables 2.3 and 4.1. Confronted with these shifts, there is a need to recognize the intrinsic connection between politics and the economy. The economy serves as the cornerstone of social development, whereas politics can exert a significant influence on economic dynamics. The primary channels of communication between nations involve personnel exchanges and international transactions. The current account (goods and services in foreign transactions) and the corresponding financial account are like two sides of a coin. This chapter, building on the groundwork from Chap. 3, examines the financial facets of this coin from a fresh standpoint—specifically, through the lens of the global flow of funds (GFF).

Using the G20 as an observation platform allows for a statistical analysis of structural changes in external debt, assets, and imbalances among G20 countries across 2018–2022. With a specific focus on CN and the US, the integration of financial network technology leads to speculations that political antagonism may impact mutual financial risks, particularly raising concerns about triggering potential debt crises. Consequently, reflecting on whether the evolving political landscape across 2018–2022 is conducive to the economic development of various nations and whether there is a risk of reverting to the Cold War dynamics becomes imperative.

The new analytical perspective (the GFF) and the methodology of statistical preparation are described in the preceding chapters. The idea originated in the 1950s, where the GFF concept extends Copeland's (1949, 1952) domestic flow of funds concept. Subsequent studies, such as Ishida (1993), established an analytical framework for international financial circulation. Additionally, Zhang (2005, 2008, 2014) proposed a theory of the GFF analysis framework and summarized the flow of funds accounts, established the GFF analysis of the data source, and constructed the observation equation model's GFF structure. Tsujimura and Tsujimura (2008, 2009) compiled a financial input–output table based on the International Monetary Fund (IMF) financial statistics to analyze the characteristics of the GFF in 1997–1998.

However, the development of statistics based on the GFF was directly inspired by the 2008 US financial crisis. Following that crisis, an international consensus was reached at the G20 Finance Ministers and Central Bank Governors Meeting in April 2009 to strengthen financial statistics and promote the systematic integrity of financial statistics. The IMF and the Financial Stability Board (FSB, 2009) requested recommendations to strengthen the flow of funds statistics and balance sheets.[1] They also called for improved data transmission methods and more detailed sectoral data to prepare financial position and flow breakdown statements for each institutional segment by its counterparties. Data sets that provide this type of information are

[1] Financial Stability Board and IMF (2009) "The Financial Crisis and Information Gaps".

called who-to-whom (W-t-W) financial statistics. The IMF began compiling international financial circulation statistics on a trial basis in 2013.[2] Aligned with the ongoing trend of statistical change, the European Central Bank has utilized the flow-of-funds data and sector account information to conduct valuable theoretical research and empirical analysis pertaining to macro imbalances, financial risks, and financial stability (2013a, 2013b). Furthermore, Zhang and Zhao (2019) aim to establish a new statistical framework for measuring the GFF based on its inherent mechanisms. It advances a previous theoretical discussion and develops a practical operational statistical matrix. Zhang and Zhu (2021) employed network theory to discuss an analytical method for the GFF and used G20 countries as the research sample with data at the end of 2018 to discuss network centrality, mutual relationships, financial risk of foreign direct investment (FDI), portfolio investment, and cross-border bank credit among the US, JP, and CN. Finally, on a from-whom-to-whom basis within a "country by country" pattern, Zhang (2022) constructed a GFF matrix (metadata) using the established GFF matrix (GFFM) table to conduct an empirical study with an econometric model and financial network analysis.

When studying international capital flows, US capital flows are the focus, and the country's increasing external liabilities are also regarded as a major financial risk. Gourinchas and Rey (2007, 2019) constructed a measure of cyclical external imbalances, emphasizing the problem of observing the US economy's external imbalance, which must be considered through the traditional "trade channel" and an unexplored "valuation channel."

Nevertheless, as the US current account deficit has persisted without change since then, this strategy is unsustainable. Consequently, while exchange rate depreciation may exert a short-term influence on the persistence of the current account deficit, it is not the primary solution for addressing unbalanced growth.

In recent years, several international organizations have expressed concerns about CN's debt issues. The IMF drew attention to CN's escalating debt-to-GDP ratio, which had tripled since the 1980s, in its Global Debt Monitor released in September 2023. Additionally, in December 2023, Moody's assigned a negative outlook rating to the Chinese government's fiscal position and downgraded CN's credit rating.[3]

Scott Davis and Zlate (2023) estimated the heterogeneous effect of the global financial cycle (GFC) on exchange rates and cross-border capital flows during the COVID-19 pandemic using weekly exchange rate and portfolio flow data for a panel of 59 advanced and emerging market economies. Their research uncovered the ability of the GFC to explain fluctuations in exchange rates, and capital flows increased dramatically during the pandemic. Significant cross-country heterogeneity exists in the response of exchange rates or capital flows to fluctuations in the GFC.

Referring to the above research results, this chapter examines the financial risks and challenges from 2018 to 2022, factoring the deteriorating political and economic relations between CN and the US, the COVID-19 pandemic, and the structural impact

[2] Errico et al. (2013, 2014).

[3] Wall Street Journal (December 7, 2023) Moody's Faces Growing Backlash Over Its Negative Outlook on China.

of the Russia–Ukraine war on the GFF. This study expands on the concept, research object, and analytical method for comprehending GFF.

A theoretical framework for analyzing GFF is established. This framework is grounded in the equilibrium relationship among savings and investment flows, foreign trade flows, external flow of funds, and external assets and liabilities. The research focuses on the G20, specifically examining the risk of debt investment in CN and the US. W-t-W data is utilized to conduct a comparative analysis across 2018–2022. This chapter unveils the structural changes among G20 countries in GFF, the imbalance between liabilities and assets, and the statistical estimation of debt risks in CN and the US. Financial network technology is employed and policy suggestions are presented afterward.

The remainder of Chapter 4 proceeds as follows. Section 4.2 investigates the structural changes within the G20. Section 4.3 uses the financial network to analyze the basic characteristics and risks in the debt market between CN and the US. Section 4.4 analyzes CN's and the US' debt securities (DS) market positions and mutual financing relationships using financial network technology. The conclusion provides an analysis summary, discusses prospects for future development, and outlines upcoming topics for study.

4.2 Changes in the Global Flow of Funds' Structure from 2018 to 2022

The following analysis uses the GFFM to examine the reciprocal dynamics of foreign financial investments between CN and the US. This exploration delves into the core of the transformative interplay mirrored in the financial relationship between the two nations. Drawing on the theoretical model of the GFFM (see Table 1.5), the balance sheet for the G20 in 2022 was assembled (Table 4.1). This table showcases a matrix of external assets and liabilities, using stock data from the IMF (2022a, 2022b, 2022c) and the Bank for International Settlements (BIS). However, because many countries lack such data, financial derivatives data are not used.

4.2.1 Matrix of Multiple Financial Instruments

Table 4.1 depicts the G20's external assets and liabilities matrix as of the end of December 2022. Each row of the matrix contains two statistical groupings, which include countries and three financial instruments for displaying the source of funds, namely direct investment (DI), portfolio investment (PI), and other investment (OI), which cover the main structural elements of external financial liabilities. Along the columns, financial assets are listed by country to show fund uses, with counterparty sectors identified for each cell. The matrix's columns delineate 25 sectors: 24 country

Table 4.1 External Asset and Liability Matrix for the G20 (as the end of 2022, USD millions)

Holder of claim (creditor)	Financial instruments	AR	AU	BR	CA	CN	FR	DE	IN	ID	IT	JP	KR	LU
Issuer of liability (debtor)														
AR	Direct investment	0	0	5897	2732	2134	1572	2085	1	0	2456	0	179	2276
	Portfolio investment	0	52	636	1153	0	436	409	0	0	1058	92	8	2992
	Other investment	0	21	2	0	0	622	0	0	0	135	0	10	29
AU	Direct investment	0	0	483	42,462	35,788	11,537	18,229	277	1561	2807	86,148	15,345	16,998
	Portfolio investment	1	0	15	24,933	18,267	19,619	32,946	4	111	6241	129,792	16,618	56,409
	Other investment	1	0	0	7038	20,494	18,897	2493	1709	474	561	16,832	1536	804
RB	Direct investment	780	0	0	17,942	3410	37,337	14,995	420	284	11,960	21,400	5444	59,337
	Portfolio investment	393	3825	0	12,297	1384	5880	5044	3	1	930	6238	7201	31,580
	Other investment	165	68	0	257	1	521	327	0	0	154	38	17	1536
CA	Direct investment	0	26,166	−1353	0	13,306	7451	20,583	742	146	5141	23,792	9403	78,180
	Portfolio investment	58	38,576	465	0	7415	38,495	95,092	39	114	5409	75,682	13,822	101,961

(continued)

Table 4.1 (continued)

Holder of claim (creditor)	Financial instruments	AR	AU	BR	CA	CN	FR	DE	IN	ID	IT	JP	KR	LU
Issuer of liability (debtor)														
	Other investment	1	1956	1571	0	15,615	5398	1136	0	82	179	47,346	577	1186
CN	Direct investment	0	2495	714	10,150	0	32,708	102,829	414	17,285	14,262	137,258	94,894	12,052
	Portfolio investment	31	23,015	10	40,956	0	14,280	14,014	216	336	2911	24,889	19,408	88,552
	Other investment	0	19,296	925	22,081	0	32,277	0	0	0	1601	0	19,484	12,891
FR	Direct investment	3	1626	1359	11,158	4814	0	122,029	144	25,990	31,640	13,167	3699	188,711
	Portfolio investment	2	32,138	313	45,135	13,868	0	404,077	393	2	169,962	206,700	25,568	487,833
	Other investment	539	11,771	3446	8956	20,764	0	203,683	4129	3060	98,604	203,311	4263	82,816
DE	Direct investment	0	0	433	8089	18,551	67,759	0	820	29	40,784	34,052	5818	213,489
	Portfolio investment	88	25,540	132	28,850	12,261	176,006	0	778	3	87,765	88,125	10,782	346,882

(continued)

Table 4.1 (continued)

Holder of claim (creditor)	Financial instruments	AR	AU	BR	CA	CN	FR	DE	IN	ID	IT	JP	KR	LU
Issuer of liability (debtor)														
	Other investment	0	262	1	370	0	52,001	0	0	0	3272	0	1254	29,376
IN	Direct investment	0	1043	60	2681	3483	26,666	24,018	0	80	6920	29,420	8595	9047
	Portfolio investment	0	9123	4	23,893	1231	12,638	8885	0	303	423	17,775	4215	56,022
	Other investment	0	2672	9	855	0	3604	0	0	0	35	0	905	152
ID	Direct investment	0	0	2	4198	24,722	1659	3104	672	0	935	35,495	13,668	1535
	Portfolio investment	1	1835	0	6089	921	1899	7510	219	0	1738	7565	1854	22,186
	Other investment	0	747	1	87	0	206	0	0	0	9	0	788	36
IT	Direct investment	0	0	527	1341	2476	85,534	48,739	256	1	0	4728	1267	102,963
	Portfolio investment	0	5526	151	8773	1433	162,500	90,616	1	2	0	55,376	1878	167,103
	Other investment	17	1379	427	217	328	138,945	33,305	63	42	0	9150	7	23,796
JP	Direct investment	0	932	207	1578	5075	18,058	12,435	117	−1	2254	0	10,604	5748

(continued)

Table 4.1 (continued)

Holder of claim (creditor)	Financial instruments	AR	AU	BR	CA	CN	FR	DE	IN	ID	IT	JP	KR	LU
Issuer of liability (debtor)														
	Portfolio investment	6	50,976	15	68,955	23,804	113,091	43,196	55	56	18,336	0	22,601	150,433
	Other investment	0	11,698	856	5829	0	35,774	0	0	0	2194	0	8897	4307
KR	Direct investment	0	540	14	1719	6674	5390	11,368	56	−278	2272	41,314	0	6884
	Portfolio investment	0	14,701	2586	19,949	12,098	15,568	10,903	71	83	3564	17,563	0	41,961
	Other investment	12	1331	418	259	26,211	1836	3391	3931	5406	19	9739	0	76
LU	Direct investment	0	0	19,314	63,268	20,555	66,041	271,375	176	−0	40,752	31,381	16,277	0
	Portfolio investment	155	14,796	4353	13,140	21,434	465,023	803,136	1638	0	709,989	108,963	34,168	0
	Other investment	41	1490	1239	4738	8135	252,691	128,577	680	71	24,477	5366	146	0
MX	Direct investment	2012	589	1114	24,193	1684	6630	18,634	243	8	4000	13,316	6164	9091
	Portfolio investment	116	2641	3254	8491	565	6607	11,366	0	0	5113	15,836	1406	30,900

(continued)

Table 4.1 (continued)

Holder of claim (creditor)	Financial instruments	AR	AU	BR	CA	CN	FR	DE	IN	ID	IT	JP	KR	LU
Issuer of liability (debtor)														
	Other investment	280	20	5	11,309	12	4431	178	0	0	64	144	27	498
NL	Direct investment	55	5759	−3085	46,299	28,302	207,981	336,252	12,852	943	35,486	138,226	13,747	485,991
	Portfolio investment	8	27,702	151	33,004	5096	289,429	261,264	304	0	82,259	88,487	9735	213,957
	Other investment	0	13,777	4748	6365	0	149,496	72,533	634	0	8144	2564	237	21,304
RU	Direct investment	0	0	0	−837	9902	16,484	17,238	336	−0	12,795	4536	5153	6197
	Portfolio investment	3	543	0	3906	135	489	4401	0	0	552	152	387	2245
	Other investment	0	52	0	58	0	22,763	0	0	0	2855	0	898	3935
SA	Direct investment	0	0	0	−2	3008	2364	1590	105	0	4432	4293	2626	15,822
	Portfolio investment	0	688	1	292	531	918	3252	3	6	1132	3463	1625	5337
	Other investment	0	70	0	0	0	9334	0	0	0	2877	17	28	496

(continued)

Table 4.1 (continued)

Holder of claim (creditor)	Financial instruments	AR	AU	BR	CA	CN	FR	DE	IN	ID	IT	JP	KR	LU
Issuer of liability (debtor)														
SG	Direct investment	0	12,999	1330	20,551	73,450	23,042	19,948	24,549	39,752	1431	105,422	33,895	99,977
	Portfolio investment	0	8621	11	8209	11,699	5791	7059	598	24,161	644	17,989	5044	27,439
	Other investment	0	43,790	21	5758	0	26,831	0	0	0	184	561	23,312	4133
ZA	Direct investment	0	486	96	1240	5742	2924	6268	524	150	2323	3643	253	4176
	Portfolio investment	4	2215	4	7216	632	2500	4585	0	0	914	3146	928	16,617
	Other investment	0	539	0	89	812	2913	2199	0	0	110	299	8	274
ES	Direct investment	406	0	3536	6270	1186	59,223	86,190	271	0	46,165	8339	1636	126,553
	Portfolio investment	12	6757	1901	9006	839	172,833	112,051	0	5	103,765	49,811	3481	112,997
	Other investment	1183	223	110	493	1018	85,594	41,246	604	155	32,278	561	15	15,347
CH	Direct investment	22	547	−9675	3584	8269	41,601	61,700	3410	2	15,259	39,009	767	350,460

(continued)

Table 4.1 (continued)

Holder of claim (creditor)	Financial instruments	AR	AU	BR	CA	CN	FR	DE	IN	ID	IT	JP	KR	LU
Issuer of liability (debtor)														
	Portfolio investment	3	15,782	1303	40,495	3771	35,989	73,690	3	3	12,646	32,101	8354	110,705
	Other investment	454	3728	223	7916	2591	120,754	24,390	502	303	2674	11,464	288	59,911
TR	Direct investment	0	0	0	−4357	3004	2359	8148	77	0	6436	0	1706	5227
	Portfolio investment	0	439	3	1547	61	1285	2895	0	0	1111	1333	393	6481
	Other investment	0	30	0	220	0	914	0	0	0	153	0	48	309
UK	Direct investment	0	124,791	6923	68,287	17,183	128,167	116,317	14,060	627	31,592	129,943	13,519	490,057
	Portfolio investment	8	65,503	555	84,936	32,305	192,992	181,547	591	114	37,929	138,393	37,315	404,103
	Other investment	257	32,760	1795	149,516	48,495	744,366	427,761	28,366	1753	23,904	340,959	7797	29,995
US	Direct investment	6582	130,822	28,182	764,921	79,172	248,753	379,534	14,993	−442	62,077	679,007	169,640	656,587
	Portfolio investment	39,192	464,645	20,477	1,416,252	217,685	369,739	608,141	7075	3157	169,390	1,694,471	423,013	1,578,515

(continued)

Table 4.1 (continued)

Holder of claim (creditor)	Financial instruments	AR	AU	BR	CA	CN	FR	DE	IN	ID	IT	JP	KR	LU
Issuer of liability (debtor)														
	Other investment	3394	81,951	43,992	424,598	20,947	313,303	83,889	18,371	3325	6893	447,702	25,845	9741
Others	Direct investment	35,921	352,131	243,294	361,318	2,382,924	388,571	401,063	34,537	18,725	174,447	364,666	188,022	1,048,553
	Portfolio investment	24	193,228	11,920	238,314	646,098	645,510	953,630	1733	2426	365,817	1,221,260	90,117	1,195,614
	Other investment	6688	32,384	132,187	293,258	822,868	307,659	619,343	68,731	72,086	61,876	250,418	91,384	387,133
Total asset of Financial Instruments	Direct investment	45,781	660,926	299,369	1,458,785	2,754,814	1,489,811	2,104,671	110,051	104,862	558,626	1,948,555	622,318	3,995,915
	Portfolio investment	40,103	1,008,867	48,259	2,145,791	1,033,532	2,749,518	3,739,708	13,725	30,884	1,789,598	4,005,202	739,920	5,258,826
	Other investment	13,032	262,015	191,976	950,267	988,291	2,331,130	1,644,451	127,720	86,757	273,252	1,346,471	187,771	690,081
Total assets		98,916	1,931,808	539,604	4,554,843	4,776,637	6,570,459	7,488,829	251,496	222,503	2,621,476	7,300,229	1,550,009	9,944,822
Difference (L > A)		39,167	240,601	405,783			831,097		1,196,280	237,565				
Total		138,082	2,172,409	945,387	4,554,843	4,776,637	7,401,556	7,488,829	1,447,776	460,068	2,621,476	7,300,229	1,550,009	9,944,822
Financial net worth		−39,167	−240,601	−405,783	561,823	530,666	−831,097	1,994,941.881	−1,196,280	−237,565	608,201	3,356,061	528,750	2,654,947.17
Reserve assets		44,585	63,583	324,704	106,908	3,306,529	242,991	294,706	562,290	137,233	225,160	1,222,571	423,164	2874

(continued)

Table 4.1 (continued)

Holder of claim (creditor)	Financial instruments	AR	AU	BR	CA	CN	FR	DE	IN	ID	IT	JP	KR	LU
Issuer of liability (debtor)														
Monetary gold		3618	3670	7585		117,235	142,563	196,293	41,259	4589	143,445	49,105	4795	131
Special drawing rights		5750	12,315	18,853	22,875	51,160	37,908	51,638	18,182	7411	28,267	59,275	14,837	2066
Reserve position in the fund		0	2590	4413	4348	10,839	7627	10,079	5215	1055	5714	10,815	3400	478
Other reserve assets		35,217	45,008	293,853	79,685	3,127,296	54,893	36,695	497,634	124,178	47,732	1,103,376	400,133	199
Adjustment item		118,157	−456,923	−715,384	−13,112	−1,305,868	−83,382	612,046	234,309	−151,308	−736,219	−1,422,693	−180,571	−2,618,925
Net Financial Position		123,575	−633,941	−796,464	655,620	2,531,328	−671,488	2,901,694	−399,681	−251,640	97,141	3,155,940	771,344	38,897

MX	NL	RU	SA	SG	ZA	ES	CH	TR	GB	US	Others	Total liability of Financial Instruments	Total liabilities	Difference (A > L)	Total
2706	17,693	26	0	800	82	19,247	3861	0	4555	12,645	7172	88,117	138,082	0	138,082
17	1172	0	0	194	1	311	316	13	1023	13,014	7057	29,957			
0	62	0	0	0	2	1089	1572	0	239	13,056	3169	20,008			
0	62,225	0	0	26,383	5840	4258	10,103	−10	93,528	173,653	48,707	656,323	2,172,409	0	2,172,409
4	27,523	0	1819	45,572	7103	2287	24,888	2	63,173	425,599	304,144	1,207,069			
4	5194	5	64	27,639	83	317	1101	6	80,176	59,218	64,371	309,017			
9305	117,347	6	6	10,117	271	45,217	12,328	13	20,310	80,963	48,065	517,257	945,387	0	945,387
2814	17,239	0	812	0	82	0	4465	9	26,506	160,210	70,275	357,189			
29	345	0	0	1	5	593	1589	0	22,577	25,011	17,707	70,941			

(continued)

Table 4.1 (continued)

MX	NL	RU	SA	SG	ZA	ES	CH	TR	GB	US	Others	Total liability of Financial Instruments	Total liabilities	Difference (A > L)	Total
871	161,631	516	1142	1930	586	10,353	31,439	4	73,029	438,766	52,850	956,678	3,993,020	561,823	4,554,843
155	28,316	0	171	23,113	388	4591	44,112	219	83,955	1,252,033	483,355	2,297,535			
958	0	15	65	15,861	71	363	2354	266	131,872	248,339	263,596	738,807			
27	58,292	312	1955	175,634	199	5364	30,241	194	25,506	126,104	1,024,050	1,872,939	4,245,970	530,666	4,776,637
59	15,901	0	7317	180,199	789	1009	10,358	14	112,643	247,214	808,304	1,612,427			
196	0	0	0	0	1373	274	4820	0	72,469	134,937	437,980	760,604			
0	170,242	561	202	1747	1272	32,152	65,107	182	109,934	112,017	129,682	1,027,437	7,401,556	0	7,401,556
210	155,883	0	10,811	0	1490	75,981	77,840	63	138,862	691,050	1,113,684	3,651,864			
183	89,071	4755	5339	32,159	519	93,334	66,968	6778	248,444	225,397	1,303,966	2,722,255			
0	382,534	3917	3482	5279	1835	19,525	71,087	2966	69,964	190,237	150,906	1,291,552	5,493,887	1,994,942	7,488,829
128	167,946	0	6513	0	1645	42,484	84,991	1	140,581	449,382	1,072,540	2,743,424			
0	26,998	0	0	0	10	9847	13,381	0	25,488	12,915	1,283,736	1,458,911			
403	37,174	283	746	79,605	848	2092	8790	166	85,610	51,553	173,803	553,085	1,447,776	0	1,447,776
13	14,681	0	795	80,611	494	378	6049	0	35,315	290,174	225,221	788,242			
0	0	0	0	0	175	24	1290	0	38,455	22,870	35,403	106,449			
157	9334	73	38	73,837	35	606	2024	105	6836	11,913	24,865	215,811	460,068	0	460,068
41	10,009	0	1593	31,577	36	283	3002	3	9215	65,034	45,568	218,179			
0	0	0	0	0	0	10	112	0	3621	9038	11,423	26,078			

(continued)

Table 4.1 (continued)

MX	NL	RU	SA	SG	ZA	ES	CH	TR	GB	US	Others	Total liability of Financial Instruments	Total liabilities	Difference (A > L)	Total
0	140,095	1295	76	−15	877	19,197	25,188	108	34,817	26,107	44,780	540,355	2,013,275	608,201	2.621,476
12	18.078	0	417	0	1564	121,893	9543	0	27,781	127,746	334,092	1,134,484			
1	8510	2287	143	340	30	20,252	18,293	3453	23,706	20,590	33,155	338,436			
0	45,368	52	197	29,343	40	469	17,351	5	12,480	77,489	29,598	269,401	3,944,167	3,356,061	7,300,229
86	44,408	0	6052	128,511	689	9316	38,278	0	147,538	1,088,862	919,622	2,874,886			
488	0	0	0	0	88	107	1224	0	175,824	326,627	225,967	799,880			
0	21,604	68	3706	22,392	1	1794	8606	6	15,626	36,655	7295	193,708	1,021,259	528,750	1,550,009
24	13,444	0	862	68,609	510	1197	15,247	196	39,203	205,816	178,608	662,764			
324	0	630	2980	7468	6	545	849	2587	16,685	44,110	35,974	164,787			
0	197,416	−13,456	8875	51,944	1575	15,338	88,551	685	285,598	605,304	265,558	2,036,529	7,289,875	2,654,947	9,944,822
977	139,891	0	8462	30,275	19,468	224,695	270,222	4	197,676	211,505	1,167,978	4,447,949			
18	13,547	2433	1399	2998	33	12,127	35,659	1375	80,630	71,928	155,599	805,397			
0	108,793	16	4	1423	6	54,055	9500	18	28,154	130,274	8053	427,973	888,352	0	888,352
0	12,536	0	1306	0	32	4611	5066	62	13,548	144,630	50,990	319,075			
0	0	0	0	49	1	2848	7493	0	11,362	87,000	15,582	141,303			
18,901	0	18,763	0	127,316	93,467	10,732	184,291	63	664,927	944,604	551,582	3,923,452	6,902,295	897,427	7,799,722

(continued)

Table 4.1 (continued)

MX	NL	RU	SA	SG	ZA	ES	CH	TR	GB	US	Others	Total liability of Financial Instruments	Total liabilities	Difference (A > L)	Total
139	0	0	1641	10,200	12.178	46,006	54,885	18	77,196	586,058	533,727	2,333,443			
28	0	657	154	0	1190	27,573	23,650	4472	157,017	45,808	105,049	645,400			
0	49,851	0	0	0	1	1865	28,505	1862	0	9637	121,294	284,819	535,395	0	535,395
22	2388	0	146	0	6	53	994	2	7783	24,438	4412	53,058			
0	0	0	0	0	6	895	15,346	0	6888	5950	137,872	197,518			
0	5968	0	0	0	1	1378	1522	−226	0	12,212	4789	59,881	300,274	217,793	518,067
1	2591	0	0	0	11	108	2013	0	5908	0	34,719	62,598			
0	0	0	0	0	8	3629	4301	0	76,904	17,949	62,182	177,795			
0	99,302	0	2346	0	826	1780	75,258	36	105,554	309,441	166,209	1,217,096	2,381,743	864,172	3,245,916
1	9730	0	1329	0	469	195	5902	1	20,210	108,784	155,874	419,759			
3	0	0	0	0	92	230	27,638	0	142,442	89,245	380,649	744,889			
0	50,598	2	197	403	0	1119	5298	0	42,109	7394	17,303	152,251	354,613	86,676	441,290
0	10,256	0	401	0	0	124	2635	4	13,531	65,153	31,875	162,741			
0	681	0	3	600	0	50	1264	1	16,728	4819	8232	39,621			
23,827	125,043	3963	−144	3891	99	0	20,306	145	104,477	35,625	79,630	736,637	2,196,432	0	2,196,432
375	35,422	0	740	6136	70	0	14,167	4	30,635	143,586	296,160	1,100,751			
1156	0	511	1526	5551	20	0	7241	1200	44,201	31,379	87,432	359,044			

(continued)

Table 4.1 (continued)

MX	NL	RU	SA	SG	ZA	ES	CH	TR	GB	US	Others	Total liability of Financial Instruments	Total liabilities	Difference (A > L)	Total
−1093	417,095	11,186	0	0	8426	9677	0	410	67,724	212,235	84,944	1,325,558	3,437,650	37,769	3,475,420
90	24,554	0	40,320	14,850	10,019	7914	0	0	85,147	600,118	260,114	1,377,971			
22	25,624	864	3245	9673	144	9185	0	2759	226,246	49,690	171,471	734,121			
0	15,877	3245	557	1718	18	7623	4118	0	4830	5761	27,677	94,023	205,314	0	205,314
70	2995	0	2480	0	20	192	1717	0	3190	20,528	18,773	65,513			
0	2322	0	0	0	7	108	2810	0	10,434	6310	22,113	45,778			
6053	641,594	0	0	0	18,135	109,623	64,889	4176	0	1,077,519	305,305	3,368,762	12,206,242	0	12,206,242
158	110,962	0	28,099	46,788	56,410	29,231	90,473	163	0	1,400,599	1,377,732	4,316,905			
555	88,210	6932	28,677	111,322	9358	33,389	174,823	22,097	0	1,062,079	1,145,408	4,520,574			
96,789	841,546	4119	6795	36,880	12,028	83,021	322,047	2810	663,369	0	367,029	5,656,262	28,635,630	0	28,635,630
22,319	547,009	0	171,513	507,794	26,864	87,234	406,766	1712	1,255,586	0	8,646,224	18,684,772			
3978	87,821	730	5223	78,269	1777	17,243	22,158	1468	1,337,720	0	1,254,258	4,294,596			
32,176	1,288,357	145,564	12,561	222,999	61,487	154,628	346,802	40,086	0	1,892,936	0	8,893,154	31,385,068	0	31,385,068
57,896	495,824	0	95,712	496,426	62,994	280,602	397,174	268	1,119,230	5,641,145	0	18,804,534			
82,284	477,601	31,545	37,195	409,505	15,005	222,767	31,169	51,503	1,697,401	678,966	0	3,687,380			
190,122	5,064,979	180,510	42,744	873,625	207,954	611,114	1,437,212	53,804	3,046,510	6,581,044	1,914,958				
85,610	1,908,756	0	389,311	1,670,856	203,332	940,694	1,571,102	2760	3,655,436	13,962,678	22,732,623				

(continued)

Table 4.1 (continued)

MX	NL	RU	SA	SG	ZA	ES	CH	TR	GB	US	Others	Total liability of Financial Instruments	Total liabilities	Difference (A > L)	Total
90,227	825,986	51,364	86,013	701,435	30,003	456,799	467,105	97,965	4,647,529	3,293,231	4,068,718				
365,959	7,799,722	231,874	518,067	3,245,916	441,290	2,008,607	3,475,420	154,529	11,349,475	23,836,953	28,716,298				
522,392		303,521				187,825		50,785	856,767	4,798,677	2,668,770				
888,352	7,799,722	535,395	518,067	3,245,916	441,290	2,196,432	3,475,420	205,314	12,206,242	28,635,630	31,385,068				
−522,392	897,427	−303,521	217,793.03	864,172.2	86,676	−187,825	37,769	−50,785	−856,767	−4,798,677	−2,668,770				
201,052	63,899	0	459,840	288,752	60,553	92,973	922,916	128,734	176,906	706,915					
6999	35,992	0	433	…	7318	16,475	60,763	45,846	18,201	474,294					
15,816	18,217	23,054	20,606	6318	5888	16,001	12,231	7331	40,738	160,537					
3538	3307	4924	3921	1548	873	3581	2287	150	7306	34,970					
174,699	6383	0	434,880	280,883	46,474	56,916	847,635	75,407	110,661	37,112					
−293,618	−192,639	303,521	29,105	−330,792	−70,532	−769,739	−163,285	−394,210	355,592	−12,080,546					
−614,958	768,687	0	706,738	822,132	76,697	−864,592	797,400	−316,261	−324,268	−16,172,307					

Data Sources IMF's CDIS, Table 6: https://data.imf.org/regular.aspx?key=61227426
IMF's CPIS, https://data.imf.org/?sk=b981b4e3-4e58-467e-9b90-9de0c3367363&sid=1424963554286
IMF's BOP/IIP: https://data.imf.org/regular.aspx?key=60587815;
BIS international banking statistics: http://stats.bis.org/statx/toc/LBS.html

sectors, all other economies (OE), total financial instruments, and total liabilities. The total assets or liabilities of all sectors equal the total assets or liabilities of the world. The matrix's columns demonstrate many countries' external assets, displaying national and regional perspectives. Each column corresponds to the sector's balance sheet; which countries/regions should appear in the matrix depends on the purpose of the analysis.

Instead of OIs, the GFFM was compiled using data from the coordinated direct investment survey (CDIS), coordinated PI survey (CPIS), and locational banking statistics (LBS). Table 4.1 depicts the relationship between debtors' cross-border liabilities (rows) and asset holders' cross-border claims (columns), which explains who traded with whom on what scale and using what financial instruments. The GFFM reveals the following structural equilibrium relationships.

First comes determining a country's external asset and liability (EAL) distribution and scale and the basic structure of its external investment position. The sources of inward financial investment to a country (debtor) can also be determined by analyzing the rows of the matrix, while the destinations of outward financial investment from a country can be identified by analyzing the columns (creditor). Concurrently, the rows in the matrix will always add up to the columns, i.e., total global assets equal total global liabilities.

Third, to reflect the actual situation of international capital in a country/region and to create the matrix table for the application analysis, countries (sectors) are set in rows and columns using the W-to-W tabulating principle. An "Other economies" sector was also created. The following is the relationship between these "Other economies" and the global total: "All other economies' liabilities (assets)" = total liabilities (assets)−liabilities (assets) of specific countries.

Fourth, the balance relationship between "total liabilities of a country = total assets of a country = the country's net financial assets" can be derived from the balance of external financial assets and liabilities, which can reveal the balance between domestic and foreign financial assets and liabilities.

Fifth, in the columns, financial assets are listed by country to show for what the funds are used, with the counterparty sectors identified for each cell. The following are the enhancements to the GFFM's updated version. This updated version[4] will allow the G20[5] to monitor its financial positions at regional, national, and cross-border levels using financial instruments based on the GFF framework.

Sixth, this improved version of the GFFM included "difference (L > A) rows" and "difference (A > L) columns" to calculate the matrix's symmetry. As such, the asymmetry in the original GFF model can be resolved by processing the data on net assets or liabilities (Zhang, 2020, 2022) to equalize the total assets and liabilities of

[4] For the first version of the GFFM, see Zhang and Zhao (2019, 535–536).

[5] As of 2020, the G20 members were Argentina (AR), AU, Brazil (BR), CA, CN, the EU, France (FR), IN, Indonesia (ID), Italy (IT), JP, Mexico (MX), Russia (RU), Saudi Arabia (SA), South Africa (ZA), Korea (KR), Turkey (TR), and the US. Singapore (SG) is a permanent guest invitee. Due to G20 restrictions, aside from FR, DE, and IT, which are also EU members, Switzerland (CH), Spain (ES), Luxembourg (LU), and the Netherlands (NL) were selected to represent the EU; therefore, the observations and analysis in this study include 24 countries and OE.

each country in the matrix. This will make it easier to perform the analysis and show the layout of the financial network. Table 4.1 is also based on the W-t-W benchmark; this matrix has the same number of rows and columns as the previous one, making it a square matrix.

Furthermore, the reserve assets item is still not included in the matrix's financial instruments in the updated GFFM. Because counterparty data on reserve assets are difficult to obtain; many countries do not publish them.

Table 4.1 explains the composition and scope of external investment in financing in greater detail, including (1) the proportion of external investment in the international financial market and its relationship with the international financial market; (2) the specific methods and composition of foreign investment in various countries; (3) the risk of external financial asset and liability imbalances; and (4) a transmission route of impacts from the onset of a financial crisis in a country/region, including a country to enable the implementation of an effective financial policy for the impacts arising from other countries. To save space, focus is on CN, the US, and JP to examine the effects of external financing such as DI, PIs, and OI (bank credit funds). Table 4.1 can also be decomposed into three matrices of financial instruments, which are shown in Tables 4.2, 4.3 and 4.4, respectively.

4.2.2 *Structural Changes in the Financial Assets and Liabilities of the G20*

Chapter 2 conducted a statistical description and network analysis of the G20's status in international capital flows for 2018. In this chapter, newly developed data from 2022 is used to perform a longitudinal comparison analysis with the 2018 data. Another significant development is the global supply chain and industrial chain fragmentation following the COVID-19 epidemic. This shift is expected to decrease the efficiency of resource allocation and drive-up global production costs. These developments inevitably altered the balance between savings and investment across various countries, consequently exerting a discernible influence on the foreign assets and liabilities held by nations. Consequently, there was an impact on international capital circulation and national balance sheets. To delve deeper, first requires examining the structural changes in the external assets and liabilities of the G20 countries.

Comparing the data in Table 4.1 with Table 2.3 (Chapter 2) shows that there has been a structural change in GFF. The first is a shrinking pool of assets. Foreign financial assets held by the G20, mainly as DI, PI, and OI, i.e., cross-border bank credit, rose from \$93.2 trillion in 2018 to \$101.3 trillion in 2022, while foreign financial liabilities rose from \$88.8 trillion in 2018 to \$98.6 trillion in 2022. As a result, the G20's net external financial assets fell from \$4.4 trillion in 2018 to \$2.7 trillion in 2022.

Second, changes occurred in the structure of the external balance sheets of the G20 countries. Between 2018 and 2022, countries with net external debt at the end

Table 4.2 Composition of bilateral investment by W-to-W (as of end-2022, USD bn.)

Creditor		China			Japan			United States			Others			Total liabilities
Debtor		DI	PI	OI	DI	PI	OI	DI	PI	OI	DI	PI	OI	
China	DI				137 (7%)			126 (1.9%)			1024 (4.1)			1873
	PI					25 (0.6%)			247 (1.8%)			808 (1.6%)		1612
	OI						0			135 (1%)			438 (5.7%)	761
Japan	DI	5 (0.18%)						77 (1.2%)			30 (0.1%)			269
	PI		24 (2.3%)						1089 (7.8%)			920 (1.8%)		2875
	OI			0						327 (2.3%)			226 (3%)	800
United States	DI	79 (2.87%)			679 (34.8%)						367 (1.5%)			5656
	PI		218 (21.1%)			1694 (42.3%)						8646 (17%)		18,685
	OI			21 (2.1%)			448 (33.2%)						1254 (16.5%)	4295
Others	DI	2671 (96.9)			1132 (58%)			6377 (97%)			23,654 (94%)			28,560
	PI		792 (76.6)			2286 (57%)			12,627 (90%)			40,352 (79.5%)		46,555
	OI			967 (97.9)			899 (66.7%)			13,501 (97%)			5694 (74.8%)	18,055

(continued)

Table 4.2 (continued)

Creditor	China			Japan			United States			Others			Total liabilities
Debtor	DI	PI	OI	DI	PI	OI	DI	PI	OI	DI	PI	OI	
Total assets	2755	1034	988	1949	4005	1346	6581	13,963	13,963	25,075	50,726	7612	129,996
Financial net worth	882	−579	228	1679	1130	547	925	−4722	9668	−3486	4171	−10,442	

Source from Table 4.1

of both periods include AR, AU, BR, IN, ID, MX, RU, ES, TR, the UK, the US, and OE. Notably, IN, RU, and the US experienced an increase in net debt, with the US net debt ratio expanding from −2.6% in 2018 to −3.7% in 2022.

Conversely, countries that maintained a net asset position from 2018 to 2022 and increased their net assets include CA, DG, IT, KR, and SG. Among them, JP underwent the most significant change, decreasing from 4.2% in 2018 to 2.58% in 2022.

Contrasting changes occurred in FR and CN during this period. FR held 0.69% of net assets, but it shifted to −0.64% of net liabilities. In contrast, CN held −0.32% of net liabilities and transitioned to 0.41% of net assets. Notably, the shift in CN's assets and liabilities in relation to the US is highly significant. In 2018, CN's net asset ratio to the US was 1.64%, but it transformed into a 5.32% net debt ratio. This change underscores the withdrawal of Chinese capital from the US, which is a consequence of the countries' economic decoupling.

4.2.3 *Composition of Bilateral Investment and Risk Between China and the US*

Table 4.2 helps examine the reciprocal changes in the international capital circulation of CN, JP, and the US across 2018–2022. Upon comparison with Table 2.4, distinctive features emerge.

A matrix focusing on CN, JP, and the US is created using Table 4.1; in Table 4.2, the rows denote financing and the columns denote what the funds are used for. Based on the W-t-W benchmark, Table 4.2 shows the composition and characteristics of mutual financial investments between CN, JP, and the US.

In Table 4.2, by the end of 2022, CN had received $137 billion from JP, $126 billion from the US, and $1,024 billion from OE through DI. Additionally, through PI, CN received $25 billion from JP, $247 billion from the US, and $808 billion from OE. This indicates that US investment through PI in CN was notably higher than JP's investment.

Furthermore, CN invested $5 billion in JP through DI, $79 billion in the US, and $2,671 billion in OE through DI. Then, through PI, CN invested $24 billion in JP, $218 billion in the US, and $792 billion in OE. The total investment stock from CN to the US in 2022 was higher than in 2018 (Table 2.4). However, due to the recent political tensions, CN's financial investment position in the US decreased by $5.332 billion when compared to 2018, while CN's investment in OE increased by $1.42 trillion.

Table 4.2 shows that CN's PI in the US accounts for 21.9% of its total PI, owing to its position as the primary holder of US Treasury bonds. CN's total investment (DI + PI + OI) in the US ranks first, accounting for 6.6% of total foreign investment, while in 2018, the figure was 10%. CN also has significant financial investment targets in the UK and FR, accounting for 2.05% and 1.78% of total foreign investment, respectively (Table 4.1). Furthermore, when comparing 2022 with 2018, the total amount of CN's investment in the US has declined, and the investment structure has changed. Among them, the PI ratio fell from 26.5% (Table 2.4) to 21.1% (Table 4.2), the DI ratio fell from 3.8% to 2.87%, and the OI ratio fell from 16.5% to 2.1%, narrowing the structure of all types of investment.

Looking at the American financing structure in the GFF (Table 4.2), the "columns" of the US show that the US investment in JP was much higher than that of CN with a focus on PI and OI. The DI, PI, and OI of US investment in CN accounted for 1.9%, 1.8%, and 1% of total US DI, PI, and OI investment, respectively, which is slightly higher than the proportion in 2018. The US investment in CN focused on DI and PI, with a smaller investment scale than JP. Regarding external investment, JP's investment in the US is much larger than CN's (Table 4.2).

Similarly, the US' DI, PI, and OI investment in CN in 2022 was higher than in 2018, with increases of $17 billion, $88 billion, and $101 billion, respectively. In other words, despite the sharp deterioration in political relations between CN and the US and despite the COVID-19 period, the scale of external financial investment between CN and the US was still higher than in 2018. However, when comparing 2022 to 2020, all types of US investment in CN have decreased significantly, particularly DI (down to −$229 billion) and PI (down to −$43 billion).

Table 4.2 depicts three characteristics of foreign investment between CN and the US. First, the forms of mutual investment between nations differ; investment is primarily in the form of PI (21.1%) and OI (2.1%), with DI accounting for only 2.87%. Second, the US has a foreign financial investment market monopoly. In comparison to the US and JP, CN's foreign investment is still relatively small; it is only 65.4% of JP's and 20% of the US.' In 2022, CN's investment in the US decreased by $5.33 billion compared with 2018. Additionally, the proportion of CN's DI, PI, and OI in the US decreased by 1.1%, 5.4% and 14.4%, respectively, but CN's investment in JP increased slightly. Apart from JP and the US, CN's investment in OE showed an upward trend, rising by 0.4%, 6.3%, and 17.6%, respectively. It shows the impact of US control on Chinese high-tech investment and reflects CN's strategic shift in outbound investment.

Third, Tables 4.2 and 2.4 show that, despite the two nations' decline in mutual investment, Chinese and US investment in OE is increasing, both in investment increment and proportion of outbound investment, indicating that both sides are looking for and adding new investment partners.

Tables 4.1 and 2.4 show that there is a corresponding relationship between asset and liability structures in GFFM. A counterparty's high financial assets, such as CN, are the inverse of the US' high debt. Regarding asset and liability linkages among countries, the GFF perspective can analyze the stability of international financial markets and the transmission of shocks to some countries. According to the total external assets and liabilities of each country at the bottom of Table 4.1, the US' external assets and liabilities at the end of 2021 were $23,837 billion and $28,635.6 billion, respectively, and net liabilities of −$4,798.7 billion. CN's foreign assets and liabilities were $4,776.6 billion and $4,246 billion, respectively, and its net assets were $530.7 billion. To comprehend financial risks, the structural situation of

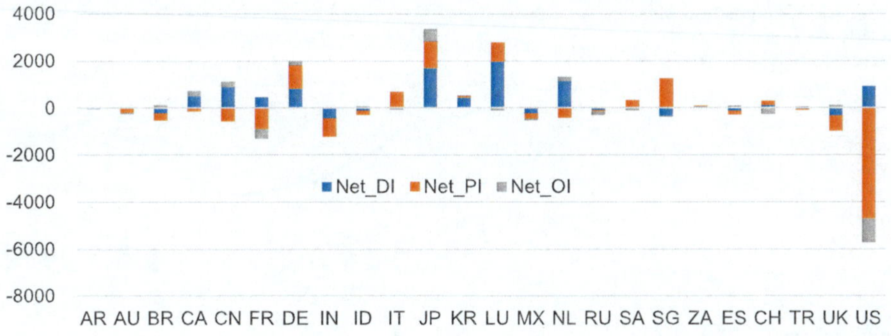

Fig. 4.1 Composition of net external investment for the G20 (as of end-2022, USD bn.) *Source* from Table 4.1

G20 countries' external assets and liabilities is illustrated in the item denoted with Difference (L > A) or (A > L) in Table 4.1. Using this data, Fig. 4.1 was generated.

Figure 4.1 shows that a country's external net assets (or net liabilities) contain DI, PI, and OI. DI largely involves investments in high-tech products, offering stability and substantial long-term economic development benefits. Whereas, PI encompasses long-term securities and short-term bonds, characterized by high yields and elevated risk attributes. The net debt of PI in the US is on a growing trend; the net debt balance of America's foreign securities investment in 2022 was −$4,722 billion, surpassing −$3,236 billion in 2018 (Table 2.4). Over the same period, CN's international portfolio position stood at −$578.9 billion. The question arises: who will be responsible for bridging this debt gap? Note that, the above analysis is based on only two data points from 2018 and 2022. A more comprehensive understanding of the trend changes in CN's and the US' EAL positions will be elucidated through an examination of a longer-term time series.

Figures 4.2 and 4.3 illustrate the net position of assets minus liabilities for DI, PI, and OI in the International Investment Position. From 2005 to 2022, CN's net position in DI and PI has consistently been negative, while that in OI has shifted to positive since 2014. Furthermore, the investment return in the Balance of Payments has consistently shown a negative trend over an extended period (see Fig. 3.9, Chap. 3). However, due to a prolonged surplus in the current account, CN's foreign exchange reserves have continued to expand, supporting the growth of its external asset position. As a result, the Net International Investment Position (NIIP) has demonstrated a persistent upward trajectory, with CN's NIIP reaching $2.53 trillion in 2022.

Nevertheless, since 2018, the international investment environment has worsened, leading to a decline in CN's foreign exports and a reduced flow of global FDI. Consequently, FDI into CN has exhibited a downward trajectory. This has significantly impacted CN's external balance, threatening the economy's stable growth. Notably, CN faces challenges due to inadequate domestic demand and persistent overcapacity. This implies that without maintaining a surplus in foreign trade and

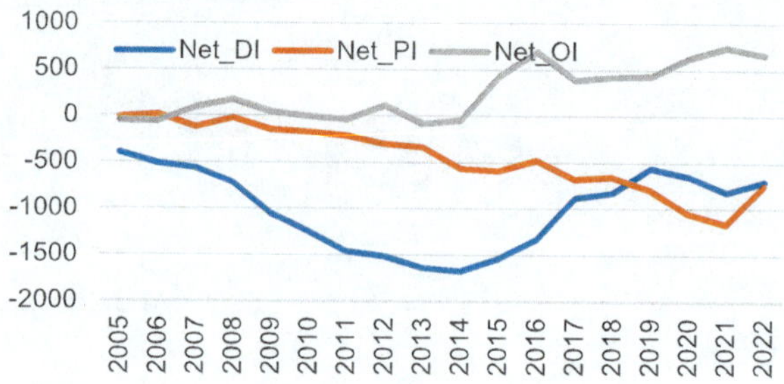

Fig. 4.2 Composition of CN's net external investment (USD bn.) *Source*: IIP (IMF, 2005–2022)

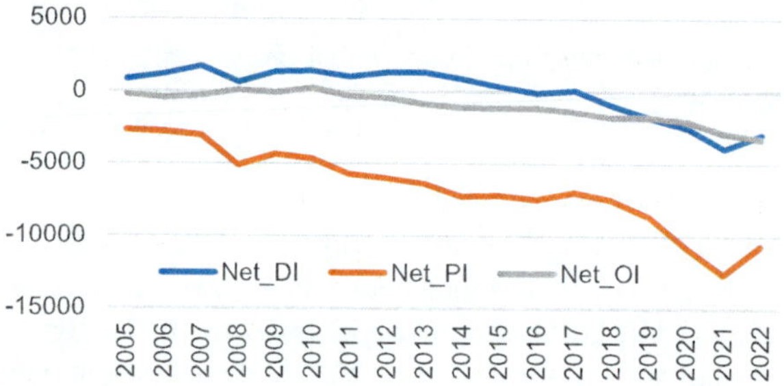

Fig. 4.3 Composition of US' net external investment (USD bn.) *Source*: IIP (IMF, 2005–2022)

with ongoing negative trends in FDI and portfolio investment, CN's economy will face a serious structural imbalance problem.

Figure 4.3 illustrates the sustained long-term decline in the net external financial position of the US. Although the Net DI has historically reflected a net asset position, it experienced a notable shift to −$976 billion starting in 2018, maintaining a negative net position of −$3 trillion through 2022. Furthermore, the extended negative growth observed in net PI and OI in the US, particularly the negative position of PI, plummeted to −$12.7 trillion by 2021, heightening debt risk.

Since 2021, the global economy has encountered two significant supply-side shocks—COVID-19 and the Russia–Ukraine war—that have contributed to heightened inflation levels. On the demand side, postpandemic, the US government implemented exceptionally loose fiscal and monetary policies, particularly emphasizing expansive fiscal measures. These policies directly affected households, augmenting their temporary income. In March 2022, the Federal Reserve embarked on a substantial initiative marked by a swift increase in interest rates and a reduction in its balance sheet. With the US NIIP standing at −$16.172 trillion at the close of 2022, potential risks are expected to escalate. Here are three key considerations to monitor.

First, the US corporate bond market; focusing on the high-yield segment commonly referred to as the junk bond market. Due to its inherently high financing costs, this market is susceptible to adverse effects if both the benchmark interest rate and risk premium increase. Such changes could pose overwhelming challenges.

Second is the US housing market, particularly the commercial real estate sector. The current 30-year mortgage rate in the US has reached 6–7%, a notably high level. Following COVID-19, there has been a recalibration of the operational model within

American enterprises. Corporate employees' efficiency working from home has not been significantly impacted. Consequently, there has been a decline in rental demand for commercial real estate, contributing to the accumulation of debt pressures in that market.

Third is the anticipated substantial rise in government debt pressures for major countries. From 2008 (see Fig. 3.7, Chap. 3), the global economy has witnessed a phenomenon known as low growth, low prices, low interest rates, and high debt. The elevated levels of debt accumulated from low growth and low interest rates, providing some relief from immediate pressure. However, the world economy's current state is characterized by sluggish economic growth coupled with increasing prices and interest rates. In this scenario, the sustainability of high debt becomes progressively untenable, and there is an expectation that the burden of servicing debt for the US and CN will notably increase.

4.3 Network Analysis of Cross-Border Debt

4.3.1 Theoretical Approach to Network Analysis

To analyze the correlation network of cross-border debt, observing the correlation between debt of different countries is necessary (Luiza, 2015). Methods of describing interconnections based on network theory have been widely used in measuring cross-border debt risk. Cross-border debt correlation-based distances, as applied to the study of stock market structures (Spelta & Araújo, 2012b), have been used in the analysis and reconstruction of geometric spaces in many fields (Acemoglu et al, 2015; Arajo & Lou, 2007). Note:

$$d_{i,j} = \sqrt{2(1 - C_{i,j})} \tag{4.1}$$

where i, j represent different countries and $C_{i,j}$ is the correlation coefficient between the claims and liabilities of cross-border debt in two different countries (or regions).

$$C_{i,j} = \frac{|\vec{s}(i)\vec{s}(j)| - |\vec{s}(i)||\vec{s}(j)|}{\sqrt{(|\vec{s}^2(i)| - |\vec{s}(j)|^2)(|\vec{s}^2(j)| - |\vec{s}(j)|^2)}} \tag{4.2}$$

The quantity in Eq. (4.1) is symmetric, nonnegative, and satisfies all the metric axioms. Therefore, it may be used to develop a geometrical analysis of the cross-border debt market structure. By setting $p(i)$ as liabilities of country (i) from another counterparty (j) at the end of a year (or quarter), and $p(j)$ as claims of country (j) to another counterparty (i), provides $\vec{p}(i) = (p_{i,1}, p_{i,2}, p_{i,3} \cdots, p_{i,j})$ and $\vec{p}(j) = (p_{1,j}, p_{2,j}, p_{3,j}, \cdots, p_{i,j})$. Using the DS' data based on each country's stocks of assets and liabilities $\vec{p}(i)$ and $\vec{p}(j)$ vis-a-vis the other reporting countries, a normalized vector is defined as:

$$\vec{Z}(i) = \frac{\vec{p}(i) - |\vec{p}(i)|}{\sqrt{n(|\vec{p}^2(i)| - |\vec{p}(i)|^2)}} \tag{4.3}$$

$$\vec{Z}(j) = \frac{\vec{p}(j) - |\vec{p}(j)|}{\sqrt{n(|\vec{p}^2(j)| - |\vec{p}(j)|^2)}} \tag{4.4}$$

where n represents the number of components (number of countries) in the vectors $\vec{p}(i)$. By inserting the normalized vector $\vec{Z}(i)$ and $\vec{Z}(j)$ into (4.2), the correlation coefficient $C_{i,j}$ is obtained, and then insert $C_{i,j}$ into (4.1) to define the distance between countries *i* and *j* using the Euclidean distance of the normalized vectors:

$$d_{i,j} = \sqrt{2(1 - C_{i,j})} = ||\vec{Z}(i) - \vec{Z}(j)|| \tag{4.5}$$

Each element of $d_{i,j}$ in the distance matrix has a range of [0, 2] because the range of the corresponding values of $C_{i,j}$ is $[-1, 1]$. If the value of $C_{i,j}$ is larger, the corresponding value of $d_{i,j}$ is smaller, indicating that the changes in the net claim between the two countries are more consistent and the correlation degree is stronger. In other words, when a country's banking sector changes, other countries/regions with small distances are more susceptible to change. When the distance between the two countries become $d_{i,j} = 0$, the degree of correlation between them reaches its peak, indicating that the net claim of the two countries changes in the same direction and proportion.

According to the network matrix $d_{i,j}$, the cross-border debt association network can be constructed. In general, the following methods can be used to depict the characteristics of cross-border debt networks.

4.3.1.1 Network Correlation Analysis

The correlation analysis of network relevance can reflect the stability, power concentration, and node equality of a relational network. More direct or indirect paths between two places improve the network correlation and strengthen the degree of correlation. However, many paths between the two regions in the relational network passing through a core area indicate that the network strongly depends on the core area and that the network correlation is poor. A change in the core area causes strong vibrations in or even paralysis of the entire network relationship. The formula for calculating the degree of correlation is:

$$NC = 1 - F/[N(N - 1)/2] \tag{4.6}$$

where F is the number of unreachable nodes in the network. In directed networks, alongside considering whether or not reachability between nodes is considered, the reachability between nodes is symmetrical. Methods such as betweenness centrality (BC), closeness centrality (CC), eccentricity, and harmonic closeness centrality

(Soramäki & Cook, 2016) can also identify the correlation among important nodes in cross-border debt networks.

To identify bilateral exposure networks, Spelta and Araújo (2012a) computed each country's CC. In a connected graph, the CC of a node i is the mean geodesic distance from i to any other node j. Formally, countries *i* and *j* are linked by their bilateral exposure ($B_{i,j}$), which is:

$$B_{i,j}(t) = B_{j,i}(t) = \frac{1}{d_{i,j}^{(3)}(t)} \tag{4.7}$$

where $d_{i,j}^{(3)}$ is the distance between countries *i* and *j* restricted to three-dimensional space and computed over a given time interval (t). If $t = 1$, $d_{i,j}$ is a two-dimensional plane showing the stock status at the end of a period. This paper chooses $t = 1$ to empirically analyze the stock of cross-border debt. Equation (4.7) indicates that a node's CC is the inverse sum of the distances (via shortest paths) from it to other nodes in the network. As with BC, CC in directed networks typically considers directed paths link direction and is only defined for connected networks (and only for strongly connected networks in the case of directed networks and paths along the direction of the links). For weight networks, the path length is defined by the sum of the link weights on the path. Because CC is based on inverse distances, nodes with higher CC are more central. In Eq. (4.7), $B_{i,j}$ represents the strength of cross-border debt association between countries *i* and *j*, matrix $B(i, j) = (B_{ij})_{N\times N}$—composed of $B_{i,j}$—can be used to represent the network matrix of the cross-border association of the countries. The network matrix determines the "edge" of the cross-border network of the countries, and each country is the "node" in the network. Together, they form the cross-border network of the countries.

4.3.1.2 Network Centrality Analysis

Methods such as BC, CC, eccentricity, and harmonic closeness centrality (Freeman et al., 1979; K. Soramäki & S. Cook, 2016) can also identify the correlation of important nodes in cross-border debt networks and the position and function of each node in the associated network.

Betweenness Centrality

BC detects the amount of influence a node has over the flow of information in a graph and is often used to find nodes that serve as a bridge from one part of a graph to another. In graph theory, BC is a measure of centrality in a graph based on the shortest paths. For every pair of vertices in a connected graph, there exists at least one shortest path between vertices such that either the number of edges that the path passes through (for unweighted graphs) or the sum of the weights of the edges (for weighted graphs) is minimized. The BC for each vertex is the number of these shortest paths that pass through the vertex.

BC finds wide application in network theory; it represents the degree to which nodes stand between each other. If the BC of this node is high, it will greatly influence the transfer of the entire graph information. The process of solving BC is divided into three steps:

1. The shortest path between each pair of nodes (s, t) is calculated, and the specific path must be obtained.
2. Whether the node is the shortest path for each node is determined.
3. Finally, judgment is accumulated and the number of nodes through which the shortest path from node s to node t is obtained. The BC of node v is given by:

$$BC(\nu) = \sum \frac{d_{st}(\nu)}{d_{st}} \tag{4.8}$$

where d_{st} is the total number of shortest paths from nodes s to t and $d_{st}(\nu)$ is the number of paths that pass through v. The denominator represents a normalization operation. Then, all the shortest paths from nodes s to t are *(n−1)(n−2)* for directed graphs and *(n−1)(n−2)/2)* for undirected graphs, indicating that is usually divided by *(n−1)(n−2)* in the analysis, where n is the number of nodes in the giant component.

Closeness Centrality

CC is used to investigate the dependence of one node on other nodes when propagating information. If one node is closer to the other, it is less dependent on others to spread information. Because the distance from one node to each point in the network is very short, this point will not be restricted by other nodes. Closeness was defined by Bavelas (1950) as the reciprocal of farness, that is:

$$C(x) = \frac{1}{\sum_y d(y, x)} \tag{4.9}$$

where d (y, x) is the distance between vertices x and y. CC is usually referred to in its normalized form, which represents the average length of the shortest paths instead of their sum. CC is generally given by the previous formula multiplied by n − 1, where n is the number of nodes in the graph. For large graphs, this difference becomes inconsequential, which eliminates the −1, resulting in:

$$C(x) = \frac{n}{\sum_y d(y, x)} \tag{4.10}$$

This adjustment allows comparisons between nodes of graphs of different sizes.

Eigenvector Centrality

In graph theory, eigenvector centrality (EC) is a measure of the influence of a node in a network. Relative scores are assigned to all nodes based on the concept that connections to high-scoring nodes contribute more to the score of the node in question than equal connections to low-scoring nodes. A high eigenvector score means that a node is connected to many nodes with high scores. The adjacency matrix can be used to find EC. EC assumes parallel duplication along walks and is based on the concept that a node's centrality depends directly on the centrality of the nodes to which it is linked. If the centrality of the ith node in a strongly connected network is denoted as x and set each node's centrality proportional to the average centrality of its neighbors, the outcome is:

$$x_i = \frac{1}{\lambda} \sum_{j=1}^{n} A_{i,j} x_j, \tag{4.11}$$

where n is the number of nodes in the network, λ is a constant, and A represents the network's (weighted or unweighted) adjacency matrix (if the adjacency matrix is weighted, moves along links with higher weights are more likely). If the vector of centralities is defined as $x = (x_1, x_2, \cdots, x_n)$, Eq. (4.11) can be rewritten as:

$$\lambda x = Ax$$

In general, many different eigenvalues λ have a nonzero eigenvector solution. However, the additional requirement that all entries in the eigenvector be nonnegative implies that only the greatest eigenvalue results in the desired centrality measure. The vth component of the related eigenvector then provides the relative centrality score of vertex v in the network. Because the eigenvector is only defined up to a common factor, only the ratios of the centralities of the vertices are well defined. To define an absolute score, one must normalize the eigenvector such that the sum of all vertices is 1 or the total number of vertices is n. Power iteration is one of many eigenvalue algorithms that can find this dominant eigenvector. Thus, the vector of centralities x is an eigenvector of the network's adjacency matrix. The Perron–Frobenius theorem states that the eigenvector of A corresponding to the largest eigenvalue has all positive entries—this eigenvector provides the nodes' ECs.

4.3.2 Debt Securities Matrix and Network for the G20

The GFFM was created based on W-t-W, and data visualization for financial network analysis requires the use of financial network technology. The IMF's CPIS gathers data on portfolio investment assets from over 86 countries and regions, presented in a matrix format with 246 rows and 84 columns. It is a global bi-annual survey of cross-border portfolio holdings by counterpart economy and by sector of holders and nonresident issuers. It shows which countries invest in a particular country, how the investments are distributed across institutional sectors, and the currency distribution of such assets. The CPIS has greater liquidity and higher risks than DI and OI (international bank credit); the CPIS statistics include equity and investment fund shares and DS.[6] To observe potential debt risks in the process of economic decoupling between CN and the US, the DS component was separated from the CPIS data source (Table 4.3).

The columns in Table 4.3 are assets and the rows are liabilities; the second-to-last row represents each country's net liabilities, and the second column on the right represents each country's net assets; lastly, the sum of the rows equals the sum of the columns. This allows quantifying the transmission of China–US bond risk shocks and exploring new policies that may be required to deal with economic decoupling.

A financial network is a system of interconnected financial institutions, such as banks, investment firms, and other financial intermediaries. The benefits of the financial network model for GFF analysis are primarily demonstrated in two areas. First, it describes the financial relations and risk exposure among countries as a whole. Second, this network describes the potential path and magnitude of shock contagion, providing a powerful analytical tool for macro-policy simulation.

According to Chap. 3, the external assets and liabilities, NIIP, net risky position, and net safe position that the US' external debt gap has been steadily increasing since 2008. Therefore, the question is whether the US' "exorbitant privilege" in international markets can be maintained despite massive shocks and rising foreign debt. As a result, a statistical test of the US' vulnerability to the abnormally high leverage ratio of external debt is required, as is discussing the countermeasures that CN should take.

The following methods are commonly used in network theory empirical research. First, a relationship model between multiple nodes is built using balance sheet data, and then the network's stability is tested by simulating the impact of a shock. Empirical research on network methods, such as Luiza (2015), has applied network analysis to the G4 economies (Euro Area, JP, the UK, and the US), and Girone et al. (2018) studied the propagation of quantity shocks in W-t-W networks. These analyses are primarily concerned with banking, and most network nodes represent the government, banks, and other institutions. This contrasts this study's network architecture, which makes the country the node.

The study of centrality, influence, sensitivity, and propagation dynamics is aided by network theory. A network is simply a graphical representation of a matrix (Soramäki & Cook, 2016), where the representation allows for a quicker interpretation of country interconnectedness. A network is made up of nodes and connecting edges, with nodes being countries and edges being asset/liability links. A link from country i to country *j* represents country i's claims (exposure) by country *j* in the financial network below. The positions of the nodes are arbitrary but their sizes are proportional to the countries' holdings of liabilities of a given type. For example, if the US is represented by a large node relating to DS exposures, this implies that the nation is a large issuer of DS. Similarly, the width of the link is proportional to the country to which each country is exposed to another.

---

[6] IMF (2023) Coordinated Portfolio Investment Survey, https://data.imf.org/regular.aspx?key=60587815.

Table 4.3 Total DS Matrix for G20 (as of end-2022, millions of USD)

Investment from	AR	AU	BR	CA	CN	FR	DE	IN	ID	IT	JP	KR	LU	MX	NL	RU
Investment in																
AR		22	7	1086	0	394	378	0	0	1053	89	2	2771	1	902	
AU	0		3	13,581	14,295	17,955	23,737	0	110	5058	109,936	9841	34,667	1	7928	
BR	231	298		3163	401	2937	1694	0	1	707	1873	5633	15,264	2774	12,097	
CA	0	19,028	15		4696	42,396	75,450	0	114	3038	51,795	6251	57,781	36	15,138	
CN	0	8780	1	1461		2749	7767	0	328	2440	15,592	5905	17,922	15	4502	
FR	0	12,639	12	10,632	9822		287,948	384	2	114,564	175,126	14,858	271,809	96	124,672	
DE	87	12,968	17	12,655	9362	112,551		754	3	67,977	68,062	3489	187,290	54	144,764	
IN	0	132	1	1578	234	586	4232		80	119	2296	257	8134	9	2309	
ID	1	375		2534	520	975	6646	0		1677	5517	649	14,329	41	7845	
IT	0	2177	7	2887	211	136,579	76,209	0	2		51,466	267	121,202		13,347	
JP	0	19,318	2	8765	18,170	95,174	19,599	0	5	13,972		6793	53,859	23	13,083	
KR	0	5610	2586	884	10,602	11,762	5215	0	83	3139	8049		14,299	5	3654	
LU	0	5911	479	6244	6508	145,421	99,439	0		16,922	25,685	4022		45	25,884	
MX	0	747	3243	7074	325	5602	10,426	0	0	5057	14,175	474	23,206		8477	
NL	0	14,819	67	9053	2796	221,740	211,786	200	0	55,136	71,924	3327	135,667	54		
RU	0	0		970	18	278	1108	0		442	58	50	1516		790	
SA	0	74	1	281	282	664	3023	0	6	1117	1036	1003	2889	1	479	
SG	0	5686	1	2492	10,568	5230	5047	448	12,806	524	11,738	3183	12,334		1472	
ZA	0	269	0	1555	294	566	3015	0	0	769	587	80	9018		6932	
ES	0	1904	1750	2403	362	147,531	89,937	0	2	99,872	44,871	1431	81,614	94	25,025	

(continued)

Table 4.3 (continued)

Investment from	AR	AU	BR	CA	CN	FR	DE	IN	ID	IT	JP	KR	LU	MX	NL	RU
Investment in																
CH	0	1280	241	3421	870	9541	8887	0	0	2436	9041	1215	19,558	0	5617	
TR	0	35	1	1078	20	707	2263	0	0	1105	852	214	5198	54	1810	
UK	1	0	29	29,322	20,207	138,054	118,516	518	113	29,625	104,239	15,896	250,718	31	37,699	
US	13,402	79,062	2693	396,182	99,351	236,396	275,732	3437	3157	96,274	996,082	105,691	650,584	12,006	159,331	
Others	15	78,696	1140	53,907	233,428	487,158	602,977	608	2110	159,772	396,419	30,017	530,249	6492	258,828	0
Total Assets	13,736	269,828	12,297	573,207	443,343	1,822,949	1,941,031	6350	18,923	682,796	2,166,505	220,549	2,521,877	21,833	882,586	
Net Liabilities	11,132	414,163	122,240	677,128	86,575	427,501		65,729	94,700	189,971				175,882	534,821	13,618
Total	24,868	683,991	134,537	1,250,335	529,917	2,250,450	1,941,031	72,078	113,623	872,767	2,166,505	220,549	2,521,877	197,715	1,417,407	13,618

Investment from	SA	SG	ZA	ES	CH	TR	UK	US	Others	Total Liabilities	Net Assets	Total
Investment in												
AR	0	28	0	270	291	1	579	10,997	12,701	24,868		24,868
AU	244	20,232	95	1924	16,764	0	31,438	175,041	438,253	683,991		683,991
BR	453	0	0	0	2270	9	16,159	24,308	91,338	134,537		134,537
CA	131	10,130	31	2764	32,104	7	57,614	524,983	622,571	1,250,335		1,250,335
CN	378	60,843	94	26	4836	39	13,270	18,931	431,501	529,917		529,917
FR	636	0	400	44,774	48,736	2	98,828	184,050	1,873,024	2,250,450		2,250,450
DE	142	0	1079	29,679	43,960	10	105,029	95,380	1,487,605	1,762,885	178,145	1,941,031

(continued)

Table 4.3 (continued)

Investment from	SA	SG	ZA	ES	CH	TR	UK	US	Others	Total Liabilities	Net Assets	Total
Investment in												
IN	417	16,437	91	0	1033	1	2626	12,866	38,608	72,078		72,078
ID	1448	11,359	0	148	2123	0	2796	27,849	67,900	113,623		113,623
IT	71	0	851	118,550	6676	1	15,401	43,710	687,506	872,767		872,767
JP	190	91,124	216	6269	16,020	0	57,317	246,834	834,683	1,252,654	913,851	2,166,505
KR	184	41,938	351	704	10,250	0	7745	24,889	119,668	205,729	14,820	220,549
LU	352	2944	295	11,762	32,270	195	38,741	61,891	547,946	696,397	1,825,480	2,521,877
MX	1099	0	0	0	4198	4	8668	71,231	112,514	197,715		197,715
NL	896	5667	319	39,171	42,833	53	41,307	160,691	1,126,470	1,417,407		1,417,407
RU	24	0	1	3	212	3	2369	3209	7797	13,618		13,618
SA		0	0	108	843	2	1058		22,832	24,842	30,843	55,686
SG	1089		26	174	3773	0	9515	37,081	174,585	226,243	479,311	705,554
ZA	143	0		9	1052	1	5845	13,917	37,629	58,596		58,596
ES	41	1304	9		10,255	1	14,395	42,781	745,644	814,430		814,430
CH	222	3261	185	1512		0	40,891	51,934	159,602	257,608	405,057	662,665
TR	2478	0	0	182	1468		1855	8126	26,293	40,402		40,402
UK	2523	17,394	1827	20,448	56,750	107		428,352	1,588,939	2,116,341		2,116,341
US	12,185	265,632	3550	47,286	184,119	734	479,950		8,451,693	9,445,149		9,445,149
Others	30,339	157,260	2805	186,867	139,830	412	329,741	1,503,227		7,560,566	5,759,013	13,319,579

(continued)

Table 4.3 (continued)

Investment from	SA	SG	ZA	ES	CH	TR	UK	US	Others	Total Liabilities	Net Assets	Total
Investment in												
Total Assets	55,686	705,554	12,227	512,630	662,665	1583	1,383,138	3,772,278	24,917,388	32,023,148		
Net Liabilities			46,368	301,800		38,819	733,203	5,672,871				
Total	55,686	705,554	58,596	814,430	662,665	40,402	2,116,341	9,445,149	24,917,388			

Source: IMF (2022), CPIS, https://data.imf.org/regular.aspx?key=60587815

As shown in Fig. 4.4, the DS Matrix in Table 4.3 is used for visualization. This is a financial network with nodes and edges. The size of each country's bond financing and investment determines the network's nodes, and the thickness of the edges indicates the amount of bond investment assets and liabilities held by each country. Cross-border links can be categorized into two types. The first type involves connections from receiving loans and other financial support (debt); the second type encompasses connections from extending credit (creditor rights). These links are measured by in-degree and out-degree values, respectively. The higher the in-degree value, the more the government industry in one country is affected by other countries' government operations. The larger the out-degree value, the stronger the ability of the country's government industry to spread its operations to other countries and the greater its influence on said operations in other countries.

According to the size of each node, the top three DS investments are the US (11.78%), LU (7.87%), and JP (6.75%). The top three DS financing countries are the US (29.49%), FR (7.03%), and the UK (6.61%). In the US, the proportion of net

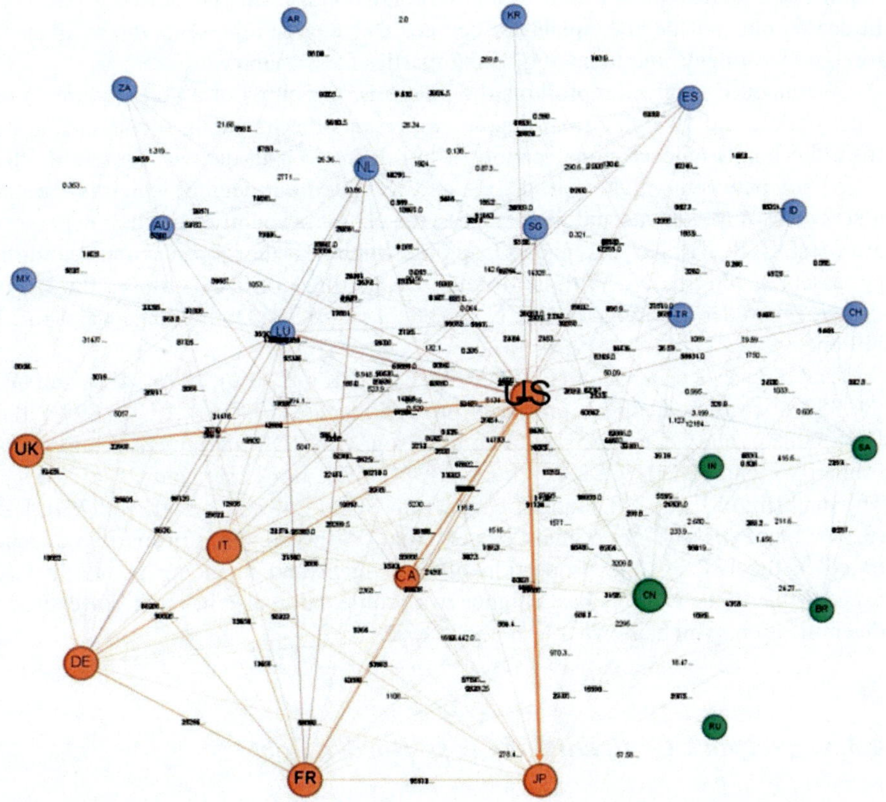

Fig. 4.4 W-t-W matrix as a network among G20 countries (as of end-2022). *Note* G7 countries are orange, related countries are blue, and BRICs countries are green

DS financing was 17.7%, which is a 1.3% increase from 2021, and the US external debt is on an increasing trend. G7 countries (orange line) and those closely related to the G7 (blue line) dominate the international bond market, while BRICS countries (green line) have relatively small investment scales.

Equations (4.8)–(4.11) were used to compute various characteristic values that reflect the G20 debt network and instead of using normalized data, raw data was used (Table 4.4). Because the processing of two different types of data were compared, the indicators calculated using the normalized data were very different from reality, which could be understood as normalized data that covered the original characteristics of the objects themselves.

The weighted in-degree and weighted out-degree in Table 4.4 indicate the debts (in-degree) and claims (out-degree) of G20 countries' cross-border debt at the end of 2022. This network connection exhibits a directional nature, characterized by both in-degree and out-degree. The in-degree represents the number of edges directed into a node, reflecting the number of countries investing in bonds from a specific country. The out-degree signifies the count of edges emanating from a node, indicating the number of countries that have invested in foreign bonds from that particular country. In-degree plus out-degree equals the degree. This degree represents the number of foreign investments and financing counterparties for a country.

"Weighted in-degree" represents the investment (claims) of a node (country) to other nodes, and "weighted out-degree" represents the financing (debt) of a node from the bonds of each node (country). "In-degree > Out-degree" represents the difference between weighted in-degree and weighted out-degree, which is the net assets of DS investments and financing. In the G20, the countries holding net assets are CH, DE, JP, KR, LU, SA, and SG, among which JP is the largest country holding net assets of foreign DS. With net foreign DS holdings of $5.67 trillion, the US is the largest holder of foreign debt. CN's net debt on foreign DS holdings was $86.57 billion.

The US invests primarily in CA (13.92%), the UK (11.36%), JP (6.54%), and OE (39.85%), while financing is primarily provided by JP (10.55%), LU (6.89%), the UK (5.08%), and OE (56.35%). In comparison, CN's DS investment destinations primarily include the US (22.41%), the UK (4.56%), JP (4.1%), and OE (52.65%). The majority of CN's DS funding came from SG (11.48%), the UK (3.57%), LU (3.38%), and OE (68.7%). Thus, CN's external DS investment is primarily concentrated in the US, but the proportion of investment and financing in OE in CN exceeds that in the US. CN has a higher risk profile, but it also has more investment diversification, which allows it to mitigate risk.

4.3.3 Network Centrality of Cross-Border Debt

The basic idea of EC is that an important node is linked to many other nodes alongside the nodes connected to it. Eigenvector analysis is particularly useful for correlation networks. The eigenvectors of a correlation matrix are orthogonal to each other, and

Table 4.4 DS Linkages and Network Centrality (as of end-2022)

Id	Indegree	Outdegree	Weighted indegree	Weighted outdegree	Indegree > Outdefree	Eigenvector centrality	Closeness centrality	Betweenness centrality
AR	7	21	13,736	24,868	−11,132	0.33	0.92	0.29
AU	22	21	269,828	683,991	−414,163	0.95	0.92	1.31
BR	22	20	12,297	134,537	−122,240	0.96	0.88	2.81
CA	24	21	573,207	1,250,335	−677,128	1.00	0.92	1.70
CH	24	21	662,665	257,608	405,057	1.00	0.92	1.70
CN	24	21	443,343	529,917	−86,575	1.00	0.92	1.70
DE	24	22	1,941,031	1,762,885	178,145	1.00	0.96	6.16
ES	21	21	512,630	814,430	−301,800	0.90	0.92	1.38
FR	24	21	1,822,949	2,250,450	−427,501	1.00	0.92	3.62
ID	22	20	18,923	113,623	−94,700	0.95	0.88	3.26
IN	7	21	6350	72,078	−65,729	0.32	0.92	0.29
IT	24	19	682,796	872,767	−189,971	1.00	0.85	0.76
JP	24	21	2,166,505	1,252,654	913,851	1.00	0.92	1.70
KR	24	21	220,549	205,729	14,820	1.00	0.92	1.70
LU	24	20	2,521,877	696,397	1,825,480	1.00	0.88	1.59
MX	20	19	21,833	197,715	−175,882	0.88	0.85	0.37
NL	24	23	882,586	1,417,407	−534,821	1.00	1.00	6.86
OE	24	23	10,951,309	5,192,296	5,759,013	1.00	1.00	6.86
RU	0	17	0	13,618	−13,618	0.00	0.77	0.00
SA	24	18	55,686	24,842	30,843	1.00	0.82	0.77
SG	15	20	705,554	226,243	479,311	0.65	0.88	1.73

(continued)

Table 4.4 (continued)

Id	Indegree	Outdegree	Weighted indegree	Weighted outdegree	Indegree > Outdefree	Eigenvector centrality	Closeness centrality	Betweenness centrality
TR	23	20	1583	40,402	−38,819	0.97	0.88	0.89
UK	24	22	1,383,138	2,116,341	−733,203	1.00	0.96	6.75
US	23	23	3,772,278	9,445,149	−5,672,871	0.95	1.00	6.30
ZA	21	19	12,227	58,596	−46,368	0.89	0.85	0.49

Table 4.5 15-order effects on G20 economies' debt security investment

Country	Δs	1	2	3	4	5	6	7	8	9	10	11	12	13	14	15	⋮	$(I-C)^{-1}*\Delta s$
AR	0.0000	−0.0011	0.0001	−0.0003	−0.0001	−0.0001	0.0000	0.0000	0.0000	0.0000	0.0000	0.0000	0.0000	0.0000	0.0000	0.0000	·	−0.0016
AU	0.0000	0.0052	−0.0061	−0.0057	−0.0023	−0.0018	−0.0011	−0.0007	−0.0005	−0.0003	−0.0002	−0.0001	−0.0001	−0.0001	0.0000	0.0000	·	−0.0139
BR	0.0000	−0.0025	−0.0015	−0.0008	−0.0006	−0.0003	−0.0002	−0.0002	−0.0001	−0.0001	0.0000	0.0000	0.0000	0.0000	0.0000	0.0000	·	−0.0064
CA	0.0000	−0.0405	0.0021	−0.0143	−0.0035	−0.0039	−0.0021	−0.0015	−0.0009	−0.0006	−0.0004	−0.0003	−0.0002	−0.0001	−0.0001	−0.0001	·	−0.0665
CN	−1.0000	0.0052	−0.0135	−0.0002	−0.0023	−0.0008	−0.0007	−0.0004	−0.0003	−0.0002	−0.0001	−0.0001	−0.0001	0.0000	0.0000	0.0000	·	−1.0136
FR	0.0000	0.0428	−0.0254	−0.0096	−0.0087	−0.0053	−0.0036	−0.0024	−0.0016	−0.0010	−0.0007	−0.0004	−0.0003	−0.0002	−0.0001	−0.0001	·	−0.0168
DE	0.0000	0.0037	−0.0240	−0.0041	−0.0079	−0.0037	−0.0029	−0.0018	−0.0012	−0.0008	−0.0005	−0.0003	−0.0002	−0.0001	−0.0001	−0.0001	·	−0.0442
IN	0.0000	−0.0007	−0.0008	−0.0004	−0.0002	−0.0002	−0.0001	−0.0001	0.0000	0.0000	0.0000	0.0000	0.0000	0.0000	0.0000	0.0000	·	−0.0027
ID	0.0000	−0.0014	−0.0006	−0.0009	−0.0004	−0.0003	−0.0002	−0.0001	−0.0001	−0.0001	0.0000	0.0000	0.0000	0.0000	0.0000	0.0000	·	−0.0040
IT	0.0000	0.0187	−0.0047	−0.0039	−0.0034	−0.0022	−0.0015	−0.0010	−0.0007	−0.0004	−0.0003	−0.0002	−0.0001	−0.0001	−0.0001	0.0000	·	0.0001
JP	1.0000	−0.0604	−0.0130	−0.0073	−0.0047	−0.0030	−0.0020	−0.0013	−0.0009	−0.0006	−0.0004	−0.0002	−0.0002	−0.0001	−0.0001	0.0000	·	0.9060
KR	0.0000	−0.0189	−0.0021	−0.0014	−0.0007	−0.0005	−0.0003	−0.0002	−0.0001	−0.0001	−0.0001	0.0000	0.0000	0.0000	0.0000	0.0000	·	−0.0244
LU	0.0000	−0.0070	−0.0031	−0.0048	−0.0028	−0.0020	−0.0013	−0.0009	−0.0006	−0.0004	−0.0002	−0.0002	−0.0001	−0.0001	0.0000	0.0000	·	−0.0236
MX	0.0000	−0.0016	0.0004	−0.0021	−0.0006	−0.0006	−0.0003	−0.0002	−0.0002	−0.0001	−0.0001	0.0000	0.0000	0.0000	0.0000	0.0000	·	−0.0056
NL	0.0000	0.0109	−0.0052	−0.0096	−0.0051	−0.0039	−0.0024	−0.0016	−0.0011	−0.0007	−0.0005	−0.0003	−0.0002	−0.0001	−0.0001	−0.0001	·	−0.0201
RU	0.0000	−0.0003	0.0000	−0.0001	−0.0001	0.0000	0.0000	0.0000	0.0000	0.0000	0.0000	0.0000	0.0000	0.0000	0.0000	0.0000	·	−0.0007
SA	0.0000	−0.0001	−0.0005	−0.0001	−0.0001	−0.0001	0.0000	0.0000	0.0000	0.0000	0.0000	0.0000	0.0000	0.0000	0.0000	0.0000	·	−0.0009
SG	0.0000	−0.0185	−0.0029	−0.0013	−0.0009	−0.0005	−0.0004	−0.0002	−0.0002	−0.0001	−0.0001	0.0000	0.0000	0.0000	0.0000	0.0000	·	−0.0252
ZA	0.0000	−0.0018	−0.0002	−0.0004	−0.0002	−0.0002	−0.0001	−0.0001	0.0000	0.0000	0.0000	0.0000	0.0000	0.0000	0.0000	0.0000	·	−0.0030
ES	0.0000	0.0155	−0.0033	−0.0043	−0.0032	−0.0022	−0.0014	−0.0010	−0.0006	−0.0004	−0.0003	−0.0002	−0.0001	−0.0001	−0.0001	0.0000	·	−0.0016

(continued)

Table 4.5 (continued)

Country	Δs	1	2	3	4	5	6	7	8	9	10	11	12	13	14	15	⋮	$(I-C)^{-1} * \Delta s$
CH	0.0000	−0.0030	−0.0025	−0.0017	−0.0010	−0.0007	−0.0004	−0.0003	−0.0002	−0.0001	−0.0001	−0.0001	0.0000	0.0000	0.0000	0.0000	·	−0.0101
TR	0.0000	−0.0005	−0.0002	−0.0003	−0.0001	−0.0001	−0.0001	0.0000	0.0000	0.0000	0.0000	0.0000	0.0000	0.0000	0.0000	0.0000	·	−0.0014
UK	0.0000	−0.0354	−0.0164	−0.0130	−0.0073	−0.0051	−0.0033	−0.0022	−0.0014	−0.0009	−0.0006	−0.0004	−0.0003	−0.0002	−0.0001	−0.0001	·	−0.0868
US	−1.0000	0.2723	−0.2240	−0.0080	−0.0441	−0.0165	−0.0146	−0.0084	−0.0059	−0.0038	−0.0025	−0.0017	−0.0011	−0.0007	−0.0005	−0.0003	·	−1.0604
Other	0.0000	−0.4167	0.0383	−0.0651	−0.0147	−0.0182	−0.0094	−0.0070	−0.0044	−0.0030	−0.0019	−0.0013	−0.0008	−0.0006	−0.0004	−0.0002	·	−0.5058

the underlying data can be projected for its eigenvectors. The eigenvectors correspond to the principal components and the eigenvalues to the proportion of the total variance explained by each principal component.

First comes performing a financial network analysis using the EC method. Equation (4.11) and Table 4.3 were used to calculate EC values (Table 4.4). The G20 countries are divided into four groups based on their EC value in the DS market at the end of 2022. The EC values for CA, CH, CN, DE, FR, IT, JP, KR, LU, NL, RU, SA, and the UK were all 1, indicating that cross-border debt is central to the G20. The next level includes AU, BR, ES, ID, TR, and the US, of which the EC was less than 1 but greater than 0.90; these countries are closely linked to the bonds and credit of other G20 countries and have a strong influence, but not the strongest. The US had an EC value of 0.96 because the country's bond financing primarily targets the G7 countries, LU, and NL, or CN, but the bond trading relationship with other countries is not very close. In the medium-level countries, MX and SG have values lower than 0.88 but greater than 0.61. AR (0.33) and IN (0.32) had the lowest values. The EC of RU was 0, which means that, due to the Russia–Ukraine war, no country has a bond investment deal with RU. Therefore, the above four levels can be roughly distinguished based on the impact of clustering in the G20 financial market, according to the EC values. However, this distinction is a little sloppy, therefore a more careful examination is required.

Using Eq. (4.10), the CC of countries with respect to cross-border banks were measured (Table 4.4). CC is calculated as the sum of the distances from a node to all other nodes. The smaller the sum, the shorter the path and the closer the node is to all other nodes. By normalizing the sum of the shortest distance between a node and other nodes, a number between (0, 1) is obtained. A larger number indicates that the node is closer to the center. For the G20 countries, the CC values of NL and the US are the only ones estimated to be 1. This indicates that CN did not yet occupy a central position in the international bond market. CN's financial openness is relatively low and less affected by the international financial crisis. However, with the opening of CN's financial system and the promotion of the RMB's internationalization, CN will strengthen its ties with other countries.

Next, is BC, which refers to the number of shortest paths in which a certain node appears between other nodes. A node with a high BC has a strong impact on the information transfer. In Table 4.4, the BC values of DE, NL, the UK, and the US are over 6.16—the highest level for G20 countries. CN's BC is 1.7, which is identical to CH, JP, KR, and SG, indicating that CN's influence in international DS is still weak but slightly stronger than that of IN, TR, and RU, which complements the results in Fig. 4.3.

4.3.4 Degree of Centrality Within the Network

The GFF's W-t-W data can be viewed as a network of interrelationships, with nodes representing countries and edges representing assets or liabilities. The amounts involved in each asset or liability relationship "weight" the network edges. Focus is on degree centrality in the network analysis to demonstrate the importance and influence of G20 countries.

In network analysis, degree centrality is the most direct metric for describing node centrality (Zhang, 2020). The greater the degree of the node, the higher the degree centrality and the more important the node. Degree centrality in an undirected graph measures the extent to which a node in the network is connected to other nodes. The degree centrality of node *i* in an undirected graph with g nodes is the total number of direct connections between *i* and other g-1 nodes, as expressed by:

$$C_D(N_i) = \sum_{j=1}^{g} x_{ij}(i \neq j) \tag{4.12}$$

where $C_D\ (N_i)$ represents the centrality of node *i*, which is used to calculate the number of direct connections between node *i* and other g–1 *j* nodes ($i \neq j$ excludes the connection between *i* and *j*, so the data in the main diagonal can be ignored), x_{ij} is the value of the cell in which the corresponding row or column in the matrix is located. $C_D\ (N_i)$ is calculated as the sum of the values of the cells in which the corresponding row or column of node *i* in the network matrix is located, that is, the sum of one country's assets (columns) or liabilities (rows), because directed relationships form a symmetric data matrix, and cells with the same rows and columns have the same value. Following Tsujimura and Tusjimura (2008), Zhang (2020)[7] proposed indicators for observing the influence coefficient (IC) and the sensitivity coefficient (SC) by financial network. However, according to network theory (Soramäki & Cook, 2016), PDI and SDI are also considered network centrality measures of a network represented by the inverse of Leontief[8] (degree centrality). PDI and SDI can be considered a type of network centrality measure, namely the degree centrality (in- and out-degree) of the weighted network represented by $(I-C)^{-1}$. This study defines in-degree as external claims and out-degree as external debts and emphasizes their connection using network theory because GFF and W-t-W provide many analytical possibilities. The degree centrality of GFF and draw network diagrams are calculated using Eq. (4.3) and based on matrix C (in Table 4.4) to obtain the inverse of Leontief by $(I-C)^{-1}$. This method is used because matrix C better represents the network of interconnections.

Table 4.3 is used to create a square matrix in Fig. 4.5. PDI and SDI are defined as the power of dispersion index (PDI) and the sensitivity of dispersion index (SDI), which both indicate the funds supplied and demanded by a country by the two different aspects of supply and demand for funds. PDI reflects a country's limits, which include the indirect effects on global financial market supply when a country increases its money supply. Because it is highly correlated with the external asset portfolio, it is best used for cross-country comparisons. When the overall demand

[7] See Zhang (2020, 312–313) for a detailed compilation method; its gist is in the Appendix.

[8] Leontief (1941).

Fig. 4.5 Degree of centrality on debt securities between G20 countries as of end-2022)[9]

for funds rises, countries with a high SDI tend to finance funds from other countries (domestic assets); so much depends on other countries' fund supply.

Figure 4.5 was created using the Appendix method and is divided into four quadrants. Moving counterclockwise, PDI and SDI are higher than average in quadrant I (greater than 1). In quadrant II, PDI is less than 1, but SDI is greater than 1. Both PDI and SDI are less than 1 in quadrant III, indicating that they are below average. In quadrant IV, PDI is greater than 1, but SDI is less than 1. The quadrant in which a country is located indicates its influence on global financial markets. Figure 4.5 depicts the G20 countries' position in international DS markets at the end of 2022. The first quadrant includes FR, DE, UK, and NL; their asset influence and liability sensitivity in the international capital market are greater than the G20 average. The US' PDI and SDI are 0.816 and 4.956, respectively, indicating that the US' influence on international bond investment will be no greater than the G20 average, but the US has the highest financing (liability) sensitivity. The financial market's capital requirements have a significant spillover effect on the US and the UK. When the international capital market's capital needs double, the capital needs of US and UK investments increase by 4.956 × and 1.38x, respectively.

CN's PDI and SDI are 1.311 and 0.598, respectively, which are in quadrant IV, and its influence is higher than that of the US and the average level of G20 countries, but the reflection degree of DS financing still lags behind the US. Compared with 2018 (see Fig. 2.3), CN's PDI and SDI in 2022 have improved. In the US, the two indicators are 1.1407 and 4.9705, and SDI in 2022 was essentially flat compared to 2018, but PDI declined. In a sign of America's growing need for external funding, securities debt has more than doubled, but AmerPDI's influence on bond investment has declined.

[9] See Appendix: (2) for the calculation method of PDI and SDI in Chap. 2 for details.

4.4 Identifying Debt Interlinkages Between China and the United States

CN and the US are facing the structural problem of growing economic decoupling, but they are still closely related in external debt. By the end of 2021, the US foreign net debt position will have reached a record high of −\$18.1 trillion,[10] whereas CN's investment in US DS continues to account for the lion's share of its total foreign securities investment, accounting for 22.41% of total foreign securities investment (Table 4.3).

4.4.1 Debt Diffusion Matrices

Based on the GFFM model from previous research (Zhang, 2020), bilateral exposures across N countries in a financial instrument k can be expressed in an n x n matrix in which the element y_{ij} denotes a claim of country i vis-à-vis country j. Therefore, the sum of each column j denotes the aggregate holdings of assets of country j in instrument k ($a_{j,k}$), and each row i denotes the aggregate holdings of liabilities of country i in instrument k ($l_{i,k}$). Aggregate assets ($a_{j,k}$) and liabilities ($l_{i,k}$) per country are observable, but bilateral exposures need to be estimated.

$$Y_k = \begin{pmatrix} y_{11} & \cdots & y_{1n} \\ \cdot & \cdot & \cdot \\ y_{n1} & \cdots & y_{nn} \end{pmatrix} \text{ with } \sum_{i=1}^{n} y_{ij} = a_{j,k} \text{ and } \sum_{j=1}^{n} y_{i,j} = l_{i,k}$$

To represent how various countries' investment behaviors react to the investment needs of others (in order to finance them), Δs is set as an exogenous variable, in which the shock itself, indicating changes in the original investment, is:

$$\Delta s = (0, \ldots, -s, 0, \ldots, 0)' \tag{4.13}$$

Using the W-t-W framework, a matrix algebra presentation of GFFM can be shown by $T = Y + \Delta s$, where T is the vector $T = \begin{pmatrix} t_1 \\ \cdot \\ t_n \end{pmatrix}$.

The elements c_{ij} is defined as the ratio of funds raised from country i to the total external financing needs of country j, that is, $c_{ij} = \frac{y_{ij}}{t_j}$. The investment ratio matrix is shown as C.

[10] BEA, Table 1.2. US NIIP at the End of the Period (March 29, 2022).

$$C = \begin{bmatrix} c_{11} & \cdots & c_{1n} \\ \cdot & \cdot & \cdot \\ c_{n1} & \cdots & c_{mn} \end{bmatrix} \tag{4.14}$$

where C is the matrix of c_{ij} determined by the form of the $n \times n$-order, and providing $y_{ij} = c_{ij} * t_j$, and the diffusion matrix is:

$$T = C * T + \Delta s \tag{4.15}$$

where $T_k = C * \Delta s$, , ($k = 0, 1, 2, ..., n$). When $k = 0$, it is called a direct effect, $k = 1$ is called an indirect first-order effect, $k = 2$ is an indirect second-order effect, and $k = n$ is an indirect n-order effect, as shown:

$$\begin{aligned} \xi_k &= T_0 + T_1 + T_2 + \cdots + T_k \\ &= T_0 + CT_0 + C^2T_0 + \cdots + C^kT_0 \\ &= (I + C + C^2 + \cdots + C^k)T_0 \end{aligned} \tag{4.16}$$

$$\text{Moreover, when } k \to \infty, \ \zeta_\infty = (1 - C)^{-1}T_0 \tag{4.17}$$

Equation (4.17) reflects the limiting effect of the n-order where $(I - C)^{-1}$ is the inverse of Leontief. Whereas Leontief considers input per unit of output, this study considers financing per unit of investment, but the overall logic is the same. The diffusion matrix elements in the model have interesting interpretations regarding well-known financial ratios. Thus, $c_{1,j}$ and $c_{2,j}$ are the ratios of financing from one country to another to the total investment of the country (also assuming that when $i = j$, $C_{i,j} = 0$, , i.e., excluding the country's own domestic PIs). The ratios $c_{1,j}$, $c_{2,j}$, $c_{3,j}$, and $c_{i,j}$ represent the mix of financing sources for a country's portfolio investment, indicating how a country relies on other countries for funding, usually by issuing treasury securities and bank debentures.

4.4.2 Shock Dynamics of the United States and China

Equation (10) is employed to assess the impact of changes in investment in one G20 country on other member countries, focusing on CN and the US. As shown in Table 4.3, the US has the largest share of DS investment in the international DS market, and during the same period, CN's DS financing accounted for only 1.05%, while its DS investments accounted for 0.5%, indicating that CN's market share remained small.

Given the large structural difference in the asset–liability ratio of DS between CN and the US, CN has long been the world's top three holders of US debt. Therefore, the impact of the debt crisis on the Chinese and US economies, including the global ripple effect, must be measured from the perspective of DS debtors. As analyzed in Chap. 3, the US' foreign liabilities are denominated in US dollars, whereas its foreign assets are denominated in foreign currencies. Using "exorbitant privilege," the US reinvests international capital raised from abroad to earn a higher return on US external assets than on its external liabilities. According to this model, a reduction (or increase) in China's investment in US bonds would have a moderate impact on US external financing.

A specific quantitative shock is examined, namely the impact of CN's reduced purchases of US treasuries on the shock of US financing, on CN itself, and on the G20 bond market, that is, a case of debt shock. The shock in unitary terms is:

ΔS =	(0	0	0	0	−1	0	0	0	0	0	1	0	0	0	0	0	0	0	0	0	0	0	−1	0)

It is assumed that the DS of: CN will be reduced by −1 unit, the US will also be reduced by −1 unit, JP will be increased by 1 unit, and the other countries will remain unchanged (with an increment of 0). Therefore, according to Eq. (4.16), using Δs and the investment ratio matrix C, the decomposition into the first 15 orders can be presented in the case of the G20 (Table 4.5), and speculate on the impact of the shock on changes in the original DS for the US and CN with $(1c)^{-1}\ \Delta s$.

Changes in investment and financing caused by shocks in Table 4.5 are governed by the set of direct and indirect relationships embedded in the W-t-W diffusion matrix, which includes intricate investment/financing paths of any order, including the 15-order one used here. Following this, a decomposition of the CN and US shocks that separates these individual n-order effects is proposed. Table 4.5 is a statistical estimation based on the asset section of Table 4.3. The limit impact effect is represented by the extreme right column.

Figures 4.6 and 4.7 were plotted using the shock effects from Table 4.5. The first-order effect of the shock is a decrease in DS investments in CN and the US and an increase in JP. CN's share of total DS investment is relatively low, at 1.38%, so a unit reduction in CN's investment has a direct impact of 0.0052, while the first indirect effect is −0.0135. The indirect effect was weak, and the positive and negative effects staggered to zero at the thirteenth order, and the maximum impact effect on CN was −1.0136 (compared to −1.0032 under the same conditions in 2021). According to the accumulated effect, CN has a longer negative shock effect; the accumulated first-order effect was −0.9948, while the accumulated indirect effect remained at −1.0113 until the 15th order. This suggests that a decrease in DS investment in CN significantly negatively affects itself (see Table 4.5 and Fig. 4.6).

Because the US has the world's largest financing market share, even if its initial investment falls by one unit, the direct effect will be 0.2723. However, the first-order indirect effect falls to −0.224 before alternating between rebound and decline until the 13th indirect effect tightens to zero. The maximum impact effect on the US was −1.0604 (compared to 0. −0.7381 under the same conditions in 2021). For a given $\Delta s\_US = -1$, the accumulated effect of the US changes significantly, except for the first indirect accumulated effect, which is relatively high (−0.7277)

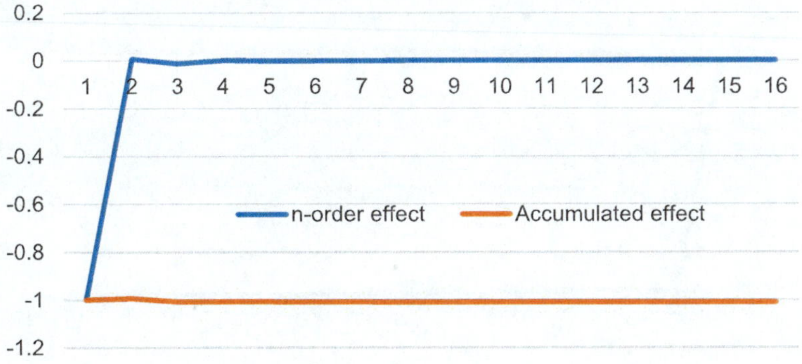

Fig. 4.6 Shock effects for CN (ΔS_CN $= -1$, ΔS_US $= -1$)

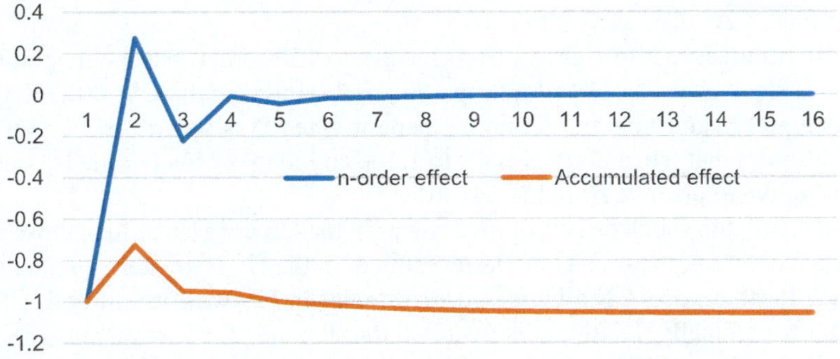

Fig. 4.7 Shock effects for the US (ΔS_US $= -1$, ΔS_CN $= -1$)

and maintains between -0.9517 and -1.0598, reflecting the strong response of the US' DS investment to the impact of risks (see Fig. 4.7).

Moreover, based on the statistical estimation results for Japanese investment in DS, it is observed that while the DS investments of CN and America are projected to decrease by one unit, the DS investment in JP is expected to increase by one unit. The limit value, encompassing both the direct and indirect nth-order effects, is calculated as 0.906 (compared to 0.9619 under the same conditions in 2021).

To test the shock and influence of CN's increased DS investment on China–US decoupling, the assumed conditions of CN's DS investment are changed, Δs_CN is set to 1, while the assumed conditions of US and JP remain unchanged. The shocks to CN are currently set at $+1$, but because CN has a small share of the DS market and has a low impact on DS investment in other countries, the first-order effects (direct effect) are 0.0052 for CN and 0.6472 for the US (Figs. 4.8 and 4.9).

However, the second-order effect (indirect effect) for CN and the US is 0.018 and 0.241, respectively. Although both CN and the US gradually drop to 0 at level 15 (Figs. 4.8 and 4.9), CN's impact effect is clearly smaller than that of the US.

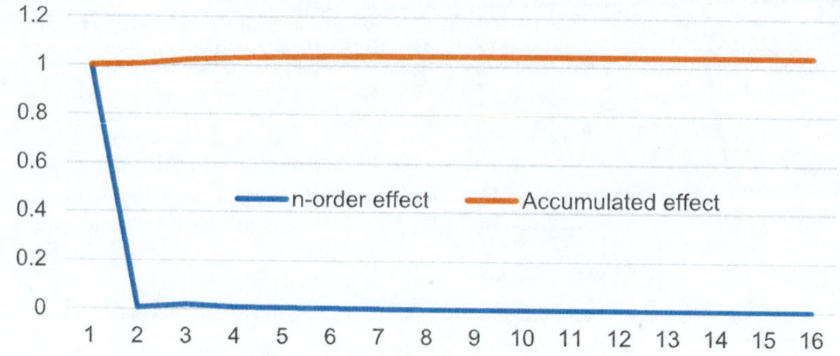

Fig. 4.8 Shock effects for CN (ΔS_CN = 1, ΔS_US = 1)

Furthermore, because Δs_CN is set to 1, the positive impact on the US will be felt sooner than if Δs_CN was set to −1.

Their cumulative effect grows from second to 15th order, but the magnitude of change is smaller in CN. CN's limiting effect is 1.0468, and the US' is 0.3708, both of which are greater than the economic benefit when D is set to −1. Furthermore, Fig. 4.9 shows that when Δs_CN is set to 1, the cumulative effect of the US changes from negative to positive from the 5th order.

As a result, the shock effects of investment in the US are greater than those in CN in 2022. Therefore, the impact of DS investment in the US is greater than that in CN in 2022. Furthermore, CN's increased investment in DS will benefit both CN and the US economically. Even if it is assumed that the US' DS investment is reduced by one unit, whereas CN and JP increase one unit to measure how the shock affects the US and CN, the US still has strong resilience to shocks due to its large share of the international financial market and close relationships with various countries regarding financial investments, and its cumulative limit effect remains the highest.

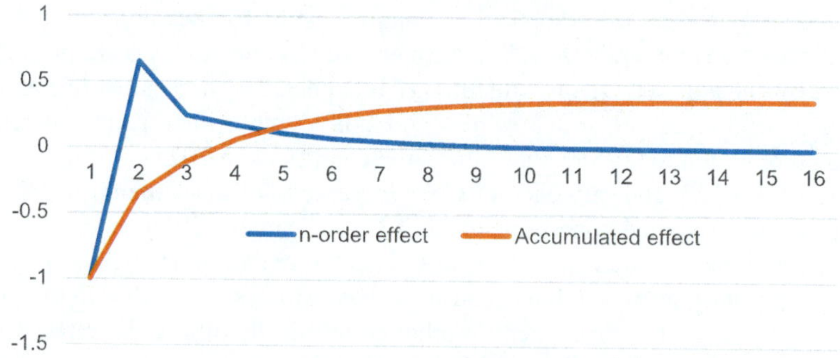

Fig. 4.9 Shock effects for the US (ΔS_US = −1, ΔS_CN = 1)

The columns in Table 4.5 show the transmission of financial risk and shock effects from one country to another. Also, when examining the shock effects of DS investment from the US to other countries under the same assumptions, there exists some meaningful results. If US DS investment decreases by −1 unit while CN and JP increase by 0 unit, the US will have a large direct negative impact or first-order effect on CN (−0.002), JP (−0.0261), CA (−0.0556), and the UK (−0.0454). The second-order effects on CN, JP, CA, and the UK are −0.0058, −0.0113, −0.0081, and −0.0167, respectively. Moreover, the cumulative effect, or limit effect, for the US is −1.2558, compared to CN (−0.0146), JA (−0.0588), CA (−0.0893), and the UK (−0.0983). CN has the smallest limit shock effect value among them, and even in Figs. 4.7 and 4.9 if a shock variable is observed, the ripple effect on CN is relatively low. In other words, CN's foreign securities investment has yielded low returns while posing low risks. The US net debt gap is nearly 1.25 × larger than expected, with effects for CN, JP, CA, and the UK. However, while 22.4% of CN's foreign bond investment is concentrated in US bonds, compared to JP, CA, and the UK, the ripple effect on CN is relatively small, with the limit loss effect being −0.0146.

4.5 Concluding Remarks

2018 to 2022 is a period of global economic and political turmoil, transitioning from economic globalization to anti-globalization. This chapter conducted a comparative analysis using the newly prepared matrix of G20 external assets and liabilities in 2022 and the same constructed matrix in 2018.

4.5.1 Structural Changes in Global Debt and Assets

Statistics describe the structural changes of the external assets and liabilities of the G20 countries. The global financial asset position and financial liabilities are still concentrated in rich advanced countries, led by the US. However, from 2018, global debt and assets have shown structural changes and are in a state of imbalance and continuous expansion, suggesting that the risk of a new global financial crisis has increased.

4.5.2 Increasing External Imbalances Between China and the United States

CN is a relatively new participant in the international capital markets, having held the world's largest amount of US debt from 2008 to 2019. However, the advantages of foreign financial investment have been limited, as the net assets of DI and PI have consistently shown negative figures over an extended period, indicating an escalation in external financial investment risks. America's net external financial position is on a long-term downward trend. Although DI had been a net asset position for a long time, the shift from a negative net position in 2018 continued to a negative net position of $3 trillion by 2022, raising the US debt risk. Moreover, the long-term negative growth of PI and OI in the US fell to −$12.7 trillion by 2021, raising debt risk. In a sign of America's growing need for external funding, net securities debt has nearly doubled since 2018, but America's influence on bond investment has declined. The huge amount of foreign debt of the US has been increasing, raising questions as to who will pay for the country's increasing foreign debt.

4.5.3 Strategic Preparation for Economic Decoupling

Comparing 2022 with 2018, CN's total investment in the US has declined while the investment structure has changed. The economic crisis caused by the coronavirus pandemic will affect the entire world, particularly vulnerable emerging markets. Foreign currency supports most of the premium on emerging market assets. If the Federal Reserve lowers interest rates, US bond yields fall dramatically. With more than $1 trillion of US debt held since 2011, CN is naturally under pressure to avoid risk. However, from 2018 to 2022, CN has shown a trend of gradually reducing its holdings of US bonds, which reflects CN's strategic preparation for decoupling.

4.5.4 New Findings from Financial Network Analysis

Through the calculation of EC, CC, BC, and degree centrality, the mutual relationship, central position, and influence of G20 countries in the securities market are known. CN's external DS investment is primarily concentrated in the US, but the proportion of investment and financing in OE in CN exceeds that in the US. CN has a higher risk profile, but it also has more investment diversification, which allows it to mitigate risk. The US had an EC value of 0.96, and its bond financing primarily targets the G7 countries, LU and NL, or CN, but its bond trading relationship with other countries is not very close.

Shock dynamics analysis demonstrates that CN's foreign securities investment has yielded low returns while posing low risks. The US net debt gap is nearly 1.25 × larger than expected, with effects for CN, JP, CA, and the UK. However, while 22.4% of CN's foreign bond investment is concentrated in US bonds, the ripple effect on CN is relatively small. The China–US DS risk simulation in Sect. 4.4 indicates that CN's foreign securities investment has resulted in low returns and low risks. Even if the US bond market crashes, the limit shock effect on CN will be minimal.

4.5.5 Future Works

Given this historical background and the dramatic changes in the global economy, here are three policy recommendations.

Owing to the deteriorating political relationship between CN and US, foreign trade relations have changed, and economic decoupling between the two countries reflected in the decline of the DI, PI, and OI investment ratios (Table 4.2), CN and the US will face greater financial risks. However, because a mutually beneficial relationship has been maintained since the 1990s, even if CN reduces the number of US bonds it holds, treasury bond purchases appear to be an unavoidable option. Therefore, to maintain a stable international trading environment, international trading rules should be perfected and adhered to with legal binding effects beyond political consciousness, rather than arbitrarily interrupt existing contracts with sanctions. Sticking to credit will provide long-term interests.

The fundamental cause of the external imbalance between CN and the US lies in the imbalance between savings and investment in the real economy. Therefore, both CN and the US need to adjust their domestic economic growth mode to balance savings and investment, and then achieve an external balance. The growing imbalances in countries' external assets and liabilities highlight the importance of macroprudential aspects. Macroprudential analysis and policy focus on the strengths and vulnerabilities of the financial system and the contagion within and between financial systems. Therefore, it is necessary to strengthen the statistical monitoring of intersectoral assets and liabilities and improve sectoral data.

Thus, CN is attempting to modify its imbalanced GFF structure to reduce its foreign reserve balance through international market transactions. However, these policy changes have been ineffective in halting the increase in foreign reserves. As a result, CN's economic structure should be adjusted by broadening domestic demand and diversifying its external financing, including the internationalization of the yuan. This would alter the China–US relationship and result in a new global economic structure. International cooperation will become even more critical.

References

Acemoglu, D., Ozdaglar, A., & Tahbaz-Salehi, A. (2015). Systemic risk and stability in financial networks. *American Economic Review, 105*(2), 564–608.

Arajo, T., & Lou, F. (2007). The geometry of crashes: A measure of the dynamics of stock market crises. *Quantitative Finance, 7*(1), 63–74.

Bavelas, A. (1950). Communication patterns in task-oriented groups. *Journal of the Acoustical Society of America, 22*(6), 725–730.

BEA. (2022). *U. S. Net International Investment Position at the End of the Period*, Table 1.2.

Bernhard Winkler, Ad van Riet, and peter Bull. (2013a). *A Flow-of-Funds Perspective on the Financial Crisis* Volume I: Money, Credit and Sectoral Balance Sheets. Palgrave Macmillan.

Bernhard Winkler, Ad van Riet, and peter Bull. (2013b). *A Flow-of-Funds Perspective on the Financial Crisis* Volume II: Macroeconomic Imbalances and Risks to Financial Stability. Palgrave Macmillan.

BIS. (2009). Global Imbalances: In Midstream? In *Reconstructing the World Economy, IMF Stuff Discussion Note*, Dec. 22. Washington: International Monetary Fund.

BIS. (2022). Locational Banking Statistics, http://stats.bis.org/statx/toc/LBS.html.

Celestino, G., Marta, R. V., & Matas, A. (2018). *Propagation of Quantity Shocks in Who-to-whom Networks', the" 35th IARIW General Conference*. Denmark.

Copeland, M. A. (1949). Social accounting for Moneyflows. *Accounting Review, 24*(3), 254–264.

Copeland, M. A. (1952). *A study of Moneyflows in the United States*. National Bureau for Economic Research Books.

Errico, L., Walton, R., Hierro, A., AbuShanab, & Amidžic, G. (2013). Global flow of funds: Mapping bilateral geographic flows. In *Proceedings 59th ISI World Statistics Congress*, 2825–2830.

Errico, L., Harutyunyan, A., Loukoianova, E., Walton, R., Korniyenko, Y., Amidžic, G., AbuShanab, H., & Shin, H. S. (2014). Mapping the shadow banking system through a global flow of funds analysis. *IMF Working*. https://www.imf.org/en/Publications/WP/Issues/2016/12/31/Mapping-the-Shadow-Banking-System-Through-a-Global-Flow-of-Funds-Analysis-41273. DC: paper P/14/10.Washington.

Financial Stability Board, and International Monetary Fund. (2009). *The Financial Crisis and Information Gaps. Report to the G—20 Finance Ministers and Central Bank Governors*. http://www.imf.org/external/np/g20/pdf/102909.pdf.

Freeman, L. C., Roeder, D., & Mulholland, R. R. (1979). Centrality in social networks: II. *Experimental Results, Social Networks, 2*(2), 119–141.

Gourinchas, P. O., Rey, H., & Sauzet, M. (2019). The international monetary and financial system, NBER WORKING PAPER SERIES", Working Paper 25782. http://www.nber.org/papers/w25782.

Gourinchas, P. O., & Rey, H. (2007). International financial adjustment. *Journal of Political Economy, 115*(4), 665–703. https://doi.org/10.1086/521966

IMF. (2014). *BPM6 Compilation Guide* vol. 144.

IMF. (2022a). Coordinated Direct Investment Survey (CDIS). https://data.imf.org/regular.aspx?key=60564262.

IMF. (2022b). Coordinated Portfolio Investment Survey (CPIS). https://data.imf.org/regular.aspx?key=60587815.

IMF. (2022c). International Investment Position. https://data.imf.org/regular.aspx?key=62805744.

IMF. (2022). *World Economic Outlook Database April 2022.*

IMF. (2023). Global debt monito.

Ishida, S. (1993). *Flow of Funds in the Japanese Economy* (in Japanese). Tokyo: Keizai Shimpo-Sha, 169–190.

Luiza, A. A. (2015). *A network analysis of sectoral accounts: Identifying sectoral interlinkages in G-4 economies, IMF Working*. Washington, DC: paper WP/15/111.

National Bureau of Statistics of China. (2022). *China Statistical Yearbook 2022, China Statistics Press.*

Newman, M. E. J. (2010). *Networks: An introduction*. Oxford University Press, 169.

Scott Davis, J., & Zlate, A. (2023). The global financial cycle and capital flows during the COVID-19 Pandemic. *European Economic Review, 156*, 104477. https://doi.org/10.1016/j.euroecorev.2023.104477.

Soramäki, K., & Cook, S. (2016). *Network theory and financial risk*. Risk Books, a Division of Incisive Media Investments Ltd.

Spelta, A., & Araújo, T. (2012a). The topology of cross-border exposures: Beyond the minimal spanning tree approach. *Physica A: Statistical Mechanics and its Applications, 391*(22), 5572–5583

Spelta, A., & Araújo, T. (2012b). *Interlinkages and structural changes in cross-border liabilities: A network approach*, Quaderni di Dipartimento, Working Paper, No. 181.

Tsujimura, K., & Tsujimura, M. (2008). *International flow-of-funds analysis: Techniques and applications*. Keio University Press, 3–59.

Tsujimura, M. (2009). The flow of funds analysis of the U.S. subprime crisis. *Input-Output Analysis, 17*, 1–2, 88–104. https://doi.org/10.11107/papaios.17.88.

Wall Street Journal. (2023). Moody's Faces Growing Backlash Over Its Negative Outlook on China. https://www.wsj.com/world/china/moodys-faces-growing-backlash-over-its-negative-outlook-on-china-87fe7ce6

Zhang, N. (2022). Measuring global flow of funds: Who-to-whom matrix and financial network. *Japanese Journal of Statistics & Data Science, 5*(1), 899–942.

Zhang, N., & Zhao, X. (2019). Measuring global flow of funds: A case study on China, Japan and the United States. *Economic Systems Research, 31*(4), 520–550. https://doi.org/10.1080/09535314.2019.1574719

Zhang, N., & Zhu, L. (2021). Global flow of funds as a network: The case study of the G20. *Japanese Journal of Monetary and Financial Economics, 9*, 21–56.

Zhang, N. (2020). *Flow of funds analysis: Innovation & development*. Springer, 137–169.

Zhang, N. (2005). The theory and practice of global flow of funds (in Japanese), Minerva Shobo Inc. (in Japanese), 75–99.

Zhang, N. (2008). Global-flow-of-funds analysis in a theoretical model-what happened in China's external flow of funds, Kyusyu University Press, quantitative analysis on contemporary. *Economic Issues*, 103–119.

Zhang, N. (2014). *The flow of funds analysis in theory and practice*. Pekin University Press (in Chinese), 1–65.

Chapter 5
A Network Analysis of the Sectoral From-Whom-To-Whom Financial Stock Matrix

Abstract This study enhances global flow of funds (GFF) statistics for assessing global financial stability at the national and cross-border sectoral levels. The investigation involves scrutinizing data sources and reconstructing the statistical framework to establish the sectoral from-whom-to-whom financial stock matrix (SFSM). The SFSM is constructed using sectoral account data, complemented by international statistics from the Coordinated Portfolio Investment Survey, International Investment Position, and Bank for International Settlements. The SFSM specifically focuses on counterparty national and cross-border exposures of sectors in China, Japan, the United Kingdom, and the United States. It is designed to create country-specific financial networks, interconnecting each country-level network based on cross-border exposures. Analytical results systematically reveal bilateral exposures among the four countries in the GFF, identifying sectoral interlinkages, characteristics of overseas investment, external shocks, and internal influences. Furthermore, this study introduces an eigenvector decomposition to analyze the effects and provided an analytical description of the shock dynamics and propagation process.

Keywords Data sources · Sectoral accounts · Balance sheet exposures · Cross-border exposures · Shock Dynamics · Financial networks

5.1 Introduction

In April 2009, the G20 Finance Ministers and Central Bank Governors Working Group on Reinforcing International Co-operation and Promoting Integrity in Financial Markets called upon the International Monetary Fund (IMF) and the Financial Stability Board (FSB) to identify information gaps and provide proposals for strengthening data collection and reporting. Consequently, in October 2009, the IMF and FSB presented 20 recommendations aimed at improving data collection

N. Zhang and Y. Zhang, *Global Flow of Funds Analysis*,
https://doi.org/10.1007/978-981-97-1029-4_5

efficiency and addressing or reducing identified gaps in four areas.[1] There is international awareness regarding information limitations, with existing data falling short in describing inherent risks in the financial system (Shrestha et al., 2012). Prior research has delved into the fundamental concept of global flow of funds (GFF) and proposed a statistical framework (Errico et al., 2013). Moreover, Errico et al. (2014) integrated sectoral account data with the Coordinated Direct Investment Survey (CDIS), Coordinated Portfolio Investment Survey (CPIS), International Investment Position (IIP), and Bank for International Settlements (BIS) statistics to analyze the shadow banking sector in the United States (US), breaking down its claims and liabilities by counterparty country and sector.

Stone (1966) and Klein (1983) outlined techniques for converting T-shaped accounts into sectoral matrices. Stone (1966) developed a financial matrix model that amalgamated the flow and stock of funds across various institutions and sectors using an input–output table. Klein (1983) suggested linking the capital flow statement with the national income account and the input–output statement with a matrix representation, culminating in a financial matrix table based on the input–output model. Tsujimura and Mizosita (2002, 2018) explored the theory and method of the flow-of-funds matrix based on "who-to-whom" (W-t-W), using flow-of-funds statistics from Japan (JP) and the US.

Zhang (IARIW-OECD conference, 2015), Zhang (2016), Zhang and Zhao (2019), and Zhang (the 36th IARIW conference, 2021) focused on three primary issues related to GFF: its definition, integration of its statistics with a system of national accounts (SNA), and exploration of data sources and approaches. They conducted research and pilot compilations of GFF statistics. By leveraging international statistical standards, data on cross-border financial exposures (CPIS, CDIS, IIP, and BIS) can be linked to domestic sectoral account data, creating a comprehensive depiction of domestic and international financial interconnections. A new challenge is to develop a GFF matrix (GFFM) that simultaneously considers risk exposures between countries and describes debt relations between counterparty sectors. The primary objective of this project is to construct a matrix facilitating the analysis of bilateral financial exposures and supporting the examination of potential sources of contagion.

In Chap. 1, the method of transforming the flow of funds table, initially a 2-dimensional account (institutional sector × transaction item), is introduced as a 3-dimensional GFF table in the form of W-t-W. Here, W-t-W signifies the flow of funds (based on flow) or debt and creditor relationships (based on stock) between countries, resulting in a country-by-country square matrix. This advancement enhances the statistical observation of more accurate international capital flows between countries and changes in debt and creditor relations. Nevertheless, for monitoring financial risk trends and preventing financial crises, a W-t-W inter-country sector matrix based on the sector of the counterparty can offer more detailed information.

---

[1] They are (i) build-up of risk in the financial sector, (ii) cross-border financial linkages, (ii) vulnerability of domestic economies to shocks, and (iv) improvement in communication of official statistics.

Two primary reasons underscore this challenge. First, the majority of countries do not furnish detailed information regarding the counterparty sector of a financial instrument issued by a specific sector, often referred to as "from-whom-to-whom" data. Second, such as the 2008 financial crisis in the United States has underscored that many risks to the global financial system stem from cross-border exposures falling under the rest of the world (ROW) sector, lacking specifications regarding the counterparty country and counterparty sector.

Some studies have employed sectoral accounts to unveil interconnections among economic agents, evaluating financial stability and systemic risk. Okuma (2013) utilized more accurate methods to calculate sectoral interlinkages in JP. Antoun de Almeida (2015) integrated sectoral account data with information from the CPIS, IIP, and BIS. The author computed bilateral exposures between financial and nonfinancial sectors across the euro area, JP, the United Kingdom (UK), and the US. However, the study did not propose a comprehensive framework for measuring GFF.

Giron et al. (2018) delved into W-t-W matrices to unveil indirect intersectoral financing and investment patterns, including exposures and risks. While the study utilized sectoral data, its focus did not extend to the interaction between and across sectors within countries. In a related vein, Hagino et al. (2019) explored the use of sectoral data for crafting financial input–output statements, and Hagino and Kim (2021) developed International SFSM tables of JP, Korea, the US, and China (CN), employing analytical applications. Zhang (2022) connected the GFFM with the sectoral account data to establish the sectoral from-whom-to-whom financial stock matrix (SFSM). The research considered the circulation of funds and debt claims among countries, aiming to extend funding operations among various sectors between countries and estimating bilateral exposure between various sectors.

This study builds upon the aforementioned research by enhancing the GFF statistical framework, integrating data sources, and introducing theories and concepts of financial networks. In this chapter, the focus lies on refining compilation methods and identifying interconnections between sectors' national tables based on the W-t-W model, drawing inspiration mainly from Zhang (2022). This enhancement involves the integration of the GFFM and the identification of sectoral interlinkages in counterparty countries.

The US, CN, and JP represent the three largest economies by GDP, and the UK holds the second largest share of international financial assets in 2021 (see Chap. 4). Accordingly, these four regions are selected as the observation samples, collectively referred to as the G-4. Despite differing economic systems, market maturities, and political structures, a GFF analysis can reveal the basic structure, mutual dependence, financial exposure risk, and homogeneity and heterogeneity of their external flow of funds. Thus, building on the theoretical enhancement of GFF statistics and developing application methods, this study emphasizes setting counterparty country sectors in the four regions. It explores new theoretical methods, considers changes in capital operation in the face of economic decoupling in CN and the US, and proposes practical countermeasures to prevent financial crises.

Propelled by advancements in data science, the utilization of financial networks for monitoring and assessing financial risks has become a prominent focus within this academic domain. Consequently, this chapter delves into the practical application of financial network theory and methodologies for observing and analyzing the SFSM, drawing notable reference value from research findings by Acemoglu et al. (2013), Billio et al. (2010), Caccioli et al. (2018), and others.

The remainder of this paper is arranged as follows. Section 5.2 discusses the integration and consistency of data sources and the balance of payments or ROW financial instruments. Section 5.3 outlines the methodology for using counterparty country sector tables and conducts an empirical analysis of the G-4. Section 5.4 examines the use of W-t-W matrices to study the international propagation dynamics of quantity shocks in investment and financing. Lastly, Sect. 5.5 deduces the application of financial networks to the SFSM.

5.2 Creating Counterparty International SFSM

The 2007–2008 financial crisis underscored the multitude of risks within the global financial system posed by cross-border exposures. Notably, in sectoral accounts data, all cross-border exposures are aggregated under the ROW sector, lacking specifications regarding counterparty countries and sectors. In response, when compiling the GFFM of the G20, the study approach extends to encompass a sectors matrix that interlinks major countries. This extension involves combining financial accounts (FA) data with IIP, CPIS, CDIS, and BIS's locational banking statistics (LBS). It connects the GFFM to the domestic sector account data, facilitating the establishment of the SFSM. This comprehensive matrix enables the measurement of cross-border exposures between sectors of major economies, connecting financial and nonfinancial sectors and providing a holistic view of domestic and cross-border financial interconnections that link to the real economy through sectoral accounts. The G-4 is used to compile the Counterparty SFSM.

5.2.1 *Data Sources for Compiling International SFSM*

We compile the financial balance sheets (FBS) of five sectors in JP, the UK, and the US using data from OECD statistics. The OECD data are sourced from the FBS of FA, compiled based on the System of National Accounts (SNA) standards (OECD, 2021). These data are included in national accounts, adhering to the classification of transaction items consistent with the Monetary and Financial Statistics Manual published by the IMF (2016). The FBS comprises 32 transaction items, categorized into eight major items: monetary gold and special drawing rights (SDRs), currency and deposits, debt securities, loans, equity and investment fund shares/units, insurance pension and standardized guarantees, financial derivatives, and employee stock options.[2] We incorporate these eight items into the balance sheets of JP, the UK, and the US, ensuring adherence to international standards.

Next, we focus on achieving alignment with international comparison standards. While the OECD data exclude CN, the People's Bank of China released the Financial Assets and Liabilities Statement (Financial Accounts) in 2022. This statement adhered to SNA standards and covered the period from 2017 to 2021. In addition, China's Center for National Balance Sheets (Li & Zhang et al., 2020) produced balance sheets spanning 2000 to 2019. For the compilation of the Counterparty SFSM, we can leverage the financial assets and liabilities data provided by the People's Bank of China for 2021. Notably, China's balance sheet strictly adheres to the framework of the SNA, categorizing all institutional units into five sectors—financial corporations (FC), nonfinancial corporations (NFC), general government (GG), household and non-profit institutions serving households (HH), and ROW. The financial transactions recorded on CN's balance sheet align with the fund flow statement compiled by the People's Bank of China, categorized into 14 items.[3]

Finally, the financial assets and liabilities tables for the G-4 are constructed using the balance sheets of the five sectors. Subsequently, leveraging the data set related to the ROW in the balance sheet, we compute the exposure risk for a country's ROW and each cross-border sector of the counterparty country. This is achieved through the analysis of the relation between the relevant ratios. To elucidate the cross-border capital flow in a sectoral W-t-W format, it is imperative to clarify the data sources and characteristics of CDIS, CPIS, BIS, and IIP. Understanding the way various sectors allocate funds and considering the international statistics published by the IMF and BIS, the ROW data in financial accounts can be directly linked and deconstructed into the counterparty sector through interrelated information channels.

(1) Direct investment is usually carried out by the non-financial corporate sector, and the corresponding information channel can be accessed through CDIS Table 3: Direct Investment Positions (Inward and Outward) as Reported by Economy and by Counterpart Economies (Mirror data). Mirror data from one economy represents the information reported by the counterpart economy. For example, if we seek data on assets with CN in banks' cross-border positions on residents of JP, and JP's LBS account does not record the corresponding data between JP and CN, resulting in missing data, we can utilize the liabilities data from CN to JP, which are recorded in banks' cross-border positions on residents of CN.[4]
(2) Portfolio investment, deposits, and loans are mainly managed by the financial sector. Relevant information channels can be accessed through the CPIS

[2] OECD. Stat (2021) Dataset: 720. Financial balance sheets, non-consolidated, SNA 2008, https://stats.oecd.org/.

[3] These include currency, deposits, loans, undiscounted bankers' acceptance bills, technical insurance reserves, interfinancial institution accounts, required and excess reserves, bonds, equity and share, security investment fund, central bank loans, foreign direct investment (FDI), changes in reserve assets, and miscellaneous (net). See The People's Bank of China (2022) Financial Assets and Liabilities Statement (Financial Accounts), http://www.pbc.gov.cn/en/3688247/3688975/4280784/index.html.

[4] See Table A6.2, https://stats.bis.org/statx/toc/LBS.html.

published by the IMF. Table 6[5] of the CPIS encompasses the country and sector of the holder, with sectors including the Central Bank, Deposit-taking Corporations except the Central Bank, Other Financial Corporations, GG, NFC, households, and NPISHs. A sectoral breakdown, such as Table 6 of the CPIS, is useful for calculating the ratio relationship and compiling the SFSM.

(3) Cross-border bank credit is accessible through the BIS Statistics Explorer, specifically the LBS—Table A6.2 with the country (residence) of the counterparty and the location of reporting bank. In BIS statistics, as interbank transactions involve buyers and sellers, missing data from one party can be obtained from the mirror data of the other party.

(4) Insurance pension and standardized guarantees are mainly held by the HH sector, and equity and investment fund shares are held by the FC, NFC, and HH sectors. This data can be obtained from Table 6 of the CPIS.

(5) Government bonds, such as treasury bonds, are generally issued through government departments and can be obtained through subsector classification in Table 6 of the CPIS.

(6) IIP reflects the stock of financial assets and liabilities of a country or region with the ROW at a specific point in time. Portfolio investment is classified by the central bank, deposit-taking corporations except the central bank, GG, and other sectors. If information for compiling the SFSM is missing or there are data structure errors, IIP data verification can be used as a reference.

The establishment of the counterparty SFSM comprises of three parts. First, the establishment of a counterparty domestic FBS for each sector. The second part involves the decomposition of the counterparty's ROW, where financial transaction items in a country's ROW, such as currency and deposits, debt securities, are allocated to the relevant sectors of the counterparty based on the ratio allocation method and the share of the country's assets and liabilities in CDIS, CPIS, and BIS. The third part involves dividing the sector proportion of other counterparties based on the assets and liabilities sheets of each domestic sector, integrating an SFSM covering the connections between each country and sector for the analysis object.

5.2.2 Compilation of FBS for the G-4

In alignment with the eight major transaction items in the FBS of the OECD, the 14 transaction items in CN's FBS are categorized and consolidated according to the OECD's statistical classification into eight items. This standardization ensures uniformity in the international comparison and classification of sectors and transaction items following the SNA. As a transitional step in the preparation of the SFSM for the G-4 economies, the initial phase involves creating their FBS based on the SNA standard (Tables 5.1, 5.2, 5.3 and 5.4).

[5] Table 6: Reported Portfolio Investment Assets by Sector of Holder, and Sector and Economy of Nonresident Issuer for Specified Economies.

Table 5.1 China's financial balance sheets (end of 2021, USD bn.)

	Assets	Liabilities	Assets	Liabilities	Assets	Liabilities	Assets	Liabilities	Assets	Liabilities
Monetary gold and SDRs	3386	0	0	0	0	0	0	0	0	3386
Currency and deposits	5890	42,745	12,375	0	5619	0	18,596	0	563	298
Debt securities	19,851	7438	70	4575	155	8443	216	0	491	327
Loans	35,913	2449	467	20,079	0	1274	0	12,196	171	553
Equity and shares	3466	6238	7909	12,232	491	0	5663	0	1582	641
IPs	0	2387	233	0	0	1625	3780	0	0	0
Financial derivatives	0	0	0	0	0	0	0	0	0	0
Other accounts receivable	6838	12,125	4893	4891	182	128	5040	247	4448	4009
Total	75,344	73,382	25,946	41,777	6447	11,469	33,295	12,443	7254	9214
Financial net worth		1962		−15,831		−5022		20,852		−1960

Source China's Center for National Balance Sheets

Source The People's Bank of China (2022) Financial Assets and Liabilities Statement
Note The yuan traded at 6.452 to the US dollar in 2021 (period average), according to the CN Statistical Yearbook 2022. IPs denotes insurance pension and standardized guarantees

Table 5.2 Japan's financial balance sheets (end of 2021, USD bn.)

	Financial corporations		Non-financial corporations		General Government		Households and NPISH		Rest of the world	
	Assets	Liabilities	Assets	Liabilities	Assets	Liabilities	Assets	Liabilities	Assets	Liabilities
Monetary gold and SDRs	49	0	0	0	74	89	0	0	62	95
Currency and deposits	7248	21,064	2952	0	1088	0	9921	0	102	248
Debt securities	12,964	2788	386	857	2373	10,442	385	0	1910	3932
Loans	14,916	6943	676	4727	174	1355	28	3264	2108	1614
Equity and shares	5631	3928	4804	9352	2089	130	2769	116	2366	4133
IPs	142	4734	37	157	0	0	4712	0	0	0
Financial derivatives	469	520	17	37	1	0	10	9	270	201
Other accounts receivable	1176	1282	2638	2522	578	458	293	142	655	938
Total	42,596	41,257	11,510	17,652	6377	12,473	18,119	3531	7471	11,131
Financial net worth		1338		−6142		−6069		14,589		−3660

Source OECD. Stat, Dataset: 720. FBS—nonconsolidated—SNA 2008

Table 5.3 United States' financial balance sheets (end of 2021, USD bn.)

	Financial corporations		Non-financial corporations		General Government		Households and NPISH		Rest of the world	
	Assets	Liabilities	Assets	Liabilities	Assets	Liabilities	Assets	Liabilities	Assets	Liabilities
Monetary gold and SDRs	11	0	0	0	164	167	0	0	161	170
Currency and deposits	5469	27,909	4138	0	1224	23	15,868	0	2169	936
Debt securities	38,261	16,153	465	7674	2152	29,254	3139	205	13,607	4336
Loans	31,930	7240	184	11,631	2563	23	1356	17,913	2897	2123
Equity and shares	44,490	48,563	12,564	78,482	615	0	63,311	0	27,907	21,842
IPs	7415	37,136	603	−2	0	5454	35,024	38	66	481
Financial derivatives	0	0	0	0	0	0	0	0	0	0
Other accounts receivable	7968	6057	16,947	18,201	1322	2203	302	475	687	289
Total	135,544	143,058	34,901	115,987	8039	37,124	118,999	18,631	47,494	30,177
Financial net worth		−7514		−81,086		−29,085		100,368		17,317
Source OECD.Stat										

Source Same as Table 5.2

Table 5.4 United Kingdom's financial balance sheets (end of 2021, USD bn.)

	Financial corporations		Non-financial corporations		General Government		Households and NPISH		Rest of the world	
	Assets	Liabilities	Assets	Liabilities	Assets	Liabilities	Assets	Liabilities	Assets	Liabilities
Monetary gold and SDRs	0	0	0	0	59	50	0	0	41	50
Currency and deposits	8695	12,975	1093	0	243	345	2736	0	4706	4153
Debt securities	5453	2536	133	594	120	3668	29	6	2787	1717
Loans	7670	3053	580	1744	318	181	25	2643	2043	3013
Equity and shares	6793	4766	1869	5307	266	0	1585	0	4886	5325
IPs	1017	6394	61	906	2	−43	6029	43	192	1
Financial derivatives	6031	6107	42	70	2	3	11	2	3272	3177
Other accounts receivable	116	89	185	298	176	157	218	134	13	29
Total	35,775	35,920	3963	8921	1185	4362	10,632	2828	17,940	17,465
Financial net worth		−145		−4958		−3176		7804		475

Source OECD.Stat

Source Same as Table 5.2

5.2.3 Establish the International SFSM

The process of compiling the sector-by-sector matrix involves two methods: one that infers the debt ratio of a transaction item between sectors (Zhang, 2020) and another that calculates financial inflow–outflow based on the input–output principle, which is deemed more direct and simpler and adopted in this study. The study emphasizes the increased exposures at both the country and cross-border levels in all governments. Notably, a significant decline can be observed in loan exposures at the cross-border level and in equity exposures at the country level (Luiza, 2015). Similar precedents, such as the US–East Asia Financial Input–Output Table (Hagino et al., 2019), further support this approach. To observe bilateral exposures at the country and cross-border levels and integrate them into the GFFM, sectoral accounts data are combined with data from the CDIS, CPIS, IIP, and BIS to calculate bilateral exposures between financial and nonfinancial sectors in three financial instruments surrounding the G-4. Establishing a counterparty country SFSM to convert the FBS prepared above into an SFSM is necessary.

To convert FBS (see Table 1.4) into an SFSM, the assets and liabilities of each sector are separated from the double-entry accounting-FBS. This involves preparing each sector's assets table (Table E) and liability table (Table R) (Zhang, 2020, 95).

As implied in Table 4.6, E denotes the financial asset matrix, R denotes the financial liability matrix, t^E denotes the aggregate of financial instruments held by each sector, and t^R denotes the aggregate of financial instruments held by each sector. Here, it is established that $t^E = t^R$. ε_j denotes the net financial liability of the jth sector, ρ_j denotes the net financial assets of the jth sector, and t denotes the total of assets or liabilities of each sector column. Each part of tables E and R is represented as a matrix, and each element of E and R matrices is indicated by e_{ij} and r_{ij}, respectively. Next, we calculate the liability coefficient b_{ij} and asset coefficient d_{ij} using e_{ij} and r_{ij}.. Subsequently, we deduce the capital inflow coefficient of Table Y (sector × sector).[6] This process results in Tables 5.5, 5.6, 5.7 and 5.8, providing practical significance to the compilation and analysis of Table Y. Additionally, conducting an analysis of the ripple effect of financial risk at a certain point in time is essential. Utilizing Tables 5.1, 5.2, 5.3 and 5.4, we compile Table Y (see Tables 5.5, 5.6, 5.7 and 5.8).

Tables 5.5, 5.6, 5.7 and 5.8 illustrate the sector-by-sector SFSM for the G-4 at the end of 2021. In these tables, rows represent assets, and columns represent liabilities. Each sector's row displays its stock of assets used in other sectors, providing insights into the sector's financing from other sectors and fund operations within each sector (diagonal elements in the matrix). If the assets of a sector exceed its liabilities, the net financial income of the sector is calculated as the net assets in the row. Conversely, if the assets of a sector are less than the liabilities, the net loss of the sectors is included as net liabilities in the column. These four SFSM tables offer the fundamental structure of domestic and external assets and liabilities in the G-4.

---

[6] For the calculation method of Table Y refer to Zhang (2020), 108–110.

Table 5.5 SFSM for China (end of 2021, USD bn.)

Liabilities	Financial corporations	Non-financial corporations	General Government	Households and NPISH	Rest of the world	Net liabilities	Total of row
Assets							
Financial corporations	22,238	26,865	9236	12,316	4689	0	75,344
Non-financial corporations	16,646	7193	137	31	1940	15,831	41,777
General Government	6047	364	4	3	29	5022	11,469
Households and NPISH	27,159	3892	1722	92	430	0	33,295
Rest of the world	1293	3464	370	1	2127	1960	9214
Net assets	1962	0	0	20,852	0		
Total of column	75,344	41,777	11,469	33,295	9214		

Source By Table 5.1 compiled by the authors

Table 5.6 SFSM for Japan (end of 2021, USD bn.)

Liabilities	Financial corporations	Non-financial corporations	General Government	Households and NPISH	Rest of the world	Net liabilities	Total of row
Assets							
Financial corporations	16,944	8120	8807	2793	5930	0	42,596
Non-financial corporations	4989	3989	536	225	1771	6142	17,652
General Government	2115	1538	1489	61	1175	6096	12,473
Households and NPISH	15,130	1783	271	31	904	0	18,119
Rest of the world	2079	2223	1370	420	1379	3688	11,160
Net assets	1338	0	0	14,589	0		
Total of column	42,596	17,652	12,473	18,119	11,160		

Source By Table 5.2 compiled by the authors

Table 5.7 SFSM for the United States (end of 2021, USD bn.)

Liabilities	Financial corporations	Non-financial corporations	General Government	Households and NPISH	Rest of the world	Net liabilities	Total of row
Assets							
Financial corporations	44,623	43,414	21,036	14,974	11,497	7514	143,058
Non-financial corporations	12,553	18,069	1687	383	2209	81,086	115,987
General Government	2758	2260	1283	1210	528	29,085	37,125
Households and NPISH	67,362	34,396	6063	671	10,507	0	118,999
Rest of the world	15,763	17,847	7055	1394	5436	0	47,494
Net assets	0	0	0	100,368	17,317		
Total of column	143,058	115,987	37,124	118,999	47,494		

Source By Table 5.3 compiled by the authors

Table 5.8 SFSM for the United Kingdom (end of 2021, USD bn.)

Liabilities	Financial corporations	Non-financial corporations	General Government	Households and NPISH	Rest of the world	Net liabilities	Total of row
Assets							
Financial corporations	17,224	4200	2671	1939	9740	145	35,920
Non-financial corporations	1700	835	129	180	1119	4958	8921
General Government	415	226	130	112	301	3176	4362
Households and NPISH	7853	1392	80	83	1224	0	10,632
Rest of the world	8728	2267	1351	514	5080	0	17,940
Net assets	0	0	0	7804	475		
Total of column	35,920	8921	4362	10,632	17,940		

Source By Table 5.4 compiled by the authors

5.2.4 Compilation of International SFSM by Counterparty (Country-Sectors)

This section focuses on the trading relation between ROW in the SFSM and FC, NFC, GG, and HH in other countries' SFSM. When determining financial transactions or debt and creditor relationships between domestic sectors and overseas entities, especially specific sectors of the counterparty, it is crucial to have accurate and standardized basic data for the counterparty's specific sectors. There should be a basic dataset reflecting FC, NFC, GG, and HH and conforming to international uniform standards. To fulfill this requirement, two methods can be employed: integration of the existing data or utilization of ratios for calculation.

We calculate the debt–bond relation between counterparty sectors by directly utilizing the W-t-W information in their source data. This method identifies links between each sector's outstanding amount of assets and each debt transaction item. Combining this method with the SFSM calculated above, three types of methods are employed to prepare the bilateral SFSM.

First, from the perspective of the nature of financial commodities, the relation between asset-holding and liability-issuing sectors is clear. For instance, financial institutions handle deposits and loans. Second, financial instruments are used, where owners can be identified from other sources. For instance, foreign deposits held by the government can be determined from GG; foreign direct investment is usually carried out by nonfinancial corporations (NFC); insurance pension, standardized guarantees, and investment trusts are usually held by HH; and financial derivatives are associated with FC. Third, for some cases, such as treasury and financial bonds, where it is impossible to specifically distinguish the counterparty, a pro-rata approach is used.

To determine the data sources and estimation methods for the sectors of the counterparty country, the following example illustrates a methodology for determining the ratio of a country's ROW sector to a counterparty's sector. Claims of sector i in Country A by sector j in Country B are calculated by multiplying Country A's foreign claims (ROW liabilities in Country A) by the share of Country B in Country A's foreign claims, share of sector i's holdings of foreign assets in Country A, and share of sector j's liabilities held by nonresidents in Country B. Data sources for calculating such claims between sectors i in JP and sector j in CN through ROW include CPIS, CDIS, LBS, and IIP.

The CPIS data categorize countries' cross-border PIs by counterparty country and instrument type (debt securities and equities) based on the sector debt position ratio of the counterparty country. The CDIS is specifically processed for transactions between the NFC sectors of the counterparty. The LBS provides information on banks' total foreign claims categorized by counterparty country, including categorization by counterparty sector (e.g., banks, private nonbanks, and public), with LBS included in the FC sector of the counterparty. The IIP dataset complements the CPIS, CDIS, and LBS datasets by providing sectoral information on countries holding foreign assets and issuing liabilities held by nonresidents. For instance, foreign claims of FC_CN vis-à-vis GG_US in debt securities are calculated as follows:

$$FC^A\_CN \rightarrow GG\_US = ROW^L_{CN} \times S_{CN\rightarrow US} \times S^A_{FC\_CN} \times S^L_{GG\_US}, \quad (5.1)$$

In the equation, ROW^L_{CN} denotes the amount of Chinese ROW sector's liabilities (the assets of CN) in debt securities sourced from the sectoral accounts data. Therefore, LBS[7] data should be used. When estimating financial assets in NFC, CDIS data should be utilized. $S_{CN\rightarrow US}$ represents the share of the US in Chinese foreign debt security claims based on the CPIS[8] data. $S^A_{FC\_CN}$ denotes the FC's share in the holdings of foreign debt securities in CN according to the IIP data. $S^L_{ROW\_US}$ denotes the GG's share in the US liabilities in debt securities held by nonresidents, according to the IIP dataset. Notably, these datasets are conceptually consistent with each other, and their external claims compiled by country and instrument are almost equal.

Using Eq. (5.1) and relevant data, we compile an international SFSM with counterparty country-sectors (Table 5.9).

To complement Tables 5.5, 5.6, 5.7 and 5.8, the rows and columns in Table 5.9 are defined: the columns represent liabilities, and the rows represent assets.[9] While Tables 5.5, 5.6, 5.7 and 5.8 are W-t-W tables indicating the credit relation between the assets and liabilities of domestic counterparties sectors for each country, Table 5.9 provides more detailed sector-to-sector information. A column categorizes a sector's assets by counterparty, revealing both the use of financial assets among domestic sectors and the creditor's rights held by various sectors of countries and cross-border sectors of other countries. ROW in the bottom row refers to financial investments (creditors) by counterparty country sectors in countries other than the target country. The total assets of ROW are calculated by summing up the total assets of ROW in all the G-4 economies. A row categorizes a sector's liabilities by counterparty, revealing the financial liabilities between domestic sectors and the liabilities held by counterparty countries and cross-border sectors. ROW in the last column refers to the financial liabilities of sectors to counterparty sectors other than those listed in the table. Thus, the total liabilities of the ROW sector are calculated by summing up the total liabilities of ROW in all G-4 economies.

Table 5.9 utilizes the debt and claims relation between the domestic sectors of the G-4 at the end of 2021 and the bilateral risk exposure of one country to another to construct the financial network of a specific country. It demonstrates how sectoral account data can be harmonized with CDIS, CPIS, LBS, and IIP data to derive information regarding cross-border risk exposure at each country level. Table 5.9 follows the matrix table of Stone-mode, focusing on observing the situation and effect of financing counterparties of various sectors (liabilities). Table 5.10 presents

[7] To avoid double counting, the claims, that is, loans and deposits of CN to the US in Table A6.2-S banks" cross-border positions on residents of CN in the LBS account, are subtracted from the claims of FC by ROW in SFSM (see Table 7).

[8] CPIS: Table 6, Reported Portfolio Investment Assets by Sector of Holder, and Sector and Economy of Nonresident Issuer for Specified Economies, December 2018.

[9] This is the arrangement of rows and columns which designed by Stone's formula (see Chap. 3).

Table 5.9 International SFSM with sectoral data (at the end of 2021, USD bn. by Stone-mode)

Liablities	FC_JP	NFC_JP	GG_JP	HH_JP	FC_CN	NFC_CN	GG_CN	HH_CN	FC_US	NFC_US	GG_US	HH_US	FC_UK	NFC_UK	GG_UK	HH_UK	ROW
Assets																	
FC_JP	16,944	8120	8807	2793	11	30	3	0	1389	1025	405	80	410	33	20	8	2561
NFC_JP	4989	3989	536	225	1	146	0	0	264	952	118	23	149	189	23	9	45
GG_JP	2115	1538	1489	61	0	0	0	0	47	53	21	4	20	5	3	1	1021
HH_JP	15,130	1783	271	31	3	7	1	0	151	171	68	13	17	4	3	1	476
FC_CN	27	29	18	5	22,238	26,865	9236	12,316	590	639	253	50	137	32	19	7	2884
NFC_CN	14	17	9	3	16,646	7193	137	31	179	280	80	16	387	107	60	23	765
GG_CN	0	0	0	0	6047	364	4	3	3	3	1	0	0	0	0	0	20
HH_CN	2	3	2	0	27,159	3892	1722	92	34	38	15	3	20	5	3	1	304
FC_US	540	339	209	64	199	153	16	0	44,623	43,414	21,036	14,974	2022	176	105	40	8111
NFC_US	234	369	154	47	32	203	9	0	12,553	18,069	1687	383	59	1169	4	0	175
GG_US	0	0	0	0	0	0	0	0	2758	2260	1283	1210	0	0	0	0	567
HH_US	457	488	301	92	9	24	3	0	67,362	34,396	6063	671	17	4	3	1	9128
FC_UK	125	133	82	25	193	190	20	0	2014	1312	519	102	17,224	4200	2671	1939	5023
NFC_UK	0	13	0	0	142	395	41	0	183	833	82	16	1700	835	129	180	5
GG_UK	0	0	0	0	3	9	1	0	37	42	17	3	415	226	130	112	189
HH_UK	10	11	7	2	21	56	6	0	158	178	71	14	7853	1392	80	83	692
ROW	655	805	579	178	676	2,245	269	1	10,715	12,320	5407	1068	5458	533	1103	421	

Source Tables 5.5, 5.6, 5.7 and 5.8 from dataset with 720 financial balance sheets of OECD Statistics; CPIS's Table 5.6, CDIS's Table 5.3, and IIP's Table 5.5 published by the IMF; LBS's Table A6.2 of BIS

Table 5.10 International SFSM with sectoral data (at the end of 2021, USD bn. by Klein-mode)

Assets	FC_JP	NFC_JP	GG_JP	HH_JP	FC_CN	NFC_CN	GG_CN	HH_CN	FC_US	NFC_US	GG_US	HH_US	FC_UK	NFC_UK	GG_UK	HH_UK	ROW
Liablities																	
FC_JP	16,944	4989	2115	15,130	27	14	0	2	540	234	0	457	125	0	0	10	655
NFC_JP	8120	3989	1538	1783	29	17	0	3	339	369	0	488	133	13	0	11	805
GG_JP	8807	536	1489	271	18	9	0	2	209	154	0	301	82	0	0	7	579
HH_JP	2793	225	61	31	5	3	0	0	64	47	0	92	25	0	0	2	178
FC_CN	11	1	0	3	22,238	16,646	6047	27,159	199	32	0	9	193	142	3	21	676
NFC_CN	30	146	0	7	26,865	7193	364	3892	153	203	0	24	190	395	9	56	2245
GG_CN	3	0	0	1	9236	137	4	1722	16	9	0	3	20	41	1	6	269
HH_CN	0	0	0	0	12,316	31	3	92	0	0	0	0	0	0	0	0	1
FC_US	1389	264	47	151	590	179	3	34	44,623	12,553	2758	67,362	2014	183	37	158	10,715
NFC_US	1025	952	53	171	639	280	3	38	43,414	18,069	2260	34,396	1312	833	42	178	12,320
GG_US	405	118	21	68	253	80	1	15	21,036	1687	1283	6063	519	82	17	71	5407
HH_US	80	23	4	13	50	16	0	3	14,974	383	1210	671	102	16	3	14	1068

(continued)

Table 5.10 (continued)

Assets	FC_JP	NFC_JP	GG_JP	HH_JP	FC_CN	NFC_CN	GG_CN	HH_CN	FC_US	NFC_US	GG_US	HH_US	FC_UK	NFC_UK	GG_UK	HH_UK	ROW
FC_UK	410	149	20	17	137	387	0	20	2022	59	0	17	17,224	1700	415	7853	5458
NFC_UK	33	189	5	4	32	107	0	5	176	1169	0	4	4200	835	226	1392	533
GG_UK	20	23	3	3	19	60	0	3	105	4	0	3	2671	129	130	80	1103
HH_UK	8	9	1	1	7	23	0	1	40	0	0	1	1939	180	112	83	421
ROW	2561	45	1021	476	2884	765	20	304	8111	175	567	9128	5023	5	189	692	

the Klein-mode and operates by fund supply (assets). See Chap. 2 for details about Stone-mode and Klein-mode.

5.3 Statistical Descriptive Analysis with the SFSM

Before employing SFSM for impact analysis and financial network assessment, it is imperative to conduct a statistical description using the balance sheet of the G-4 and the sector table with W-t-W form. This step is crucial for gaining insights into the fundamental characteristics of each G-4 sector and identifying interlinkages among sectors.

5.3.1 Characteristics of the Assets and Liabilities in the Sectors of G-4

Table 5.5 illustrates that, at the end of 2021, CN's stock of financial assets and liabilities surpasses that of JP and the UK but is lower than that of the US. The total financial assets of domestic sectors amount to $141.031 trillion, and the total financial liabilities reach $139.071 trillion, resulting in net external assets of $1.960 trillion,[10] constituting 7.91% of CN's total financial assets. The HH and FC sectors exhibit a net surplus of funds, holding net financial assets of $20.852 trillion and $1.962 trillion, respectively. By contrast, the NFC and GG sectors are net debtors, holding net liabilities of $15.831 trillion and $5.022 trillion, respectively. Notably, comparing the data to the period before the COVID-19 outbreak in 2019,[11] CN's FC sector has undergone a transformation, shifting from a net debt of $490 billion to a net fund surplus of $1.962 trillion. Simultaneously, GG has transitioned from a net fund surplus of $13.786 trillion to a net debt of $5.022 trillion.

Examining the net assets in the row of JP's SFSM (Table 5.6), the FC sector displays a net financial surplus of $1.338 trillion. The NFC and GG sectors have net financial liabilities of $6.142 trillion and $6.096 trillion, respectively. The HH sector emerges as the largest holder of financial net surplus at $14.589 trillion. In total, domestic sectors have total financial assets amounting to $78.602 trillion and total financial liabilities of $74.914 trillion, with net external assets of $3.688 trillion, constituting 18.8% of JP's total financial assets. Compared with values in 2019,[12] the asset and liability positions of JP's various sectors have remained basically stable, with no structural changes.

[10] The ROW sector in the FA is designed from a foreign standpoint, so its assets and liabilities show an opposite relationship when observed from the domestic standpoint, and its assets are the liabilities of the domestic.

[11] See Table 7–9 of Zhang (2022).

[12] See Table 7–9 of Zhang (2022).

Table 5.7 reveals that the US holds a significantly higher stock of financial assets and liabilities compared with other economies. As of the end of 2021, domestic sectors have total financial assets amounting to $297.484 trillion and total financial liabilities amounting to $314.801 trillion. The net external liabilities amount to $17.317 trillion, constituting 12.8% of the total US financial debt. Within these sectors, the FC sector is the most indebted sector, with total debt reaching $143.058 trillion. The majority of its financing comes from HH, constituting 47.1% of the total financing, with a net financial debt of $7.514 trillion. The NFC and GG sectors have net financial debts of $81.086 trillion and $29.085 trillion, respectively. The HH sector stands as the largest holder of net financial assets at $100.368 trillion in the G-4. When compared to the values in 2019, the financial corporations (FC), non-financial corporations (NFC), general government (GG), and external net debt of the United States have increased by 0.7 times, 17.6 times, 5.7 times, and 1.7 times, respectively.

In Table 5.8, the UK's financial transaction system has relatively developed, with the scale of external investment and financing with the ROW sector from FC being higher than those of CN and JP. The scale of financial assets held is slightly lower than those of CN, JP, and the US. The domestic sector holds financial assets and liabilities of $51.555 trillion and $52.030 trillion, respectively, with net external liabilities of $475 billion, constituting 5.43% of the total UK financial debt. The net liabilities amount to $145 billion for FC, $4.958 trillion for NFC, and $3.176 trillion for GG, respectively. The net financial assets held by HH amount to $7.804 Strillion.

Tables 5.5, 5.6, 5.7 and 5.8 provide insights into the financial stock structure of the G-4. First, the financial stock structure of CN and JP appears to be fundamentally identical. The NFC and GG sectors exhibit net debt positions, and the FC and HH sectors demonstrate fund surpluses and hold substantial net external assets. The financial stock structures of the US and the UK are similar. The FC, NFC, and GG sectors have financial net debt positions, but HH maintains a fund surplus and holds substantial net external liabilities. Second, the scale of financial assets and liabilities in the US is the largest, in line with the financing capacity and operational scale of the country's HH. The net financial assets held by the HH sector of the US are 6.9 times those of JP, 4.8 times those of CN, and 12.9 × those of the UK. Third, from the perspective of external financial investment, although JP's scale of foreign investment is lower than those of the US and the UK, its net foreign financial position is the highest. Concurrently, the US net external debt has hit a record high of $17.317 trillion.

Figure 5.1 displays the individual time series evolution of each sector's assets and liabilities across 2017–2021,[13] revealing three significant features. First, CN's sectors have undergone structural changes in assets and liabilities. Since 2017, amidst strained political and economic relations between CN and the US, HH_CN has reduced consumption, leading to increased savings and a rise in net assets. GG_CN has transitioned from positive to negative net assets. NFC_CN has experienced a

[13] As of the end of November 2023, China has not released the FBS data for 2022, so the time series of FBS data for various sectors in China can only be from 2017 to 2021.

continuous increase in net debt. Since 2020, CN's NFC sector, alongside GG and HH, has exhibited a tendency toward balance sheet recession. Efforts to deleverage and tighten off-budget borrowing before the pandemic have failed to bring the deterioration of net financial worth to a halt. Second, NFC's net liabilities and HH's net assets in the US have shown an increasing trend, and GG has an unprecedented huge net debt, highlighting an increased vulnerability of the FC sector to the GG sector. Third, the sectoral balance sheets of JP and the UK remain structurally stable.

To analyze the risk associated with the net position of foreign assets and liabilities in each country, Fig. 5.2 illustrates their net foreign financial assets. The graph underscores the contraction of CN's net external assets and the expansion of the US net external liabilities. It indicates a significant increase in the US net foreign liabilities, revealing that the balance of its external liabilities can no longer be solely covered by capital credit from other countries. Meanwhile, JP's external financial investments have been relatively successful, yielding significant returns with net financial assets amounting to \$3.07 trillion annually. Finally, the UK has maintained a net external financial debt of \$317.1 billion per year.

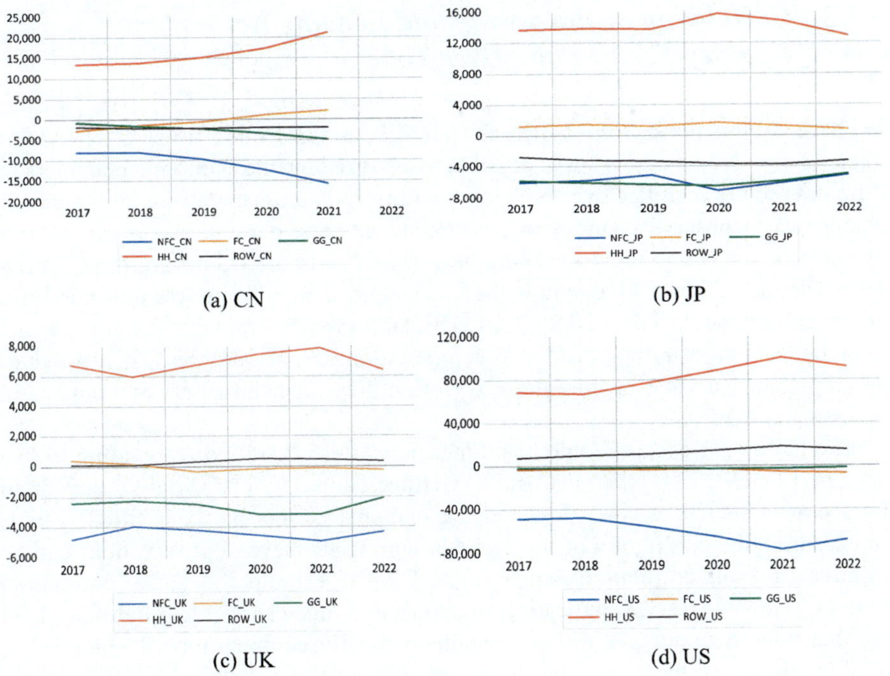

Fig. 5.1 Evolution of FBS (Net worth in USD bn.). *Source* The People's Bank of China (2022) Financial Assets and Liabilities Statement; 720 FBS of OECD Statistics

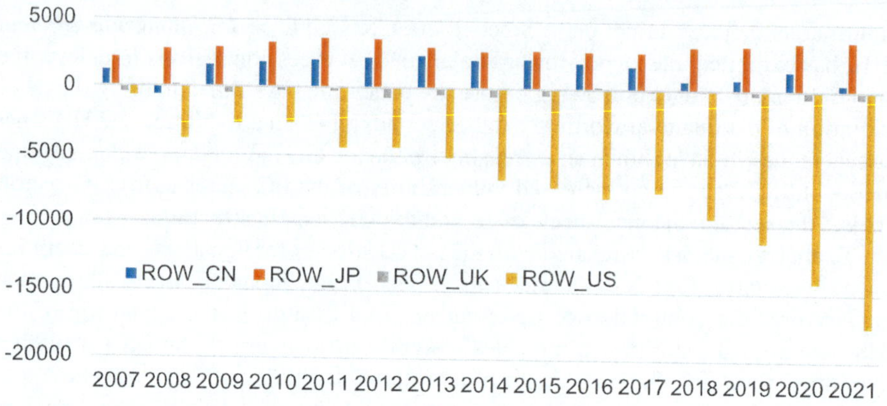

Fig. 5.2 Evolution of External FBS (Net worth in USD bn.). *Source* The People's Bank of China (2022) Financial Assets and Liabilities Statement. Li, Y. and Y. J. Zhang (2020) China's National Balance Sheet 2020, China Social Sciences Press. OECD. Stat, Dataset: 720. FBS—nonconsolidated—SNA 2008

5.3.2 Correlation of Borrowing and Lending Across Country-Sector Pairs Over Time

We have compiled a sectoral matrix for CN, US, and JP from 2018 to 2021 (Annex Tables 5.1, 5.2, 5.3 and 5.4). It is possible to analyze how the economic behaviors of different sectors correlate over time, both within and across countries. This analysis enables us to observe changes in correlation patterns, particularly in the context of political and economic decoupling between CN and the US during the COVID-19 pandemic. Table 5.11 displays the correlation matrix for assets and liabilities across country-sectors over 2018–2021. This table provides valuable insights into the interconnectedness among different sectors within CN, JP, and the US. It provides a comprehensive view of how these sectors utilize financial assets and engage in financing activities.

The matrix for each variable has dimensions of 15 × 15, corresponding to five sectors (FC, GG, HH, NFC, and ROW) in three countries. The variables are sorted by sectors rather than countries, and the diagonal of the matrix is filled with 1, representing the correlation of the variable with itself. Several observations can be inferred from the correlation matrix. First, FC_CN exhibits a negative correlation with FC_US and HH_US, indicating an opposite change in assets and liabilities held by both sides. Second, GG_CN acts countercyclically, compensating the increase in lending elsewhere by increasing its budget deficit. The correlation of government lending with FC, HH, and NFC's lending is negative. Third, NFC, HH, and GG sectors seem to increase or decrease lending simultaneously within a sector across the G-3. Cross-border correlations within these sectors are fairly positive. Fourth, the data suggest the presence of different pull and push factors for capital flows. The

Table 5.11 Correlation of assets and liabilities across country-sector pairs over time (correlations 2018–2021)

	FC_CN	FC_JP	FC_US	GG_CN	GG_JP	GG_US	HH_CN	HH_JP	HH_US	NFC_CN	NFC_JP	NFC_US	ROW_CN	ROW_JP	ROW_US
FC_CN	1.00	−0.93	−0.60	−0.82	0.00	0.00	0.81	−0.85	−0.93	0.70	0.45	0.03	−0.78	0.00	0.00
FC_JP	−0.93	1.00	0.83	−0.92	0.31	−0.52	0.72	0.98	0.85	0.98	0.94	0.53	0.00	−0.34	0.00
FC_US	−0.60	0.83	1.00	−0.93	0.58	0.74	0.93	0.68	0.95	0.37	−0.64	0.99	0.00	0.00	0.46
GG_CN	−0.82	−0.92	−0.93	1.00	0.00	0.00	1.00	−0.94	−0.16	−0.97	0.31	−0.84	−0.82	0.00	0.00
GG_JP	0.00	0.31	0.58	0.00	1.00	0.24	0.00	−0.03	−0.19	0.00	−0.32	0.70	0.00	0.85	0.00
GG_US	0.00	−0.52	0.74	0.00	0.24	1.00	0.00	−0.56	−0.95	0.00	0.97	0.99	0.00	0.00	−0.12
HH_CN	0.81	0.72	0.93	1.00	0.00	0.00	1.00	0.54	0.83	−1.00	0.89	0.59	0.04	0.00	0.00
HH_JP	−0.85	0.98	0.68	−0.94	−0.03	−0.56	0.54	1.00	0.84	0.92	−0.39	0.51	0.00	0.91	0.00
HH_US	−0.93	0.85	0.95	−0.16	−0.19	−0.95	0.83	0.84	1.00	−0.47	−0.84	0.56	0.00	0.00	0.95
NFC_CN	0.70	0.98	0.37	−0.97	0.00	0.00	−1.00	0.92	−0.47	1.00	0.96	0.74	0.05	0.00	0.00
NFC_JP	0.45	0.94	−0.64	0.31	−0.32	0.97	0.89	−0.39	−0.84	0.96	1.00	0.48	0.00	0.29	0.00
NFC_US	0.03	0.53	0.99	−0.84	0.70	0.99	0.59	0.51	0.56	0.74	0.48	1.00	0.00	0.00	−0.99
ROW_CN	−0.78	0.00	0.00	−0.82	0.00	0.00	0.04	0.00	0.00	0.05	0.00	0.00	0.00	0.00	0.00

(continued)

Table 5.11 (continued)

	FC_CN	FC_JP	FC_US	GG_CN	GG_JP	GG_US	HH_CN	HH_JP	HH_US	NFC_CN	NFC_JP	NFC_US	ROW_CN	ROW_JP	ROW_US
ROW_JP	0.00	−0.34	0.00	0.00	0.85	0.00	0.00	0.91	0.00	0.00	0.29	0.00	0.00	0.00	0.00
ROW_US	0.00	0.00	0.46	0.00	0.00	−0.12	0.00	0.00	0.95	0.00	0.00	−0.99	0.00	0.00	0.00

Source Table 5.10 and Annex Tables 5.1, 5.2, 5.3 and 5.4

acquisition of assets and the occurrence of liabilities in the ROW sector are highly negatively correlated across countries.

5.3.3 *Dynamic Structure Analysis for the Sectors of CN, JP, and the US*

The power of dispersion index (PDI) and sensitivity of dispersion index (SDI)[14] for each sector in the financial asset-liability matrix are observed from the sectoral perspective. PDI reflects the capital ripple effect of the entire system resulting from the increase of unit capital supply from a certain sector. SDI focuses on the capital ripple effect of a certain sector when the capital demand of the whole system increases by one unit.

A sector's PDI is calculated as the ratio of the column sum of the Jth column of the Leontief inverse[15] matrix to the column mean. SDI is determined as the ratio of the row sum of the ith row to the row mean. We use d_j^p to define the PDI of sector j and use d_i^s to define the SDI of sector i as follows:

$$d_j^p = \frac{\sum_{i=1}^{n} c_{ij}}{\frac{1}{n}\sum_{j=1}^{n}\sum_{i=1}^{n} c_{ij}} \quad d_i^s = \frac{\sum_{j=1}^{n} c_{ij}}{\frac{1}{n}\sum_{i=1}^{n}\sum_{j=1}^{n} c_{ij}} \tag{5.2}$$

In the equation, c_{ij} denotes Leontief inverse matrix $(I - C)^{-1}$ elements. C denotes the input coefficient matrix of matrix Y, corresponding to the direct consumption coefficient in the Input–Output Model. When $PDI > 1$, the impact of fund raising (supply) on other sectors is higher than the overall average level. When $PDI < 1$ it is lower. The higher the PDI, the greater the impact of capital fluctuations on the capital market. Similarly, the size of SDI reflects differences in the perception of the overall average fund raising (supply) among different sectors.

C can be substituted into the input coefficient matrix with Stone-type C^S and Klein-type C^K, respectively and Stone-formula PDI^S and SDI^S and Klein-formula PDI^K and SDI^K can be solved. PDI^S denotes the total amount of capital supply in the economy as a whole (i.e., indirect and direct) when unit capital raising increases in a sector. PDI^K denotes the ripple effect on the total demand for capital in the economy when a sector increases the supply of capital per unit. SDI^S denotes the ripple effect of a sector that can increase the fund supply when the fund raising of all sectors increases per unit. SDI^K denotes the ripple effect on a sector's amount

[14] It was Rasmussen (1956) who invented the dispersion indices for the input–output analysis. While the PDI is the mean-normalized column sum, the SDI is the mean-normalized row sum of the Leontief inverse.

[15] Leontief (1953, 1963).

of capital demand when the capital supply of all sectors increases by one unit. Both coefficients can partly reflect risk bearing and dispersion. For instance, if the influence of a sector's capital supply PDI^K and the sensitivity of capital raising SDI^S are large, the potential debt risks of other sectors have increased, gathering to this sector through inter-sectoral transmission.

Using the prepared asset–liability matrix, we calculate the time series of PDI^S and SDI^S and PDI^K and SDI^K (Figs. 5.3 and 5.4). Figure 5.3 illustrates the PDI scatter diagram for each sector from 2018 to 2021. The horizontal axis PDI^S denotes the liability influence coefficient, and the vertical axis PDI^K denotes the asset influence coefficient. The intersection's coordinates are (1, 1). Beginning at the upper-right corner, each sector's PDI is divided counterclockwise into four quadrants (I, II, III, and IV), indicating the basic ranking of each sector's influence in the SFSM system.

Figure 5.3 indicates the distribution of each sector across the quadrants. From the perspective of stock, FC_CN's influence on fund raising (supply) is significantly higher than those of other sectors during 2018–2021. Moreover, HH_CN and HH_JP PDI^K>1, PDI^S< 1, indicate that their influence on increasing fund raising is lower than the average. The influence of fund supply is greater, and the impact of HH_US, ROW_US, and GG_CN on fund supply during 2018–2019 is marginally greater than their influence during 2020–2021. ROW_JP and GG_JP are in quadrant IV during 2018–2021, implying that their impact of liabilities increases and that of assets decreases. NFC_US and GG_US (2018–2019), mainly in quadrant IV, $PDI^K < 1$,

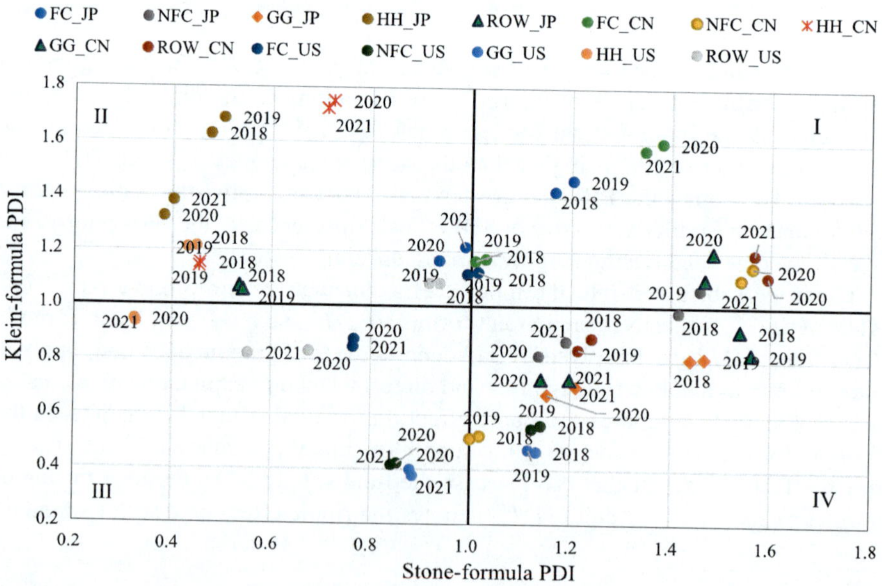

Fig. 5.3 Scatter diagram of the power-of-dispersion index for SFSM. *Source* Annex Tables 1–4 compiled by the authors

$PDI^S > 1$, just contrary to HH_CN and HH_JP, have a greater influence on fund circulation of the entire society when fund demand increases, whereas fund supply has a lower influence than that in other sectors.

In Fig. 5.4, the horizontal axis SDI^S denotes the liability induction coefficient. The vertical axis SDI^K denotes the asset induction coefficient. The figure demonstrates that all sectors' fund raising and supply have a large ripple effect on the FCs for CN, JP, and the US. NFC sectors are the main borrowers, and the basic distribution is in quadrants I and II, rendering $SDI^S < SDI^K$. The liability sensitivity is lower than asset sensitivity. When the fund supply of the whole society increases, the sensitivity of the NFC sectors are greater than the ripple effect of the fundraising for the whole society. The GG and ROW sectors are located in quadrant III, far from the coordinate intersection point (1, 1), implying that the sensitivity of their social financing is low and that they are not the primary beneficiaries of the financial market. Meanwhile, all HH sectors are in quadrant IV, $SDI^K < 1$, $SDI^S > 1$. When the fund raising of the entire society increases, HHs have a strong sense of the change in capital supply.

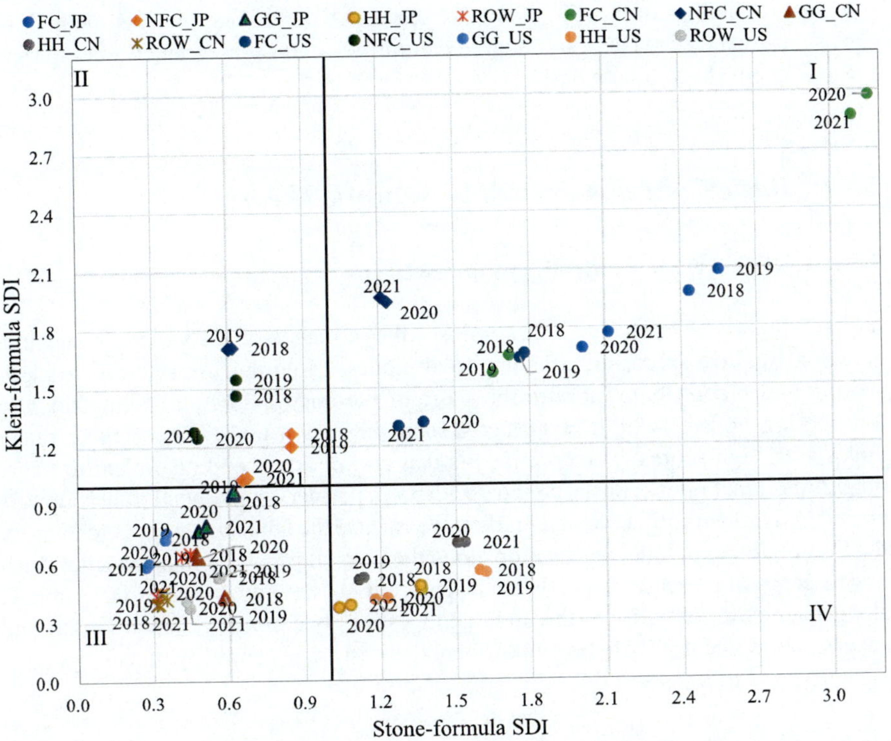

Fig. 5.4 Scatter diagram of the SDI for SFSM. *Source* Annex Tables 1–4 compiled by the authors

5.4 Financial Network Analysis for the SFSM

Network theory provides tools for the study of centrality, influence, sensitivity, and propagation dynamics. A network is merely a graphical representation of a matrix (Soramäki & Cook, 2016), allowing for a faster interpretation. A network comprises nodes and edges connecting them. In a network, nodes are international sectors of SFSM and the edges—the links between nodes—are asset/liability links. Nodes in the financial network denote different countries, and a link from country i to j denotes country i's claims (exposure) by country j. To facilitate the identification of systemically important countries, the node sizes are proportional to the countries' holdings of liabilities of a given type. For instance, if FC_US is denoted by a large node depicting exposures in debt securities, it implies that FC_US is a large issuer of debt securities. Likewise, the width of the link is also proportional to the size of each sector's exposure to another sector. Because networks are constructed to assess financial stability, instead of drawing a link proportional to the absolute value of a bilateral claim, its width is based on the creditor sector's capacity to absorb the potential loss represented by the claim. The smaller the sector, the less able it is to absorb the loss of a claim. Next, we discuss two network analysis methods—eigenvector centrality (EC) and degree centrality—and then conduct an empirical analysis of SFSM as a network.

5.4.1 Basic Concepts Related to Network Theory

5.4.1.1 Eigenvector Centrality in the Network

In graph theory, EC (Zhang, 2020) assesses the influence of the various nodes in a network. Relative scores are assigned to all nodes, given that connections to high-scoring nodes contribute more to the score of the node than equal connections to low-scoring nodes. A high eigenvector score implies a node connected to many nodes with high scores. The core idea is that an important node is linked to many other important nodes. Thus, a country with more partners in financial transactions is considered more important in the market. We can use the adjacency matrix to identify the EC, assuming parallel duplication along the links. It is based on the concept that a node's centrality directly depends on the centrality of the nodes whereto it is linked. If we denote the centrality of the ith node in a strongly connected network as x_i and set each node's centrality to be proportional to the average centrality of its neighbors, we derive the following:

$$x_i = \frac{1}{\rho} \sum_{j=1}^{n} A_{i,j} x_j, \tag{5.3}$$

In the equation, n denotes the number of nodes in the network, ρ denotes a constant, and A represents the network's (weighted or unweighted) adjacency matrix. Notably, if the adjacency matrix is weighted, moves along links with higher weights are more likely.

Many different eigenvalues ρ have a nonzero eigenvector solution. However, the additional requirement that all of the entries in the eigenvector be nonnegative implies that only the greatest eigenvalue results in the desired centrality measure. The vth component of the related eigenvector then provides the relative centrality score of vertex v. Because the eigenvector is only defined up to a common factor, only the ratios of the centralities of the vertices are well defined. To define an absolute score, it is essential to normalize the eigenvector such that the sum over all vertices is 1 or the total number of vertices is n. The power iteration may be used to find this dominant eigenvector. Thus, the vector of centralities x is an eigenvector of the network's adjacency matrix. According to the Perron–Frobenius theorem (Loriana Pelizzon Newman, 2010), the eigenvector of A corresponding to the largest eigenvalue has all positive entries. This eigenvector provides the nodes' ECs.

5.4.1.2 Degree of Centrality Within the Network

The GFF's W-t-W data can be viewed as a network of interrelationships in which the nodes are countries and the edges are assets or liabilities. The edges in the network are "weighted" by the amounts involved in every asset or liability relationship. Given space constraints, we focus on degree centrality in the network analysis to illustrate the importance and influence of the sectors.

Degree centrality employs the most direct metric to describe node centrality in network analysis (Zhang, 2020). The greater a node's degree, the higher its degree centrality and the more important the node. In an undirected graph, degree centrality assesses the extent to which a node is associated with other nodes. For an undirected graph with g nodes, the degree centrality of node i is the total number of direct connections between i and other g-1 nodes, which is expressed as follows:

$$C_D(N_i) = \sum_{j=1}^{g} x_{ij}(i \neq j) \tag{5.4}$$

In the equation, C_D (N_i) denotes the centrality of node i ($i \neq j$ excludes the connection between i and j, so the data in the main diagonal can be ignored), x_{ij} denotes the value of the cell in which the corresponding row or column in the matrix is located. C_D (N_i) denotes the sum of assets (columns) or liabilities (rows) of one country to another, considering that undirected relationships form a symmetric data matrix. Cells with the same rows and columns have the same value. Tsujimura and Mizoshita (2002) proposed indicators to observe PDI and SDI. According to network theory (Soramäki & Cook, 2016), PDI and SDI are considered network centrality

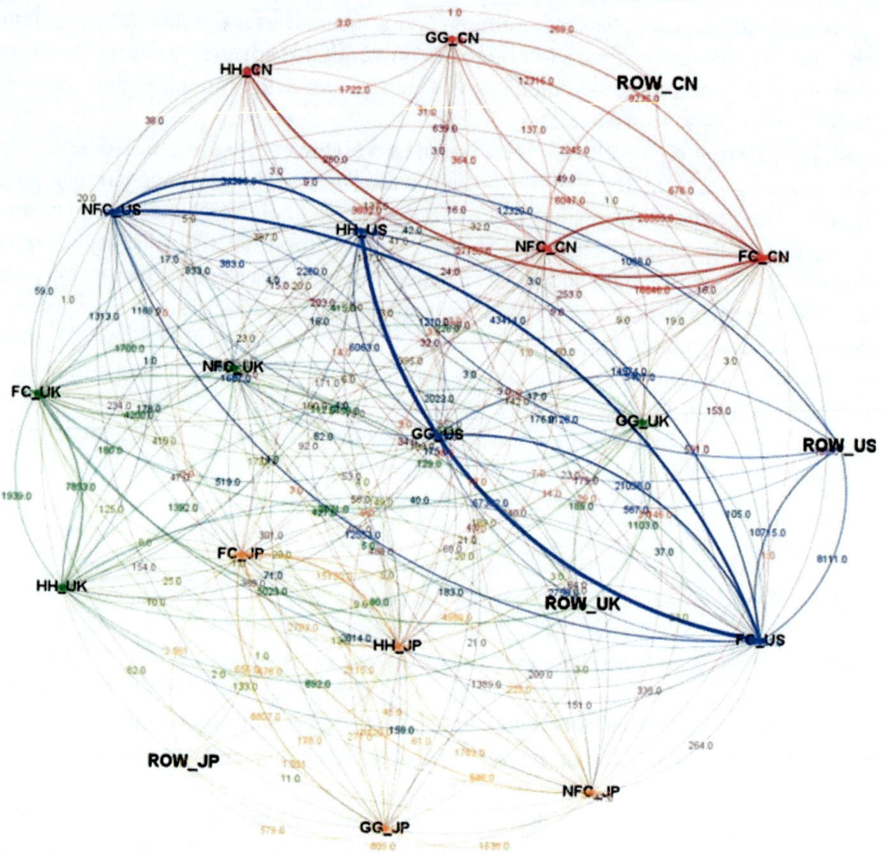

Fig. 5.5 Cross-border exposure networks in the sectors of G-4 (as of the end of 2021)

measures of a network, represented by the inverse of Leontief[16] (degree centrality). Moreover, PDI and SDI can be regarded as a certain network centrality measure, that is, degree centrality (in- and out-degree) of the weighted network represented by $(I - C)^{-1}$. We define in-degree as external claims and out-degree as external debts. Using Eq. (5.3) and based on matrix C (in Table 5.9) to obtain the inverse of Leontief by $(I - C)^{-1}$, we calculate the degree centrality of SFSM and draw the network diagrams (Figs. 5.5, 5.6 and 5.7). We consider this approach because matrix C denotes the network of interconnections better.

[16] Leontief (1941).

Table 5.12 International sector linkages and network centrality (at the end of 2021)

Id	ROW_US	GG_US	ROW_UK	ROW_CN	ROW_JP	GG_CN	GG_UK	NFC_UK	GG_JP	NFC_US
Eigenvector centrality	0.2275	0.2453	0.2505	0.2517	0.2695	0.4088	0.6792	0.7481	0.7682	0.8758
Id	NFC_JP	HH_CN	FC_US	HH_US	FC_UK	HH_UK	FC_JP	HH_JP	FC_CN	NFC_CN
Eigenvector centrality	0.9185	0.9286	0.9471	0.9471	0.9487	0.9487	0.9500	0.9500	1	1

5.4.2 *Network Correlation of the Sectors of G-4*

The W-t-W data is a network of interrelationships. EC computes the centrality for a node based on the centrality of its neighbors. If we define the vector of centralities $x = (x_1, x_2, \ldots, x_n)$, the EC for node i is obtained as follows:

$$Ax = \rho x \tag{5.5}$$

In the equation, A denotes the adjacency matrix of Fig. 5.5 with eigenvalue ρ. By virtue of the Perron–Frobenius theorem, there is a unique and positive solution if ρ is the largest eigenvalue associated with the eigenvector of the adjacency matrix A.

First, we use the EC method to conduct a financial network analysis. The EC values are calculated based on Eq. (5.5) and Table 5.10. The results are presented in Table 5.12. The sectors of G-4 countries are divided into four categories by EC value. The EC values for FC_CN and NFC_CN are close to 1, denoting the central position of cross-border exposures and having a higher contribution. The next level includes HH_JP, FC_JP, HH_UK, FC_UK, HH_US, FC_US, HH_CN, and NFC_JP, the EC values for which are lower than 0.95 but higher than 0.9. The lower-level sectors include NFC_US, GG_JP, NFC_UK, GG_UK, and GG_CN, with EC values being less than 0.9 and greater than 0.4. The EC value of GG_US and ROW sectors are low,[17] at about 0.25, indicating sectors poor in centrality and low in contribution. Therefore, according to the EC values, the above four levels can be roughly distinguished according to the impact of clustering with counterparties in the G-4 financial market.

The SFSM data (Table 5.10) are used to establish the network matrix (Fig. 5.5). Figure 5.5 presents a network diagram that indicates the relation between the financial positions of the sectors based on W-t-W. In addition, the colors of nodes are divided into red (CN), orange (JP), green (the UK), and blue (the US). Nodes represent sectors, and edges represent different degrees of mutual holding of claims and debts.

[17] The data for the ROW sector in Table 5.9 refer to the amount of financial assets or liabilities held with other economies (excluding the G4), so the ROW sectors in the G-4 network have a lower EC value.

Regarding EC of network nodes and the thickness of network edges, FC_US, HH_US, FC_CN, FC_JP, and FC_UK have more influence than others (Fig. 5.5). Additionally, the width of the links indicates that the scale of financial investment of these sectors are larger than those of other sectors. Furthermore, they have a strong influence on the creditor's rights (weighted in-degree) and debts (weighted out-degree) of other sectors. According to the order of the width of the links of the edges (degree centrality (financial investments)), they are arranged as follows: FC_US, HH_US, FC_CN, FC_JP, FC_UK, NFC_US, HH_CN, ROW_US, NFC_CN, HH_JP, NFC_JP, HH_UK, GG_US, ROW_UK, GG_CN, GG_JP, NFC_UK, ROW_CN, ROW_JP, and GG_UK.

Figure 5.5 represents the investment amount. In the global financial market, FC_US still holds the largest share of financial liabilities, with 23.5% of the global market, worth \$143.059 trillion. The sector also holds \$136.021 trillion in financial claims, constituting 22.2% of the global market. Accordingly, the net debt of the FC_US sector is \$7.038 trillion. Figure 5.5 indicates that FC_US has raised \$1389 billion, \$264 billion, \$47 billion, and \$151 billion, respectively, from FC_JP, NFC_JP, GG_JP, and HH_JP; \$591 billion, \$179 billion, \$3 billion, and \$34 billion from FC_CN, NFC_CN, GG_CN, and GG_CN; and \$2014 billion, \$183 billion, \$37 billion, and \$158 billion from FC_UK, NFC_UK, GG_UK, and HH_UK. Similarly, we can observe the position of financial assets or liabilities held by a sector against a sector in another country.

In the claims and debts of ROW[18] to other domestic sectors, the US and the UK occupy a larger proportion than CN and JP. The external claims of the US and the UK constitute 2.93% and 0.96% of the global total, while CN and JP constitute 0.65% and 0.67%, respectively. The external debt of the US and the UK constitute 4.82% and 1.23% of the global total, while CN and JP constitute 0.52% and 0.36%, respectively (see Table 9).

5.4.3 *The Network Analysis of the G-4 by the SFSM*

Table 5.12 and Fig. 5.5 indicate the bilateral risk exposure between the cross-border sectors of the G-4. From this, we build the financial network. Then, we connect each country-sector level network through cross-border exposures to achieve financial network visualization (Fig. 5.5). As a preliminary attempt, we conduct the following two aspects of the analysis.

5.4.3.1 Observing Bilateral Exposures Within Countries

Table 5.12 and Fig. 5.5 reveal that the national sectors of the G-4 hold the creditor's rights and debts of their counterparties. These nodes are larger with wider edges.

[18] The ROW sector here means external investment and financing in addition to G-4.

Thus, we can understand the situation of bilateral fund operations of the domestic sectors of the G-4. The largest exposures at the country level are from FC_CN by NFC_CN, FC_JP by HH_JP, FC_UK by HH_UK, and FC_US by HH_US. Here, we focus on the FC sector for analysis.

The ranking of the size of the FC node is as follows: the US, CN, and JP. FC_US holds financial assets worth $136.021 trillion, and it applies its assets to NFC_US, GG_US, and HH_US, accounting for 31.92%, 15.47%, and 11.01% of the total assets, respectively. However, the internal fund use of FC_US constitutes the largest proportion of its total assets (32.81%), while its total liabilities are $143.059 trillion, with 8.77% from NFC_US, 1.93% from GG_US, and 47.094% from HH_US.

FC_CN holds $75.344 trillion in financial assets with NFC, GG, and HH sectors, constituting 35.66%, 12.26%, and 16.35% of their assets, respectively. However, the total debt of FC_CN is $73.38 trillion, and the financing proportion from NFC_CN is 22.68%, 8.24% from GG_CN, and 37.01% from HH_CN.

FC_JP holds financial assets of $42.64 trillion, providing strong investment activities to NFC_JP, GG_JP, and HH_JP, constituting 19.04%, 20.66%, and 6.55% of assets, respectively. However, the total debt of FC_JP is $41.243 trillion. The proportion of financing from NFC_JP is 12.1%, from GG_JP is 5.12%, and from HH_JP is 36.69%.

FC_UK holds financial assets of $35.774 trillion, providing strong investment activities to NFC_UK, GG_UK, and HH_UK, constituting 11.74%, 7.47%, and 5.42% of assets, respectively. However, the total debt of FC_UK is $35.89 trillion, and the proportion of financing from NFC_UK is 4.74%, from GG_UK is 1.16%, and from HH_UK is 21.88%.

The above analysis shows that the high exposures of FC_US and FC_CN are mainly concentrated in their NFC sectors, whereas the larger exposures from FC_JP are by GG_JP. Regarding fund-raisers, the main fund-raiser of FC_US, FC_CN, FC_JP, and FC_UK is the HH sector. However, from the perspective of net financial position, FC_US and FC_UK are in a state of net debt, with deficits of $7.038 trillion and $114 billion, respectively, whereas FC_JP and FC_CN hold a net financial position of $1.397 trillion and $1.964 trillion.

5.4.3.2 Bilateral Cross-Border Exposure

As shown in Fig. 5.5, because the edges of the cross-border exposures are much smaller than those of national exposures, another reference base for the width of the links is used for cross-border links to visualize the differences in exposures to different countries. We focus on cross-border exposure of the FC, NFC, and ROW sectors between the G-4.

First, we observe the characteristics of overseas investment from a macro perspective. Figure 5.5 shows that the US has the biggest exposure ($17.98 trillion), followed by the UK ($5908 billion), JP ($4103 billion), and CN ($3973 billion), which are all lower than the outbound financial investment of the G-4 in 2019,[19] showing the impact of COVID-19. However, in the US, 8.8% of the FC sector's assets are used by ROW, and the financing proportion from ROW is 11.02%.The UK has the highest share, showing the traditional advantage of outbound financial investment; 27.22% of the FC sector's assets are used by ROW, with 24.33% being its proportion of financing. In JP, 14.02% of the FC sector's assets are used by the ROW sector with 5% being the proportion of financing. However, in CN, 6.22% of the FC sector's assets are applied to ROW with only 1.76% raised from the sector. This means that CN's FC sector still has much work in opening overseas markets and expanding financing.

Regarding cross-border exposures, JP's FC sector in the US is greater than that in CN. These exposures amount to 3.26%, 2.4%, 0.95%, and 0.19% of FC_JP's total assets, respectively, whereas FC_JP's exposure to similar sectors in CN only accounts for 0.03%, 0.07%, 0.01%, and 0%, respectively. The degree of closeness centrality in the JP–US financial network is higher than that in the CN–US relationship.

Although the scale of the risk exposure of FC_CN is less than FC_JP, the risk exposure of FC_CN to the FC_US is also greater than FC_CN to FC_JP. The risk exposure of FC_CN to the US' FC, NFC, GG, and HH sectors accounts for 0.78%, 0.85%, 0.34%, and 0.07% of the total assets of FC_CN, respectively. However, the exposure of FC_CN to similar sectors in JP only accounts for 0.04%, 0.04%, 0.02%, and 0.01% of FC_CN's total assets, respectively.

For cross-border exposures, NFC_US has larger vulnerabilities from CN and JP's other sectors (FC, NFC, GG, and HH) because NFC_US holds the largest exposures with $80.841 trillion, having a bigger node than the others' NFC sectors. The funds used by NFC_US to FC_CN, NFC_CN, GG_CN, and HH_CN account for 0.09%, 0.58%, 0.03%, and 0.0001% of its total assets held, respectively. However, the NFC_US' financing from FC_CN, NFC_CN, GG_CN, and HH_CN accounts for 0.41%, 0.13%, 0.002%, and 0.023% of its total financing, respectively. The funds used by NFC_US in FC_JP, NFC_JP, GG_JP, and HH_JP account for 0.67%, 1.05%, 0.44%, and 0.14% of its assets, respectively. However, the NFC_US' financing from FC_JP, NFC_JP, GG_JP, and HH_JP accounts for 0.88%, 0.82%, 0.045%, and 0.15% of its debts, respectively. This implies that the cross-border exposure of the NFC_US sector to JP's other sectors is larger than that to CN's sectors.

5.4.3.3 The Degree Centrality Between the Sectors of G-4

Next, we use degree centrality to observe the position of the G-4 in the financial network. Tsujimura (2002) introduced input–output structure analysis to the asset–liability matrix derived from the JP FBS. The literature proposed PDI and SDI calculation methods. Zhang (2020) introduced PDI and SDI into financial network analysis based on degree centrality theory to measure network centrality. PDI and SDI can be regarded as nodes in the W-t-W network and seen as a certain network centrality

[19] Nan Zhang (2022).

measure: degree centrality (in-degree and out-degree) of the weighted network represented by $(I - C)^{-1}$. PDI is a relative indicator of the amount of funds supplied to international markets, including indirect effects, when a country increases its use of funds. If direct funds are supplied to a country holding external net debt, PDI will be small. Conversely, if countries with financing channels provide funds supply, PDI will be large. However, from the perspective of fund demand, when the global fund demand increases, a country's SDI is relatively lower when it obtains direct financing from other countries' banks. Conversely, when the country obtains indirect financing from international markets or regional banks, its SDI increases. Therefore, the size of PDI largely depends on the country's asset portfolio, whereas the size of SDI largely depends on other countries' liability portfolios. To facilitate comparison of the position of financial investment between G-4 countries, we use the data in Tables 5.9, 5.10 to calculate degree centrality among sectors and draw Figs. 5.6 and 5.7, which depict the position of PDI and SDI between the sectors of bilateral national exposures.

Figure 5.6 is divided into four quadrants, moving anticlockwise. Liability influence index (PDI^S) and asset influence index (PDI^K) in the upper-right quadrant are higher than average (greater than 1). In Quadrant II, PDI^S is less than 1, but PDI^K is greater than 1. In Quadrant III, both PDI^S and PDI^K are less than 1 (below average). In Quadrant IV, PDI^S is greater than 1, but PDI^K is less than 1. The quadrant in which a given sector lies indicates its influence tendencies in G-4.

Although there are no prominent influential sectors in Fig. 5.6, the distribution of Quadrant I sectors indicate that the asset influence and liability influence of these six sectors in international capital markets are slightly higher than the G-4 average. Quadrant II is the region where debt influence is lower than average, but asset influence is higher than average. Quadrant III is where the impact of both assets and liabilities is below average. Surprisingly, all five U.S. sectors fall into this quadrant.

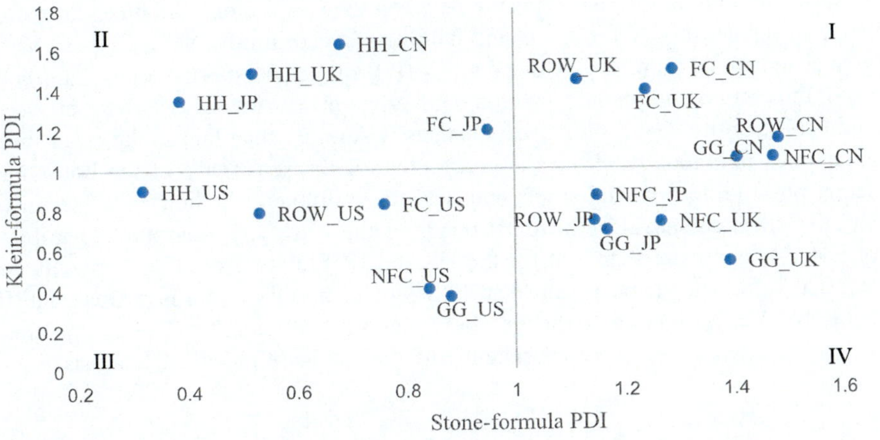

Fig. 5.6 Degree centrality on bilateral sectors exposures by PDI (at the end of 2021). *Source* International SFSM (Tables 5.9 and 5.10) compiled by the authors

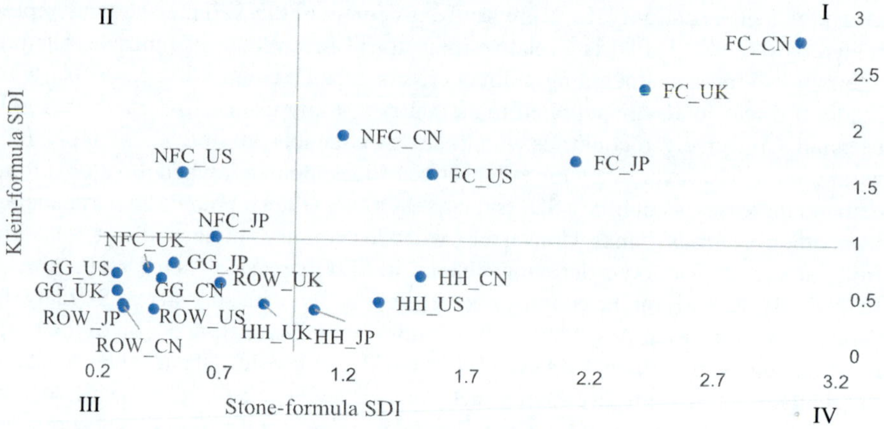

Fig. 5.7 Degree centrality on bilateral sectors exposures by SDI (at the end of 2021). *Source* International SFSM (Tables 5.9 and 5.10) compiled by the authors

Being in Quadrant IV means the sector has above-average liability influence but below-average asset influence.

Figure 5.7 indicates the distribution of liability sensitivity index (SDI^S) and asset sensitivity index (SDI^K). From the distribution of Quadrants I–III, there is a linear state of approximately positive correlation between the two indexes. Sectors in Quadrant I mean that when the capital demand or capital supply changes in the international market, it ultimately depends on the corresponding sectors to a large extent. The government and ROW sectors of the G-4 are distributed in Quadrant III. In addition, NFC_US and NFC_JP are distributed in Quadrant II, while HH_CN, HH_US, and HH_JP are distributed in Quadrant IV.

When looking at the risk exposure between the G-4 sectors, the financial sector has a strong sensitivity for assets and liabilities. Accordingly, NFC_CN also has a greater sensitivity than NFC_US and NFC_JP. From the perspective on the ability of capital operation and planning with the government sector, the sensitivity of assets and liabilities of GG_JP on market changes is stronger than that in the other three economies individually. From the perspective of the sensitivity of the household sector on holding financial assets and liabilities, although the household sector of CN, the US, and JP are all in the fourth quadrant, the sensitivity of assets and liabilities of HH_CN is greater than that for the US and JP. Finally, from the perspective of external fundraising capacity, the distribution of each ROW sector is in Quadrant III, but ROW_UK is higher than that in other economies individually. These projections based on network theory are consistent with the results of the above analysis.

5.5 Shock Dynamics and Propagation Across the SFSM

This section develops centrality measures on the W-t-W matrix, which directly represents the net of interlinks. In particular, focus is on FC_CN, FC_JP, FC_US, and FC_UK EC and on capturing direct and indirect links because these sectors are integral in the network, which are in Quadrant I in Figs. 5.6 and 5.7. Therefore, we propose a decomposition of shocks into n-order effects on the basis of an "inverse of Leontief" representation of the W-t-W matrices and carry out an empirical analysis on quantity shock changes.

5.5.1 *A Theoretical Model for Estimating Bilateral Exposures*

Based on the GFFM model from Zhang (2020), bilateral exposures across N countries in a financial instrument k can be expressed in an n x n matrix in which the element y_{ij} denotes a claim of country i vis-à-vis country j. So, the sum of each column j denotes the aggregate SFSM holdings of assets of country j in instrument k ($a_{j,k}$), and the sum of each row i denotes the aggregate holdings of liabilities of country i in instrument k ($l_{i,k}$). Aggregate assets ($a_{j,k}$) and liabilities ($l_{i,k}$) per country are observable but bilateral exposures need estimating.

$$Y_k = \begin{pmatrix} y_{11} & \cdots & y_{1n} \\ \cdot & \cdot & \cdot \\ y_{n1} & \cdots & y_{nn} \end{pmatrix} \text{ with } \sum_{i=1}^{n} y_{ij} = a_{j,k} \text{ and } \sum_{j=1}^{n} y_{i,j} = l_{i,k}$$

To represent how the investment behaviors of various countries react to the investment needs of others (in order to finance them), Δs is set as an exogenous variable, which is the shock itself, indicating changes in the original investment, its transpose vector, and can be represented as follows:

$$\Delta s = (0, \ldots, -s, 0, \ldots, 0) \tag{5.6}$$

By the W-t-W framework, a matrix algebra presentation of GFFM can be shown by $T = Y + \Delta s$.

where T is the vector $T = \begin{pmatrix} t_1 \\ \cdot \\ t_n \end{pmatrix}$.

We then define the elements c_{ij} as the ratio of funds raised from country i to the total external financing needs of country j, that is, $c_{ij} = \frac{y_{ij}}{t_j}$.

C is the matrix of c_{ij} determined by the form of the $n \times n$ order, and we can get $y_{ij} = c_{ij} * t_j$, and the diffusion matrix is represented as follows:

$$T = C * T + \Delta s \tag{5.7}$$

In the equation, $T_k = C * \Delta s$, (k = 0, 1, 2, …, n). When $k = 0$ is a direct effect, $k = 1$ is an indirect first-order effect, $k = 2$ is an indirect second-order effect, and $k =$ n is an indirect n-order effect. On the basis of the power series representation of the inverse of Leontief, the total change in investment produced by a shock is as follows:

$$\begin{aligned} \xi_k &= T_0 + T_1 + T_2 + \cdots + T_k \\ &= T_0 + CT_0 + C^2T_0 + \cdots + C^kT_0 \\ &= (I + C + C^2 + \cdots + C^k)T_0 \end{aligned} \tag{5.8}$$

When

$$k \to \infty, \quad \xi_\infty = (I - C)^{-1}T_0 \tag{5.9}$$

Equation (5.9) reflects the limiting effect of n-order where $(I - C)^{-1}$ is the inverse of Leontief. While Leontief deals with input per unit of output, we consider financing per unit of investment, but the overall logic behind the two representations is identical. The elements in the diffusion matrix in our model have interesting interpretations for well-known financial ratios. Thus, $c_{1,j}$ and $c_{2,j}$ are the ratios of financing from one country to another to the total investment of the country (also assuming that when $i = j, C_{i,j} = 0$, i.e., excluding the country's own domestic portfolio investments). The ratios $c_{1,j}, c_{2,j}, c_{3,j}$, and $c_{i,j}$ represent the mix of financing sources for a country's portfolio investment, indicating how a country relies on other countries for funding, usually by issuing treasury securities and bank debentures.

5.5.2 *Shock Dynamics Between the Sectors of G-4*

This section uses the shock dynamics model (Eq. 5.7) to measure the impact of investment changes in W-t-W networks between the G-4 sectors. According to Table 5.10, we consider it to be more effective to measure the effect of a shock between FC_CN, FC_JP, FC_US, and FC_UK from the perspective of sectoral investments to measure financial risk. Therefore, we use the table to calculate the investment ratio matrix shown in Table 12. Additionally, according to the positions of the G-4 in the first quadrant of Figs. 5.6 and 5.7 by the Stone-model with regard to the vector Δs put in $T_k = C * \Delta s$, in terms of the transpose vector of Δs above, we have our shock in unitary terms as:

Δs =	(1	0	0	0	-1	1	0	0	0	1	1	0	0	0	1	1	0	0	0	1)′

(5.10)

This is based on the assumption that the Row_CN will be reduced by −1 unit, FC and ROW sectors of the G-4 will be increased by +1 unit each, and the financing of other sectors of countries are assumed to be unchanged.

Alternatively, if we focus on the supply of funds, which use Klein-model, Δs' transpose vector also can be represented as follows:

$$\Delta s = (1\ 0\ 0\ 0\ 1\ 1\ 0\ 0\ 0\ 1\ -1\ 0\ 0\ 0\ -1\ 1\ 0\ 0\ 0\ 1)' \tag{5.11}$$

Here, FC_US and Row_US will be reduced by −1 unit, financial and ROW sectors of CN, JP, and the UK will be increased by +1 unit each, and the financial investment of other sectors of countries are assumed to be unchanged. Therefore, according to Eq. (5.8), using Δs and the investment ratio matrix C (calculated with Tables 5.9 and 5.10), we can present the decomposition into the 15-orders in the G-4 sectors, as shown in Tables 5.13 and 5.14. We can speculate on the impact of the shock on changes in the original financial investments for the G-4.

The shock effects shown in Table 5.12 can be decomposed into four parts: (i) the shock itself, the vector of Δs' indicating the changes in investment positions, (ii) the investment effort needed to finance the change based on the vector $T_0 = C * \Delta s$, (iii) the investment effort needed to finance the investment change based on the vector $C^2 T_0, C^3 T_0, \ldots C^{15} T_0$, and (iv) infinite *n*-order investment efforts based on the vector $(I - C)^{-1} \Delta s$.

In Table 5.13, the changes in investment and financing triggered by shocks are governed by the set of direct and indirect relationships embedded in the W-t-W diffusion matrix, including intricate investment/financing paths of any order, even beyond the fifteenth order referred to in this example. Next, we propose a decomposition of the shocks to the G-4 sectors of that separating these individual *n*-order effects.

We plot Figs. 5.8, 5.9, 5.10 and 5.11 using Table 5.13. The shock itself, or the first-order effect, consists of a reduction in external fund inflow in CN and financial liability increase in the G-4. Since FC_US has the largest global market share of financing, assuming an increase in original financing by one unit, its direct effect will be shown with 0.6584. It quickly recovers to 0.1359 of its original position after the six-order indirect effects, and the shock effect declines gradually after the 9th-order, tightening to zero effect by the 15th order. From a cumulative effect perspective, it exhibits a significant impact function, ranging from 2.1341 for the second order to 3.2116 for the eighth order. Subsequently, the positive shock starts to diminish but gradually climbs to 3.4155 at the 15th order (Fig. 5.8). The overall limit value for FC_US, encompassing both primary and indirect effects, stands at 3.4516 (see Table 5.13).

Figure 5.9 shows that under the influence of positive shocks in FC_US and constraints from other sectors in CN, JP, and the UK, GG_US experiences relatively minor positive shock effects. The first-order effect is 0.0385, which decreases to 0.0038 and gradually converges to zero thereafter. The comprehensive limit effect is 0.1311 (see Table 5.13), which is lower than the shock effects on the government sectors of CN, JP, and the UK.

When we assume that FC_CN increases by 1 and ROW_CN decreases by −1, while both FC and ROW sectors in JP, the US, and the UK increase by +1, and the change of other sectors is 0, the change of FC_CN is indicated in Fig. 5.10. The direct effect of FC_CN is −0.422, and its negative effect changes from the first-order to the

Table 5.13 The 15-order effects on financial financing for all sectors of G-4 (Stone-model)

Sectors	Δs	1	2	3	4	5	6	7	8	9	10	11	12	13	14	15	⋮	$(I–C)^{-1}*\Delta s$
FC_JP	**1**	1.0429	0.7921	0.574	0.4311	0.3238	0.2436	0.1833	0.138	0.104	0.0783	0.0591	0.0445	0.0336	0.0254	0.0192		5.1528
NFC_JP	0	0.134	0.1801	0.15	0.114	0.0859	0.0646	0.0487	0.0367	0.0276	0.0208	0.0157	0.0119	0.009	0.0068	0.0051		0.9269
GG_JP	0	0.2993	0.1056	0.0798	0.06	0.0453	0.034	0.0256	0.0193	0.0145	0.0109	0.0082	0.0062	0.0047	0.0035	0.0027		0.7281
HH_JP	0	0.4724	0.3943	0.3082	0.2254	0.1695	0.1273	0.0958	0.0721	0.0543	0.0409	0.0308	0.0232	0.0175	0.0132	0.01		2.086
ROW_JP	**1**	0.0154	0.0407	0.0292	0.0224	0.0168	0.0127	0.0095	0.0072	0.0054	0.0041	0.0031	0.0023	0.0017	0.0013	0.001		1.1757
FC_CN	**1**	−0.422	0.0843	−0.0873	0.0012	−0.016	0.0016	0.002	0.0062	0.007	0.0078	0.0078	0.0075	0.007	0.0064	0.0057		0.654
NFC_CN	0	0.0406	−0.067	0.0183	−0.0076	0.0059	0.0027	0.0049	0.0045	0.0046	0.0043	0.0039	0.0036	0.0032	0.0028	0.0024		0.0409
GG_CN	0	0.0752	−0.0334	0.0062	−0.0068	0	−0.0012	0.0002	0.0002	0.0005	0.0006	0.0007	0.0007	0.0006	0.0006	0.0005		0.048
HH_CN	0	0.2848	−0.1345	0.0196	−0.0285	−0.0009	−0.0049	0.0009	0.0014	0.0029	0.0032	0.0034	0.0034	0.0032	0.003	0.0027		0.1767
ROW_CN	**−1**	0.009	0.0002	−0.0036	0.0003	−0.0006	0.0002	0.0001	0.0003	0.0003	0.0003	0.0003	0.0003	0.0003	0.0002	0.0002		−0.9908
FC_US	1	0.6584	0.4757	0.3343	0.2467	0.1824	0.1359	0.1017	0.0764	0.0576	0.0436	0.033	0.0251	0.0191	0.0145	0.0111		3.4516
NFC_US	0	0.1012	0.1005	0.0807	0.0597	0.0445	0.0332	0.0249	0.0187	0.0141	0.0107	0.0081	0.0062	0.0047	0.0036	0.0027		0.5225
GG_US	0	0.0385	0.0255	0.0175	0.0127	0.0093	0.0069	0.0051	0.0038	0.0029	0.0022	0.0016	0.0012	0.0009	0.0007	0.0005		0.1311
HH_US	0	0.7915	0.3991	0.3012	0.2166	0.1598	0.1184	0.0882	0.066	0.0496	0.0374	0.0283	0.0214	0.0162	0.0124	0.0094		2.346

(continued)

Table 5.13 (continued)

Sectors	Δs	1	2	3	4	5	6	7	8	9	10	11	12	13	14	15	⋮	$(I–C)^{-1}*\Delta s$
ROW_ US	**1**	0.0749	0.0728	0.0536	0.0389	0.0286	0.0212	0.0158	0.0118	0.0089	0.0067	0.005	0.0038	0.0029	0.0022	0.0017		1.3542
FC_ UK	**1**	1.1677	0.7835	0.6225	0.4753	0.3664	0.2816	0.2165	0.1664	0.1279	0.0983	0.0755	0.058	0.0446	0.0343	0.0263		5.6328
NFC_ UK	0	0.0512	0.0683	0.0501	0.0395	0.0302	0.0233	0.0179	0.0138	0.0106	0.0082	0.0063	0.0049	0.0037	0.0029	0.0022		0.3408
GG_ UK	0	0.037	0.0233	0.0197	0.0149	0.0115	0.0088	0.0068	0.0052	0.004	0.0031	0.0024	0.0018	0.0014	0.0011	0.0008		0.1447
HH_ UK	0	0.3125	0.2822	0.2044	0.1598	0.1225	0.0943	0.0725	0.0558	0.0429	0.0329	0.0253	0.0195	0.015	0.0115	0.0088		1.4894
ROW_ UK	1	0.1521	0.2024	0.1403	0.1108	0.0848	0.0653	0.0502	0.0386	0.0297	0.0228	0.0175	0.0135	0.0104	0.008	0.0061		1.9728

Note Extrapolation based on data from Table 5.9

Table 5.14 The 15-order effects on financial investment for all sectors of G-4 (Klein-model)

Sectors	Δs	1	2	3	4	5	6	7	8	9	10	11	12	13	14	15	⋮	$(I-C)^{-1}*\Delta s$
FC_JP	**1**	0.5572	0.4918	0.3505	0.261	0.1934	0.144	0.1074	0.0803	0.0602	0.0452	0.034	0.0256	0.0193	0.0146	0.0111		3.4311
NFC_JP	0	0.3883	0.2609	0.1945	0.1438	0.1069	0.0797	0.0596	0.0446	0.0335	0.0252	0.019	0.0143	0.0108	0.0082	0.0062		1.4156
GG_JP	0	0.3487	0.1797	0.1423	0.1039	0.0773	0.0575	0.0429	0.0321	0.0241	0.0181	0.0136	0.0102	0.0077	0.0059	0.0044		1.0827
HH_JP	0	0.1091	0.0463	0.0398	0.0286	0.0214	0.0159	0.0119	0.0089	0.0067	0.005	0.0038	0.0028	0.0021	0.0016	0.0012		0.3091
ROW_JP	**1**	0.0601	0.0659	0.0461	0.0342	0.0253	0.0188	0.014	0.0104	0.0078	0.0058	0.0044	0.0033	0.0025	0.0019	0.0014		1.3065
FC_CN	**1**	0.4696	0.7576	0.4956	0.4674	0.3691	0.3151	0.2592	0.2165	0.1793	0.1489	0.1234	0.1023	0.0846	0.07	0.0579		5.3897
NFC_CN	0	0.9265	0.3914	0.3852	0.2939	0.2541	0.2075	0.1737	0.1436	0.1193	0.0988	0.0818	0.0677	0.056	0.0463	0.0382		3.4638
GG_CN	0	0.1908	0.0733	0.1021	0.0721	0.0652	0.0524	0.0443	0.0366	0.0305	0.0253	0.021	0.0174	0.0144	0.0119	0.0098		0.8135
HH_CN	0	0.1637	0.078	0.1244	0.0817	0.0769	0.0608	0.0519	0.0427	0.0356	0.0295	0.0245	0.0203	0.0168	0.0139	0.0115		0.8865
ROW_CN	**1**	0.0383	0.0368	0.037	0.0273	0.0241	0.0196	0.0165	0.0137	0.0114	0.0094	0.0078	0.0065	0.0054	0.0044	0.0037		1.2793
FC_US	**−1**	−0.5785	−0.2618	−0.0902	−0.0006	0.0411	0.0568	0.0591	0.055	0.0482	0.0408	0.0337	0.0275	0.0222	0.0178	0.0142		−1.4606
NFC_US	0	−0.6519	−0.2455	−0.0761	0.0079	0.0466	0.0604	0.0615	0.0565	0.0492	0.0415	0.0343	0.0279	0.0225	0.018	0.0143		−0.4782
GG_US	0	−0.3029	−0.0893	−0.0326	−0.0019	0.0129	0.0185	0.0196	0.0183	0.0161	0.0137	0.0113	0.0093	0.0075	0.006	0.0048		−0.2704
HH_US	0	−0.1355	−0.0689	−0.0284	−0.008	0.0025	0.0071	0.0086	0.0086	0.0078	0.0067	0.0057	0.0047	0.0038	0.003	0.0024		−0.1706

(continued)

Table 5.14 (continued)

Sectors	Δs	1	2	3	4	5	6	7	8	9	10	11	12	13	14	15	⋮	$(I-C)^{-1}*\Delta s$
ROW_ US	**−1**	−0.0567	−0.0488	−0.0219	−0.0079	−0.0007	0.0028	0.0041	0.0044	0.0041	0.0036	0.0031	0.0026	0.0021	0.0017	0.0014		−1.1007
FC_ UK	**1**	1.2035	0.8278	0.6511	0.4971	0.3833	0.2957	0.2286	0.177	0.1373	0.1065	0.0828	0.0644	0.0501	0.0391	0.0305		5.885
NFC_ UK	0	0.188	0.1944	0.1474	0.1151	0.0891	0.0691	0.0536	0.0417	0.0324	0.0252	0.0196	0.0153	0.0119	0.0093	0.0072		1.0452
GG_ UK	0	0.2212	0.1222	0.0971	0.0737	0.0567	0.0437	0.0338	0.0261	0.0202	0.0157	0.0122	0.0095	0.0074	0.0057	0.0045		0.7658
HH_ UK	0	0.1101	0.084	0.0633	0.0487	0.0374	0.0288	0.0222	0.0172	0.0133	0.0103	0.008	0.0062	0.0049	0.0038	0.0029		0.4719
ROW_ UK	1	0.14	0.1853	0.1267	0.0995	0.076	0.0586	0.0452	0.0349	0.0271	0.021	0.0163	0.0126	0.0098	0.0077	0.006		1.8881

Note Extrapolation based on data from Table 5.10

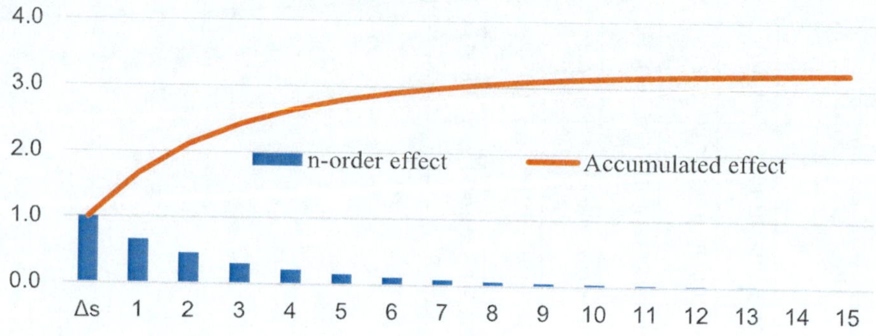

Fig. 5.8 Shock effects on liability-side for the FC_US

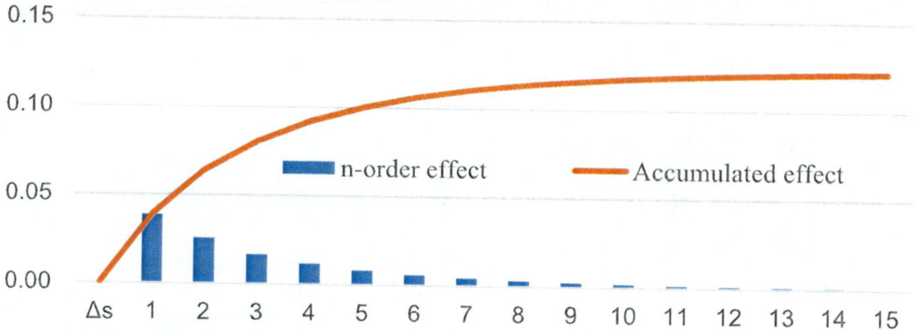

Fig. 5.9 Shock effects on liability-side for the GG_US

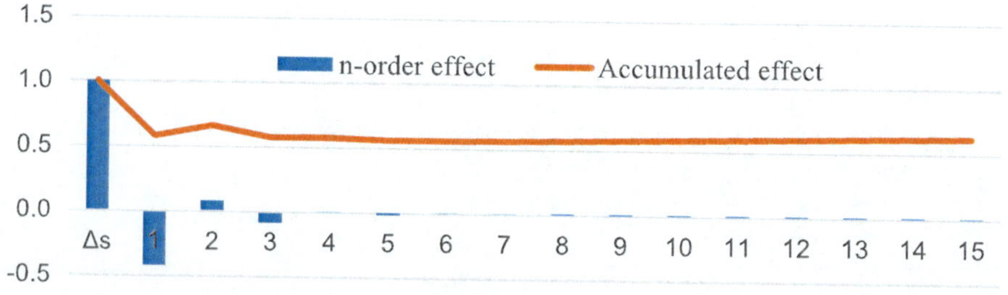

Fig. 5.10 Shock effects on liability-side for FC_CN

6th-order to a lower positive effect of 0.0016. This low positive effect continues until 15th order, with a limit effect of 0.654. Moreover, the cumulative effect of hedging negative and positive effects in order 15 is 0.619, which is significantly lower than that of the FC sector in JP, the US, and the UK (see Table 5.13). FC_CN is less able to cope with debt shocks because of the decrease in overseas financing.

Given the aforementioned circumstances, GG_CN experiences a significant positive impact of 0.0752 from the first effect, but it exhibits a negative influence of −0.0334 in the second instance, gradually diminishing to −0.0012 by the sixth order.

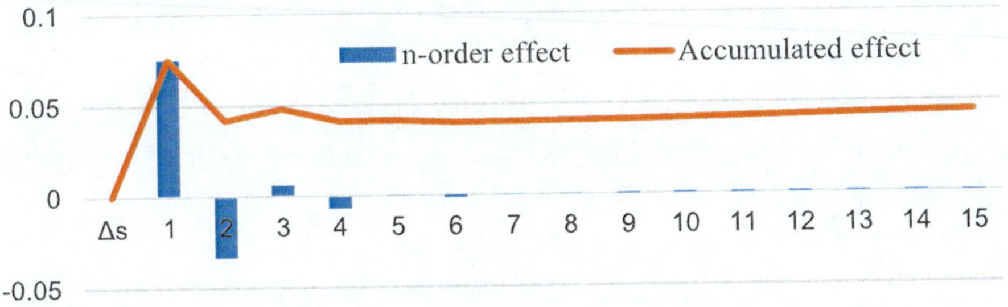

Fig. 5.11 Shock effects on liability-side for GG_CN

Thereafter, although the effect turns positive, it continues to show a low effect and the final limit effect is only 0.048. Moreover, the final possible cumulative impact are 0.0446 (GG_CN), 0.7197 (GG_JP), 0.1294 (GG_US), and 0.1419 (GG_UK) (see Table 5.13). As a result, we know that GG_CN is less able to respond to debt shocks than the other GG sectors.

Next, we focus on the supply of funds, which uses Klein-model, to observe n-order effects on financial investment for all sectors of the G-4 (see Eq. 5.11 and Table 5.14).

The shocks to PC_US and ROW_US are set at −1, while FC_CN, FC_JP, and FC_UK are set at +1. Since FC_US has the largest global market share of financial investments, assuming that its original investment falls by one unit, its direct effect will be shown with −0.5785, which is only lower than that of FC_UK. But it quickly recovers to +0.0411 of its original position after the five-order indirect effects, and the shock effect declines gradually after the 10th order, tightening to zero effect by the 15th order (Fig. 5.12). The combined limit value of the US including the original shock and indirect impacts is −1.4606, while FC_CN's limit effect is 5.3897, the FC_UK's limit effect is 5.885 (see Table 5.14). From the perspective of the accumulated effect, it has a strong recovery function, which start turning from the eighth order, the negative shock began to abate and slowly recover −1.5148 by the 15th order, and the cumulative effects of FC_CN, FC_JP, and FC_ UK, are 5.1167, 3.3956, and 5.7747 (see Fig. 5.12), respectively.

The data related to the direct, indirect, and cumulative effects of shocks on FC_US all exceed the figures from 2019.[20] This indicates a rise in global financial risk and heightened financial pressure stemming from the impact of COVID-19, including significant global political and economic changes.

Figure 5.13 shows that when the investment of FC_US and ROW_US declines by −1 unit, the direct impact effect on GG_US is −0.3029, and the negative effect lasts until the 5th order, when it turns into a lower level of positive effect, but the limit effect including the direct effect and indirect effect is −0.2704. The cumulative impact on GG_US from the first order to the fifteenth order is −0.2888, coming at the lowest level compared to the other GG sectors in CN (0.7671), JP (1.0684), and the UK (0.7496) (see Table 5.14). It is worth noting that, whether assessed in terms

[20] Zhang (2022).

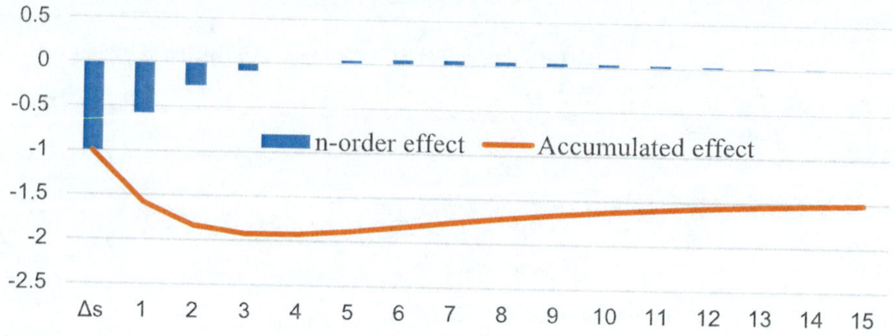

Fig. 5.12 Shock effects on asset-side for the FC_US

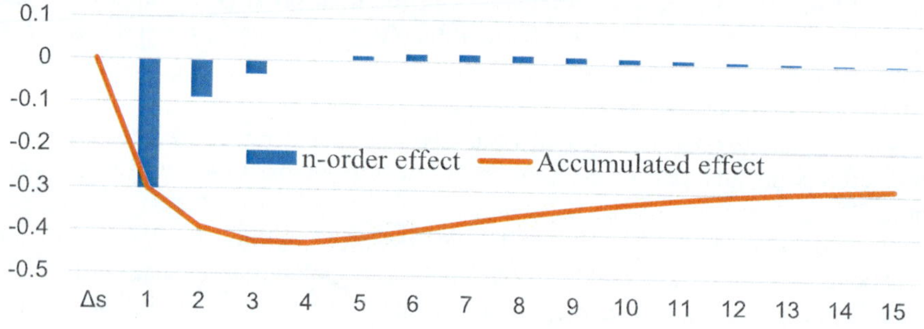

Fig. 5.13 Shock effects on asset-side for the GG_US

of asset reduction or liability increase, the effect on GG_US' financing operations remains minimal, and lower than other sectors.

We observe the shocks on FC_CN and GG_CN when the fund supply of the FC_US and ROW_US decreases by −1, while GG_CN and ROW_CN, JP, and the UK increase the fund supply by +1. As shown in Table 5.14 and Fig. 5.14, the direct impact on FC_CN is 0.4696, while on FC_JP and FC_UK are 0.5572 and 1.2035, respectively. In this regard, due to the close political and economic relations between the UK and the US, Table 5.14 shows that even if the investment in FC_US decreases by −1 unit, it still has a more positive impact on the UK than on CN.

However, when considering the second-order effect, CN, JP, and the UK returned values of 0.7576, 0.4918, and 0.8278, respectively. The positive effect of FC_CN decreases from the 8th order to the 15th order, which is 0.05791 and its limit effect is 5.3897. By comparison, FC_JP and FC_UK are 3.4311 and 5.885, respectively. Therefore, we know that at the end of 2021, the financial investment changes of FC_CN have some impact on the other FC sectors.

Under the impact of the above constraints, the direct effect of GG_CN is 0.1908, the second-order indirect effect is 0.0733, and then continues to decline to the 15th-order effect of 0.0098; the limit effect which includes direct effect and indirect effect is 0.8135. From the inferred direct effects and indirect effects and limit effects of all

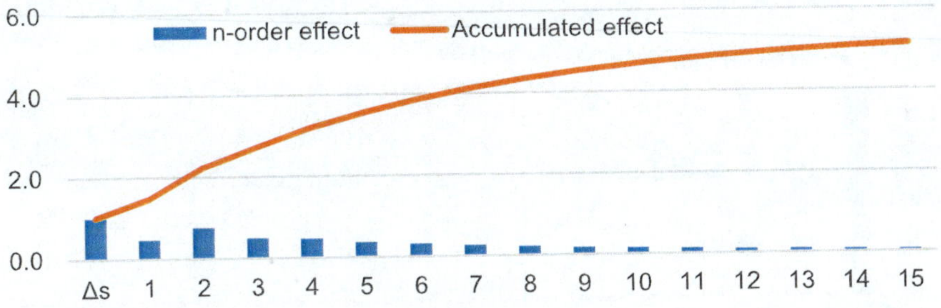

Fig. 5.14 Shock effects on asset-side for the FC_CN

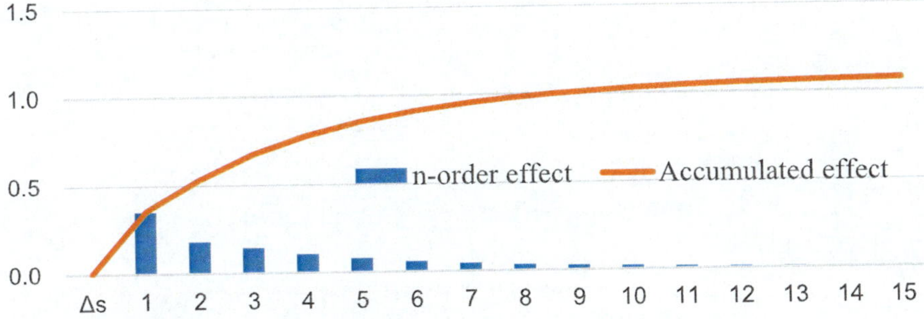

Fig. 5.15 Shock effects on asset-side for the GG_CN

levels, the positive impact of GG_CN on the impact is higher than that of the US and the UK, but lower than JP (see Table 5.14). Moreover, the accumulative effect of GG_CN is 0.7572 (see Fig. 5.15), compared t0 1.0684 (JP), −0.2888 (US), and 0.7496 (UK).

Under the above assumption, the direct impact of the liability side (see Table 5.13) on the NFC sectors are 0.0406 (CN), 0.134 (JP), 0.1012 (US), and 0.0512 (UK); the cumulative effects are 0.0271, 0.9108, 0.5136, and 0.3333, respectively. The direct impact of the increase in assets side (see Table 5.14) on the NFC sectors are 0.9265 (CN), 0.3883 (JP), −0.6519 (US), and 0.188 (UK); the final possible cumulative impact are 3.2841, 1.3956, −0.533, and 1.019, respectively. At the end of 2021, by the perspective of financing, the cumulative limit effect on NFC_CN is slightly lower than the other NFC sectors, but from the perspective of assets, the cumulative limit effect on NFC_CN is the highest.

Returning to the financial risks associated with the influence of foreign capital raising and utilization in both CN and the US, the direct effect brought by ROW_CN's reduction of external financial assets including US Treasury bonds is 0.009, but it turns to −0.0036 in the third order. The negative indirect impact has a 5th order negative impact on CN's use of external financial assets, and then turns to positive, but the limit effect including the direct effect and indirect effect is −0.9908 (see

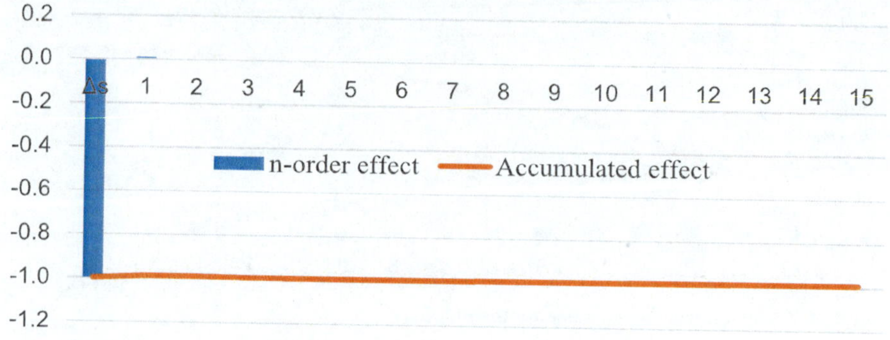

Fig. 5.16 Shock effects on liability-side for the ROW_CN

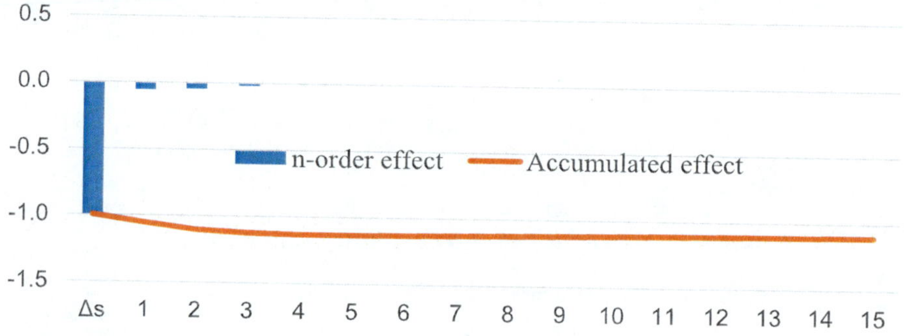

Fig. 5.17 Shock effects on asset-side for the ROW_US

Table 5.13). The cumulative effect including from first order until 15th order is –0.9921 (see Fig. 5.16).

From the inferred results of Table 5.14, the direct effect of reducing ROW_US external financing by one unit is −0.0567, where the indirect negative effect will last to the 5th order, including the limit effect at the 15th order which is −1.1007, and the total cumulative effect is −1.106 (Fig. 5.17). Through this comparison, we can observe that the reduction in the US' foreign financing poses a more significant financial risk impact compared to CN's utilization of foreign assets.

5.5.3 *Shock Propagation Across the SFSM*

To focus on the impact of changes in the G-4 sectors, we adapt the 20-order matrix in Table 5.9 to the 20-order matrix C, and to reflect the shock of a sector's investment changes on other sectors, the diffusion matrix C as an operator on the vector can be calculated.

We put $V, \rho as$ the matrix of eigenvectors and diagonal matrix of eigenvalues of C, which we assume as diagonalizable, as in our example so that $C = V * \rho * V^{-1}$ and $C^n = V * \rho^n * V^{-1}$ (see Meyer, 2000). This allows a representation of the n-effects:

$$C^{n-1} * \Delta s = V * \rho^{n-1} * (V^{-1} * \Delta s) \quad (5.12)$$

The vector $(V^{-1} * \Delta s) = \begin{pmatrix} e_1 \\ e_2 \\ e_3 \\ e_4 \\ \vdots \\ e_{20} \end{pmatrix}$ contains the components of the shock vector Δs expressed in the base of eigenvectors, where Δs (see Eq. 5.10). It is assumed that ROW_CN will be reduced by 1 unit, ROW_JP, ROW_US, and ROW_UK will be increased by 1 unit, and other sectors will be assumed unchanged with a zero increment.

The matrix of eigenvectors V and the diagonal matrix ρ of eigenvalues can set as:

$$V = \begin{pmatrix} v_{1,1} & v_{1,2} & v_{1,3} & v_{1,4} & \cdots & v_{1,20} \\ v_{2,1} & v_{2,2} & v_{2,3} & v_{2,4} & \cdots & v_{2,20} \\ v_{3,1} & v_{3,2} & v_{3,3} & v_{3,4} & \cdots & v_{3,20} \\ v_{4,1} & v_{4,2} & v_{4,3} & v_{4,4} & \cdots & v_{4,20} \\ \vdots & \vdots & \vdots & \vdots & \cdots & \vdots \\ v_{20,1} & v_{20,2} & v_{20,3} & v_{20,4} & \cdots & v_{20,20} \end{pmatrix} \text{ and } \rho = \begin{pmatrix} \rho_1 & 0 & 0 & 0 & 0 & 0 \\ 0 & \rho_2 & 0 & 0 & 0 & 0 \\ 0 & 0 & \rho_3 & 0 & 0 & 0 \\ 0 & 0 & 0 & \rho_4 & 0 & 0 \\ 0 & 0 & 0 & 0 & \cdots & 0 \\ 0 & 0 & 0 & 0 & 0 & \rho_{20} \end{pmatrix} \text{ in our}$$

20×20.

example, Eq. (5.11) can also be expressed as:

$$C_i^{n-1} * \Delta s = \rho_1^{n-1} * e_1 * \begin{pmatrix} v_{1,1} \\ v_{2,1} \\ v_{3,1} \\ v_{4,1} \\ v_{\vdots,1} \\ v_{20,1} \end{pmatrix} + \rho_2^{n-1} * e_2 * \begin{pmatrix} v_{1,2} \\ v_{2,2} \\ v_{3,2} \\ v_{4,2} \\ v_{\vdots,2} \\ v_{20,2} \end{pmatrix}$$

$$+ \rho_3^{n-1} * e_3 * \begin{pmatrix} v_{1,3} \\ v_{2,3} \\ v_{3,3} \\ v_{4,3} \\ v_{\vdots,3} \\ v_{20,3} \end{pmatrix} + \cdots + \rho_{20}^{n-1} * e_{20} * \begin{pmatrix} v_{1,20} \\ v_{2,20} \\ v_{3,20} \\ v_{4,20} \\ v_{\vdots,20} \\ v_{20,20} \end{pmatrix} \quad (5.13)$$

The individual n-order effects are expressed as a linear combination of the eigenvectors of the diffusion matrix, denoted by λ_i. By utilizing the 20-order diffusion matrix C, calculated from Table 5.9, we determine the corresponding eigenvalues ρ with the components of the shock in the base formed by the eigenvectors V. The transpose eigenvalues ρ denoted as $\rho\prime$, is:

ρ'=(	0.8210	0.7638	0.7304	0.6008	-0.3814	-0.1793	-0.1440	-0.1220	0.0922	0.0922	0.1078	0.0913	0.0362	-0.0527	0.0027	-0.0289	-0.0289	-0.0250	-0.0198	-0.0219	)

The matrix of eigenvectors V and Inverse matrix of eigenvector V^{-1} are calculated by Table 5.9, and using $V^{-1} * \Delta s = E$ we can get the vector E, the transpose of matrix E is denoted as E^T as:

E^T=(	-0.2803	-2.0361	0.7279	-0.3732	-1.2808	-1.0778	1.2175	-0.4578	-2.0861	-2.0861	-1.1709	-0.6328	-0.1322	1.2446	-0.6579	-4.5388	-4.5388	12.7952	5.8458	0.1962)

This way, we can get the decomposition of the impact on ROW_CN when CN's foreign asset utilization decreases, that is, the eigenvector decomposition of (n > 1)-order effects for ROW_CN (Table 5.15).

This presentation would allow us better understanding the features that govern the propagation effects and link them to network centrality, including perform dimensionality reduction to simplify the presentation of the shock dynamics. Figure 5.18 and 5.19 show the decomposition of the effects on ROW_CN and ROW_US for n > 1 (indirect effects).

According to Eq. (5.12), using the eigenvalues and eigenvectors, we can analyze and decompose the shock propagation on the US and CN. The shock propagation equation can be made for each eigenvalue as below. Table 5.15 demonstrates that some of the eigenvalues in observing the external impact on ROW_CN are very small. For the convenience of observation, we adopt a dimensionality reduction method and only retain λ_1, λ_2, λ_5, λ_{11} λ_{14}. It allows better understanding of the features that govern the propagation effects and link them to network centrality, including perform dimensionality reduction to simplify the presentation of the shock dynamics.

We first calculated the shock propagation on CN using Eq. (5.13). Figure 5.18 shows the decomposition of the effects on CN for n > 1 (indirect effects). The power after the shock is decomposed into a persistent negative sub-effect (gray line (λ_2), and four sign-oscillating sub-effects, which are green (λ_{14}), red (λ_1), orange (λ_5), and blue (λ_{11}), inducing the alternation of positive and negative effects described. The nature of the signs as oscillating or not depends on the sign of the corresponding eigenvalue, those with positive value ($\rho_1 = 0.821$ in ROW_CN) delivering a constant sign contribution which depends on the sign of the product of the component of the shock in the eigenbase ($E_5 = -1.2808$ in ROW_CN), and the sector component in the eigenvector associated to the eigenvalue ($V_{14} = -0.6885$), delivering a small negative sign path.

The size of the sub-effects depends on the corresponding module of the eigenvalues,[21] the components of the eigenvectors, and the components of the shock.

[21] Eigenvalues, eigenvectors, and shocks in the base of eigenvectors are in general complex numbers if we allow for diffusion matrices that are diagonalizable in the complex plane. This case pertains to

Table 5.15 The eigenvector decomposition of (n > 1)-order effects for ROW_CN

FC_JP		1	2	3	4	5	6	7	8	9	10	11	12	13	14	15
	λ1	0.0072	0.0059	0.0049	0.004	0.0033	0.0027	0.0022	0.0018	0.0015	0.0012	0.001	0.0008	0.0007	0.0006	0.0005
NFC_JP	λ2	−0.0102	−0.0078	−0.006	−0.0046	−0.0035	−0.0027	−0.002	−0.0016	−0.0012	−0.0009	−0.0007	−0.0005	−0.0004	−0.0003	−0.0002
GG_JP	λ3	0.0001	0.0001	0.0001	0	0	0	0	0	0	0	0	0	0	0	0
HH_JP	λ4	−0.0001	−0.0001	0	0	0	0	0	0	0	0	0	0	0	0	0
ROW_JP	λ5	−0.017	0.0065	−0.0025	0.0009	−0.0004	0.0001	−0.0001	0	0	0	0	0	0	0	0
FC_CN	λ6	0.0008	−0.0001	0	0	0	0	0	0	0	0	0	0	0	0	0
NFC_CN	λ7	0.0001	0	0	0	0	0	0	0	0	0	0	0	0	0	0
GG_CN	λ8	0	0	0	0	0	0	0	0	0	0	0	0	0	0	0
HH_CN	λ9	0	0	0	0	0	0	0	0	0	0	0	0	0	0	0
ROW_CN	λ10	0	0	0	0	0	0	0	0	0	0	0	0	0	0	0
FC_US	λ11	−0.0181	−0.002	−0.0002	0	0	0	0	0	0	0	0	0	0	0	0
NFC_US	λ12	0.0007	0.0001	0	0	0	0	0	0	0	0	0	0	0	0	0
GG_US	λ13	0	0	0	0	0	0	0	0	0	0	0	0	0	0	0
HH_US	λ14	0.0451	−0.0024	0.0001	0	0	0	0	0	0	0	0	0	0	0	0

(continued)

Table 5.15 (continued)

FC_JP		1	2	3	4	5	6	7	8	9	10	11	12	13	14	15
	λ1	0.0072	0.0059	0.0049	0.004	0.0033	0.0027	0.0022	0.0018	0.0015	0.0012	0.001	0.0008	0.0007	0.0006	0.0005
ROW_US	λ15	0	0	0	0	0	0	0	0	0	0	0	0	0	0	0
FC_UK	λ16	−0.0003	0	0	0	0	0	0	0	0	0	0	0	0	0	0
NFC_UK	λ17	−0.0003	0	0	0	0	0	0	0	0	0	0	0	0	0	0
GG_UK	λ18	0.0011	0	0	0	0	0	0	0	0	0	0	0	0	0	0
HH_UK	λ19	0	0	0	0	0	0	0	0	0	0	0	0	0	0	0
ROW_UK	λ20	0.0001	0	0	0	0	0	0	0	0	0	0	0	0	0	0

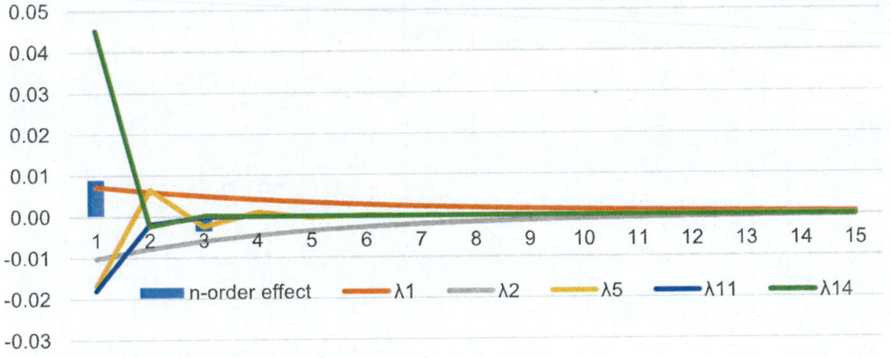

Fig. 5.18 Shock propagation on ROW_CN

Therefore, the sub-effects linked to the eigenvalue are extremely small and disappearing fast for ROW_CN, to the extent that they can be totally ignored. Combined with Fig. 5.18 and EC shown in Table 5.11, when CN's foreign assets decrease the amount of shocks are small and very limited for CN. Among them, FC_JP (red line in Fig. 5.18) has a positive effect on ROW_CN and NFC_JP (gray line) has a negative effect, and the effects of these two terms continue to the 8th order; while FC_US (blue line) has a negative effect on CN in the short term, HH_US (green line) has a positive effect in the short term, but they only last to the third stage at the end of 2021. In addition, the largest factor in the effect λ_5 (yellow line) comes from E_5, which the component of the shock in the eigenbase reflects the short-term negative effect of the combined action of various sectors of the G-4 on ROW_CN.

Taking the same approach with the data from Table 5.10, to calculate the shock propagation on ROW_US, the transpose eigenvalues λ denoted as λ' and the transpose of matrix $E = V^{-1} * \Delta s$ is denoted as E^T, shown below:

ρ'= (	0.8210	0.7638	0.7304	0.6008	-0.3814	-0.1793	-0.1440	-0.1220	0.0922	0.0922	0.1078	0.0913	0.0362	-0.0527	0.0027	-0.0289	-0.0289	-0.0250	-0.0219	-0.0198	)
E^T= (	1.5167	1.7642	0.5891	2.7008	-0.9666	-0.8311	-1.0481	-0.3778	0.0403	0.0403	1.4254	0.9550	0.9240	3.0916	-0.2750	6.1660	6.1660	-5.9791	-1.1762	-12.0927	)

By Eq. (5.13), we also can get the eigenvector decomposition of (n > 1)-order effects for ROW_US (Table 5.16).

Using Table 5.16, we plotted Fig. 5.19 to show the shock propagation on ROW_US that we can know that the persistence of the n-order effects depends on the module of the eigenvalue. Thus, the sub-effects linked to the eigenvalue 0.7638 (orange (λ_2) line in Fig. 5.19) and 0.6008 (yellow (λ_4) line) show the significant sub-effects before 8th order. But after that, the associated sub-effects are extremely small and disappear quickly, to the extent that they can be totally ignored. In addition, λ_{20} (red line) and λ_{18} (green line) have a short-term impact on ROW_US, where it tends to stop at

economic analysis. When the eigenvalue or eigenvector exhibits a very small imaginary component, considering only the real part of the complex number does not significantly impact the accuracy of prediction. Therefore, this paper exclusively focuses on the real component of the complex number.

Table 5.16 The eigenvector decomposition of (n > 1)-order effects for ROW_US

	1	2	3	4	5	6	7	8	9	10	11	12	13	14	15
λ1	0.0078	0.0064	0.0052	0.0043	0.0035	0.0029	0.0024	0.002	0.0016	0.0013	0.0011	0.0009	0.0007	0.0006	0.0005
λ2	0.0412	0.0314	0.024	0.0183	0.014	0.0107	0.0082	0.0062	0.0048	0.0036	0.0028	0.0021	0.0016	0.0012	0.0009
λ3	0.0032	0.0023	0.0017	0.0012	0.0009	0.0007	0.0005	0.0004	0.0003	0.0002	0.0001	0.0001	0.0001	0.0001	0
λ4	−0.1466	−0.0881	−0.0529	−0.0318	−0.0191	−0.0115	−0.0069	−0.0041	−0.0025	−0.0015	−0.0009	−0.0005	−0.0003	−0.0002	−0.0001
λ5	−0.0003	0.0001	−0.0001	0	0	0	0	0	0	0	0	0	0	0	0
λ6	0.0035	−0.0006	0.0001	0	0	0	0	0	0	0	0	0	0	0	0
λ7	−0.0017	0.0002	0	0	0	0	0	0	0	0	0	0	0	0	0
λ8	−0.0009	0.0001	0	0	0	0	0	0	0	0	0	0	0	0	0
λ9	0	0	0	0	0	0	0	0	0	0	0	0	0	0	0
λ10	0	0	0	0	0	0	0	0	0	0	0	0	0	0	0
λ11	−0.0025	−0.0003	0	0	0	0	0	0	0	0	0	0	0	0	0
λ12	0.0083	0.0008	0.0001	0	0	0	0	0	0	0	0	0	0	0	0
λ13	−0.0033	−0.0001	0	0	0	0	0	0	0	0	0	0	0	0	0
λ14	0.0027	−0.0001	0	0	0	0	0	0	0	0	0	0	0	0	0
λ15	0	0	0	0	0	0	0	0	0	0	0	0	0	0	0
λ16	0.0169	−0.0005	0	0	0	0	0	0	0	0	0	0	0	0	0
λ17	0.0169	−0.0005	0	0	0	0	0	0	0	0	0	0	0	0	0
λ18	0.0135	−0.0003	0	0	0	0	0	0	0	0	0	0	0	0	0
λ19	−0.0008	0	0	0	0	0	0	0	0	0	0	0	0	0	0
λ20	−0.0196	0.0004	0	0	0	0	0	0	0	0	0	0	0	0	0

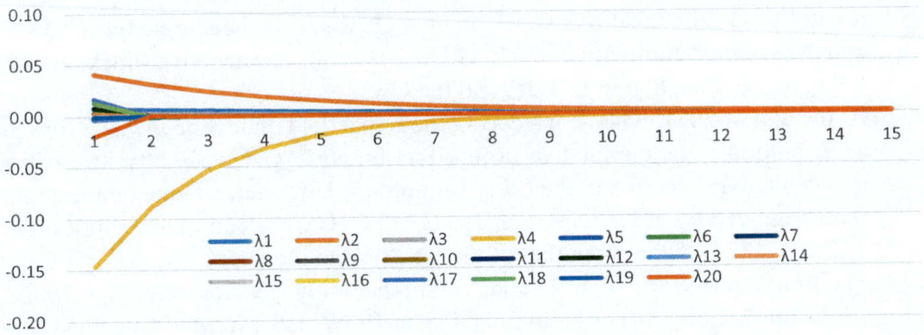

Fig. 5.19 Shock propagation on ROW_US

the second order. Therefore, we know that NFC_JP and HH_JP have a large shock on ROW_US, where NFC_JP is a positive impact and HH_JP is negative. Also, the short-term effect on ROW_US by ROW_UK and GG_UK only last until the second-order. Where E_{20} for λ_{20} is -12.0927 and E_{18} for λ_{18} is -5.9791, the component of the shock in the eigenbase also reflects the short-term positive and negative effects of the combined action of various sectors of the G-4 on ROW_US.

5.6 Concluding Remarks

This study presents a new statistical approach to measure the GFF and establishes the SFSM based on the statistic system of the GFF. It also discusses the data sources needed to establish the SFSM and the integration of the dataset. Sectors of G-4 statistical matrix based on W-t-W are established through empirical analysis and the analysis method of SFSM is discussed.

Regarding GFF as a network, the established GFFM and SFSM are both square matrices. By denoting each country and sector as nodes and the scale of bilateral debt as the edge of the network, network analysis can be conducted using network theory. The results of the network analysis are as follows.

(1) The columns in Table 5.13 show the transmission of financial risk and shock effects in a country's sector on other country sectors. On the liability side, if we assume that the FC sectors all increase their financing by +1 unit, the external financing of JP, the US, and the UK also increases by +1 unit, and only when CN's external financing is reduced by −1 unit due to the change in international environment, FC_CN is less able to cope with debt shocks because of the decrease in overseas financing, and the first-order effect (direct effect) on the financing of GG_CN will decrease by 0.0752. From this, we know that GG_CN is also less able to respond to debt shocks than the other GG sectors.

(2) In Table 5.14, if we assume that the FC and ROW sectors of CN, JP, and the UK both increase assets by +1 unit, while the same sectors of the US both decrease

assets by −1 unit, the direct shock on GG_US assets will decrease by −0.3029, and the eventual cumulative shock will be −0.2888. The negative shock on GG_US is higher than that in CN, JP, and the UK.

(3) At the end of 2021, when considering the impact of the decline in CN's foreign asset holdings, the cumulative limit effect on NFC_CN is slightly lower than that of the NFC sectors in the other economies. However, when examining the risks triggered by the reduction in assets in FC_US, the cumulative limit effect on NFC_CN is higher than that of the US, JP, and the UK.

(4) The limit effect of ROW_CN underweight including US Treasuries is −0.9908, while the limiting effect of the decline in ROW_US external financing is −1.1007. This comparison demonstrates that the reduction in America's foreign financing poses a more significant financial risk impact compared to CN's utilization of foreign assets.

The shock propagation in the ROW sectors of CN and the US foreign financial investment is inferred as follows.

CN's foreign assets decrease, the amount of shocks are small and very limited for CN. Among them, the negative effect of NFC_JP and the positive effect of FC_JP are longer, lasting from the first to the eighth, while the shock of FC_US and GG_US only maintains the second-order effect.

NFC_JP and HH_JP have a large shock on ROW_US, where NFC_JP is a positive impact and HH_JP is negative. The shock effect on ROW_US persists to the 8th-order, short-term effect on ROW_US by ROW_UK and GG_UK, and then will no longer have a shock effect.

Also, the component of the shock in the eigenbase reflects the short-term positive and negative effects of the combined action of various G-4 sectors on ROW_CN and ROW_US.

Through the above statistical speculation, deduction, and analysis, we reach the following conclusions. First, we discussed the preparation and application of counterparty matrix by country level, that is, GFFM in Chaps. 1–4. The GFFM meets the needs of GFF data by employing the W-t-W benchmark. However, it is not possible to provide more detailed financial information of bilateral exposures between financial and nonfinancial sectors in different financial instruments within and across countries to observe the impact channel of bilateral exposure. Therefore, we construct the theoretical framework of the GFFM and establish a practical GFFM to further develop an SFSM to identify sectoral interlinkages using the G-4 and put forward the basic concept, data source, and compilation method for building the SFSM.

Second, this study is the first to compare sectoral financial exposures across the G-4 economies using the financial network. By comparing the shock dynamics of financial operations on sectors of the US and CN, we can see the following two points. (A) The changes in the liabilities of the US financial sector impact the domestic sector and other countries than that of CN; (B) the reduction of US financial investment and external debt has had a greater impact on the GG sectors of JP and UK.

Third, while the US' debt risk leads to the decline of its relative position in the world economy, CN is facing a balance sheet recession. The decoupling of the

Chinese and American economies seems inevitable. The challenge now is to navigate this separation in an organized fashion, minimizing its economic impact, preventing further conflicts, and preserving the possibility of future historic cooperation.

Fourth, the proposed statistical framework can be used to decompose effects caused by quantity shocks of any nature such as central bank quantitative easing affecting the volume of assets held by the relevant sectors, with shock affecting the distribution of stock value, and price shocks.

This study has some limitations, which can be addressed in future studies. First, the accuracy of the GFF table, especially in processing reserve data, needs improving. The data of reserve assets are not included in the GFFM because of the mismatch of data sources, but put the the data of reserves assets in SFSM. CPIS, CDIS, and LBS have their own information system, all of which can be carried out on the basis of the W-t-W matrix. However, the data of reserves are from IIP and cannot be carried out similarly. Therefore, the integration and matching of data systems needs strengthening.

The second is to enhance the function of the SFSM. The BSA and external-sector matrices could potentially be extended to flow data to identify changes in transactions and other changes in the volume of an asset/liability. This could be a rather challenging task given that the flow data would need to be decomposed by the contributing country.

Third, the financial network analysis method, new approaches, and the network theory need expanding. This includes the development of centrality measures of GFF that directly represent the interlinks, especially EC and capturing direct and indirect links with financial instruments.

Fourth, future research should tackle the lack of temporal dimension in the description of the propagation process. While large structural changes in the W-t-W relationships would be difficult to model, very short-term changes might be studied on the basis of the literature on eigenvalue perturbations, such as Bauer-Fike Theorem (see for instance Wei et al., 2006), from which boundaries for time paths might be derived.

Finally, continuously improving financial accounts and GFF statistics is needed, alongside reducing data gaps including obtaining data on special purpose entities activities (SPEs[22]). The future work should be directed towards improving the methodological framework for compiling GFF matrices, trying to extend the GFFM to flow data, further developing the methods for the GFF analysis.

Appendix A

See Tables A.1, A.2, A.3 and A.4.

[22] IMF (2016); Carlos Sanchez-Munoz, Artak Harutyunyan, Padma S Hurree Gobin (2022).

Table A.1 International SFSM with sectoral data (at the end of 2018, USD bn.)

Liabilities / Assets	FC_CN	NFC_CN	GG_CN	HH_CN	ROW_CN	FC_JP	NFC_JP	GG_JP	HH_JP	ROW_JP	FC_US	NFC_US	GG_US	HH_US	ROW_US
Assets															
FC_CN	21,868	23,188	4559	7907	3447	31	39	24	6	0	573	442	257	41	0
NFC_CN	10,301	2542	83	0	601	8	12	7	2	0	144	121	69	11	0
GG_CN	7382	10,908	46	0	105	1	1	1	0	0	16	13	8	1	0
HH_CN	22,551	20,012	163	83	270	2	3	2	0	0	42	35	20	3	0
ROW_CN	704	2642	162	243	0	0	0	0	0	0	0	0	0	0	0
FC_JP	4	17	1	2	0	13,916	7408	9177	2579	2339	979	503	293	47	0
NFC_JP	2	130	1	1	0	4014	3035	410	141	1608	318	749	152	24	0
GG_JP	0	0	0	0	0	1701	1721	638	93	1483	73	60	35	6	0
HH_JP	1	3	0	0	0	14,483	2229	286	36	252	60	49	29	5	0
ROW_JP	0	0	0	0	0	998	1265	772	192	0	0	0	0	0	0
FC_US	147	93	5	8	0	462	331	202	50	0	33,150	28,986	14,141	13,199	8168
NFC_US	20	187	5	7	0	216	274	167	42	0	7115	12,468	1212	322	3143
GG_US	0	0	0	0	0	0	0	0	0	0	1985	1590	845	996	561
HH_US	12	48	3	4	0	329	418	255	63	0	49,005	19,976	7349	491	6035
ROW_US	0	0	0	0	0	0	0	0	0	0	9475	7620	4714	751	0

Table A.2 International SFSM with sectoral data (at the end of 2019, USD bn.)

Liabilities	FC_CN	NFC_CN	GG_CN	HH_CN	ROW_CN	FC_JP	NFC_JP	GG_JP	HH_JP	ROW_JP	FC_US	NFC_US	GG_US	HH_US	ROW_US
Assets															
Assets															
FC_CN	20,865	23,588	4929	8591	3360	33	34	25	7	0	517	439	199	48	0
NFC_CN	10,300	2542	86	0	556	10	13	8	2	0	144	200	60	14	0
GG_CN	7406	11,594	51	0	123	1	1	1	0	0	19	17	8	2	0
HH_CN	23,825	22,060	162	67	326	3	3	2	1	0	47	43	19	5	0
ROW_CN	531	2578	189	245	0	0	0	0	0	0	0	0	0	0	0
FC_JP	5	21	1	2	0	15,054	7352	9358	2725	2450	1115	620	281	67	0
NFC_JP	2	135	0	1	0	4355	2901	442	181	1255	306	798	127	30	0
GG_JP	0	0	0	0	0	1780	1655	604	89	1466	91	83	38	9	0
HH_JP	1	3	0	0	0	14,745	1692	289	33	263	66	60	27	7	0
ROW_JP	0	0	0	0	0	758	860	733	198	0	0	0	0	0	0
FC_US	155	134	9	12	0	544	337	249	67	0	35,413	33,174	15,041	13,278	9469
NFC_US	21	208	6	8	0	244	382	184	50	0	8091	15,226	1299	391	2561
GG_US	0	0	0	0	0	0	0	0	0	0	2047	1797	894	999	581
HH_US	14	58	4	5	0	377	385	284	77	0	54,320	23,643	7465	528	7177
ROW_US	0	0	0	0	0	0	0	0	0	0	10,072	9051	4365	1044	0

Table A.3 International SFSM with sectoral data (at the end of 2020, USD bn.)

Liabilities Assets	FC_CN	NFC_CN	GG_CN	HH_CN	ROW_CN	FC_JP	NFC_JP	GG_JP	HH_JP	ROW_JP	FC_US	NFC_US	GG_US	HH_US	ROW_US
Liabilities															
FC_CN	19,600	23,159	7453	10,301	3325	26	30	17	5	0	352	338	152	30	0
NFC_CN	14,898	5959	123	30	1229	13	17	8	3	0	151	239	71	14	0
GG_CN	5408	349	3	3	20	0	0	0	0	0	3	3	1	0	0
HH_CN	22,697	2910	1467	82	258	2	2	1	0	0	25	26	12	2	0
ROW_CN	879	2257	241	2	0	0	0	0	0	0	0	0	0	0	0
FC_JP	12	30	3	0	0	17,824	8936	9661	3012	3336	1435	962	432	85	0
NFC_JP	2	144	1	0	0	5319	4452	622	236	404	260	836	123	24	0
GG_JP	0	0	0	0	0	2382	1679	1514	67	978	74	78	35	7	0
HH_JP	3	7	1	0	0	16,293	2041	316	34	496	156	164	74	14	0
ROW_JP	0	0	0	0	0	840	1116	704	220	0	0	0	0	0	0
FC_US	217	186	18	0	0	645	464	261	82	0	39,339	37,615	19,049	13,770	11,274
NFC_US	35	210	8	0	0	247	417	161	50	0	10,232	15,671	1558	354	2100
GG_US	0	0	0	0	0	0	0	0	0	0	3626	2084	1038	1130	680
HH_US	24	58	6	0	0	403	465	262	82	0	59,803	28,465	6965	638	7769
ROW_US	0	0	0	0	0	0	0	0	0	0	10,800	11,309	5372	1056	0

Table A.4 International SFSM with sectoral data (at the end of 2021, USD bn.)

Liabilities	FC_CN	NFC_CN	GG_CN	HH_CN	ROW_CN	FC_JP	NFC_JP	GG_JP	HH_JP	ROW_JP	FC_US	NFC_US	GG_US	HH_US	ROW_US
Assets															
Liabilities															
FC_CN	22,238	26,865	9236	12,316	3079	27	29	18	5	0	590	639	253	50	0
NFC_CN	16,646	7193	137	31	1342	14	17	9	3	0	179	280	80	16	0
GG_CN	6047	364	4	3	21	0	0	0	0	0	3	3	1	0	0
HH_CN	27,159	3892	1722	92	333	2	3	2	0	0	34	38	15	3	0
ROW_CN	1039	2901	338	1	0	0	0	0	0	0	0	0	0	0	0
FC_JP	11	30	3	0	0	16,944	8120	8807	2793	2988	1389	1025	405	80	0
NFC_JP	1	147	0	0	0	4989	3989	536	225	266	264	952	118	23	0
GG_JP	0	0	0	0	0	2115	1538	1489	61	1051	47	53	21	4	0
HH_JP	3	7	1	0	0	15,130	1783	271	31	490	151	171	68	13	0
ROW_JP	0	0	0	0	0	805	978	678	208		0	0	0	0	0
FC_US	199	153	16	0	0	540	339	209	64	0	44,623	43,414	21,036	14,974	9977
NFC_US	32	203	9	0	0	234	369	154	47	0	12,553	18,069	1687	383	1161
GG_US	0	0	0	0	0	0	0	0	0	0	2758	2260	1283	1210	528
HH_US	9	24	3	0	0	457	488	301	92	0	67,362	34,396	6063	671	9133
ROW_US	0	0	0	0	0	0	0	0	0	0	13,106	14,686	6094	1204	0

References

Acemoglu, D., Ozdaglar, A., & Tahbaz-Salehi, A. (2013). System risk and stability in financial networs, NBER working paper series, http://www.nber.org/papers/w18727. Accessed 25 June 2023.

Antoun de Almeida, L. (2015). *A Network Analysis of Sectoral Accounts: Identifying Sectoral Interlinkages in G-4 Economies.* IMF Working Paper WP/15/111, Washington DC.

Billio, M., Getmansky, M., & Lo, A. W. (2010). Econometric measures of systemic risk in the finance and insurance sectors, NBER working paper series, http://www.nber.org/papers/w16223. Accessed 25 June 2023.

Caccioli, F., Barucca, P.,& Kobayashi, T. (2018). Network models of financial systemic risk: A review. *Journal Computer Social Science, 1*, 81–114.

Errico, L., Walton, R., Hierro, A., AbuShanab, H., & Amidžic, G. (2013). global flow of funds: Mapping bilateral geographic flows. *Proceedings 59th ISI World Statistics Congress*, Hong Kong, 2825–2830.

Errico, L., Harutyunyan, A., Loukoianova, E., Walton, R., Korniyenko, Y., Amidžić, G., AbuShanab, H., & Shin, H. (2014). M*apping the Shadow Banking System through a Global Flow of Funds Analysis.* IMF Working Papers WP/14/10.

European Commission, IMF, OECD, United Nations, World Bank. (2009). System of National Accounts 2008, 217–236.

Giron, C., Rodriguez Vives, M., & Matas, A. (2018). Propagation of quantity shocks in who-to-whom networks. *Paper prepared for the 35th IARIW General Conference.*

Hagino, S., & Kim, J. (2021). Compilation and analysis of international from-whom-to-whom financial stock table for Japan, Korea, the United States, and China. *Journal of Economic Structures.* https://doi.org/10.1186/s40008-021-00254-4. Accessed 20 Dec.

Hagino, S., Kim, J., & Inomata, S. (2019). Development of U.S.–East Asia Financial Input–Output Table. Institute of Developing economies, Japan External Trade Organization, http://hdl.handle.net/2344/00050809. Accessed 10 Sept 2020.

https://doi.org/10.1038/scientificamerican0963-148; reprinted as Chapter 8 of Leontief. (1966).

IMF. (2016). Monetary and Financial Statistics Manual and Compilation Guide, 55–90, IMF, Publication Services, Washington DC.

IMF. (2018). Final Report of the Task Force on Special Purpose Entities, Thirty-First Meeting of the IMF Committee on Balance of Payments Statistics, Washington DC. https://www.imf.org/external/pubs/ft/bop/2018/pdf/18-03.pdf.

IMF. (2023). World Economic Outlook Database, April 2023.

Input-output economics. Oxford University Press.

Klein, L. R. (1983). *Lectures in econometrics* (pp. 1–46). North-Holland.

Leontief, W. (1941). *The structure of American economy, 1919–1939.* Oxford University Press.

Leontief, W. W. (1953). Structural change. In W. Leontief (Ed.), *Studies in the Structure of the American Economy* (pp. 17–52). Oxford University Press, Chapter 2. Domestic Production and Foreign Trade; The American Capital Position Re-Examined on JSTOR.

Leontief, 1963 Leontief, W. W. (1963). The structure of development. *Scientific American, 209*(3), 148–166.

Li, Y., & Zhang, Y. J. (2020). China's National Balance Sheet 2020, China Social Sciences Press.

Loriana Pelizzon Newman, M.E.J (2010). *Networks: An introduction.* Oxford University Press, 169.

Meyer, C. (2000). Matrix analysis and applied linear algebra. SIAM, 489–642.

Okuma, R (2013). Sectoral Interlinkage in Balance Sheet Approach. IFC Bulletin No. 36.

Rasmussen, P. N. (1956). Studies in inter-sectoral relations. EinarHarcks Forlag.

Sanchez-Munoz, C., Harutyunyan, A., & Hurree Gobin, P. S. (2022). Special Purpose Entities: Guidelines for a Data Template, IMF, Technical Notes and Manuals No. 2022/006. https://www.imf.org/en/Publications/TNM/Issues/2022/10/03/Special-Purpose-Entities-Guidelines-for-a-Data-Template-529451.

Shrestha, M., Mink, R., & Fassler, S. (2012). *An integrated framework for financial positions and flows on a from-whom-to-whom basis: Concepts, status, and prospects.* IMF Working Paper WP/12/57 Washington DC.

Soramäki, K., & Cook, S. (2016). *Network theory and financial risk.* Risk Books, a Division of Incisive Media Investments Ltd.

OECD. Stat (2021). Dataset: 720. Financial balance sheets, non-consolidated, SNA 2008, https://stats.oecd.org/. Accessed 25 Mar 2021.

Stone, R. (1966). The social accounts from a consumer's point of view. *Review of Income and Wealth, 12*, 1–33.

The People's Bank of China. (2022). Financial Assets and Liabilities Statement (Financial Accounts). http://www.pbc.gov.cn/en/3688247/3688975/4280784/index.html. Accessed 25 Jun 2023.

Tsujimura, K., & Tsujimura, M. (2018). A flow of funds analysis of the US quantitative easing. *Economic Systems Research*, 30, 137–177.

Tsujimura, K., & Mizosita, M. (2002). *Flow-of-funds analysis: Fundamental technique and policy evaluation* (pp. 32–43). Keio University Press.

Wei, Y., Li, X., Bu, F., & Zhang, F. (2006). Relative perturbation bounds for the eigenvalues of diagonalizable and singular matrices—Application of perturbation theory for simple invariant subspaces. In *Linear Algebra and its Applications*, vol. 419 pp. 765–771.

Zhang, N. (2015). Measuring global flow of funds and integrating real and FAs: Concepts, Data Sources and Approaches. The IARIW-OECD Special Conference: "W(h)ither the SNA?".

Zhang, N. (2016). *Measuring global flow of funds: theoretical framework, data sources and approaches, contemporary works in economic sciences: Legal information* (pp. 47–60). Economics, OR and Mathematics, Kyushu University Press.

Zhang, N. (2020). *Flow of funds analysis: Innovation and development.* Springer, 283–321.

Zhang, N. (2021). Measuring global flow of funds: who-to-whom matrix and financial network. *The 36th IARIW General Conference*, https://iariw.org/wp-content/uploads/2021/07/Zhang_Paper.pdf. Accessed 30 Aug 2021

Zhang, N. (2022). (2022) Measuring global flow of funds: Who-to-whom matrix and financial network. *Japanese Journal of Statistics and Data Science, 5*, 899–942.

Zhang, N., & Zhao, X. (2019). Measuring global flow of funds: A case study on China, Japan and the United States. *Economic Systems Research, 31*(1).